DESIGN THINKING

DESIGN THINKING

New Product Development Essentials from the PDMA

Edited by

Michael G. Luchs

K. Scott Swan

Abbie Griffin

pdma
Connecting Innovators Worldwide

WILEY

Cover image: C. Wallace
Cover design: Vector Swirl © iStock.com/antishock

This book is printed on acid-free paper.

Published by John Wiley & Sons, Inc., Hoboken, New Jersey
Published simultaneously in Canada

Limit of Liability/Disclaimer of Warranty: While the publisher and author have used their best efforts in preparing this book, they make no representations or warranties with the respect to the accuracy or completeness of the contents of this book and specifically disclaim any implied warranties of merchantability or fitness for a particular purpose. No warranty may be created or extended by sales representatives or written sales materials. The advice and strategies contained herein may not be suitable for your situation. You should consult with a professional where appropriate. Neither the publisher nor the author shall be liable for damages arising herefrom.

For general information about our other products and services, please contact our Customer Care Department within the United States at (800) 762-2974, outside the United States at (317) 572-3993 or fax (317) 572-4002.

Wiley publishes in a variety of print and electronic formats and by print-on-demand. Some material included with standard print versions of this book may not be included in e-books or in print-on-demand. If this book refers to media such as a CD or DVD that is not included in the version you purchased, you may download this material at http://booksupport.wiley.com. For more information about Wiley products, visit www.wiley.com.

Library of Congress Cataloging-in-Publication Data:

Design thinking (2015)
  Design thinking : new product development essentials from the PDMA / edited by Michael G. Luchs, K. Scott Swan, Abbie Griffin.
     pages cm
  Includes bibliographical references and index.
  ISBN 978-1-118-97180-2 (cloth), 978-1-118-97182-6 (ePDF), 978-1-118-97181-9 (epub) and P010238748 (oBook)
  1. Product design. 2. Critical thinking. 3. Creative ability in business. I. Luchs, Michael, 1968– II. Swan, Scott, 1962– III. Griffin, Abbie. IV. Product Development & Management Association. V. Title.
  TS171.D4695 2015
  658.5′752—dc23
                                                                                    2015021569

V10015358_111019

CONTENTS

6 BOOSTING CREATIVITY IN IDEA GENERATION USING DESIGN HEURISTICS71

Colleen M. Seifert
Richard Gonzalez
Seda Yilmaz
Shanna Daly

9 THE ROLE OF DESIGN IN EARLY-STAGE VENTURES: HOW TO HELP START-UPS UNDERSTAND AND APPLY DESIGN PROCESSES TO NEW PRODUCT DEVELOPMENT   125

J. D. Albert

10 DESIGN THINKING FOR NON-DESIGNERS: A GUIDE FOR TEAM TRAINING AND IMPLEMENTATION   143

Victor P. Seidel
Sebastian K. Fixson

11 DEVELOPING DESIGN THINKING: GE HEALTHCARE'S MENLO INNOVATION MODEL   157

Sarah J. S. Wilner

14 STRATEGICALLY EMBEDDING DESIGN THINKING IN THE FIRM 205

Pietro Micheli
Helen Perks

PART III: DESIGN THINKING FOR SPECIFIC CONTEXTS 221

15 DESIGNING SERVICES THAT SING AND DANCE 223

Marina Candi
Ahmad Beltagui

18 BUSINESS MODEL DESIGN 265

John Aceti
Tony Singarayar

19 LEAN START-UP IN LARGE ENTERPRISES USING HUMAN-CENTERED DESIGN THINKING: A NEW APPROACH FOR DEVELOPING TRANSFORMATIONAL AND DISRUPTIVE INNOVATIONS 281

Peter Koen

PART IV: CONSUMER RESPONSES AND VALUES 301

20 CONSUMER RESPONSE TO PRODUCT FORM 303

Mariëlle E. H. Creusen

21 DRIVERS OF DIVERSITY IN CONSUMERS' AESTHETIC RESPONSE TO PRODUCT DESIGN 319

Adèle Gruen

ABOUT THE EDITORS

Dr. Michael G. Luchs is an Associate Professor and is the Director of the Innovation and Design Studio at the College of William & Mary's Raymond A. Mason School of Business. He earned his Ph.D. from the University of Texas at Austin in 2008. Prior to earning his Ph.D., Dr. Luchs worked for over a decade as a consultant and executive. As a Principal with the consulting firm Pittiglio, Rabin, Todd & McGrath (now PwC), Dr. Luchs worked with a broad client base to improve their product development and marketing practices and performance. Clients included Fortune 500 companies as well as SMEs in the consumer packaged goods, consumer durables, computing & telecommunications equipment, and telecommunications services industries. In addition to his consulting experience, Dr. Luchs worked in industry as the Sr. VP of Marketing for Labtec Inc. (now Logitech) and as a Product Manager for Black & Decker Power Tools.

K. Scott Swan is a Professor of International Business, Design, and Marketing at The College of William & Mary. Prof. Swan was awarded a Fulbright and named the 2015–2016 Hall Chair for Entrepreneurship in Central Europe at WU (Vienna, Austria) and The University of Bratislava (Slovakia). His latest publications are: *Innovation and Product Management: A Holistic and Practical Approach to Uncertainty Reduction* (Springer Science & Business Media) October 2014 and *A Review of Marketing Research on Product Design with Directions for Future Research* (forthcoming in JPIM). Dr. Swan has lectured internationally at University of Applied Science Upper Austria (Wels), Corvinus University in Budapest, MCI in Innsbruck, Tsinghua University in Beijing, Aoyama Gakuin University in Tokyo, WHU in Koblenz, Germany, and the Vienna Business School (WU) in Austria. His research interests have led to publications in journals such as *Strategic Management Journal, Journal of International Management, Journal of International Business Studies, Management International Review, Journal of Business Research*, and *The Journal of Product Innovation Management*. He serves on the editorial boards of *JPIM* and *The Design Journal*.

Abbie Griffin holds the Royal L. Garff Presidential Chair in Marketing at the David Eccles School of Business at the University of Utah, where she teaches the 1st-year core MBA Marketing Management course. Professor Griffin obtained her B.S. ChE from Purdue University, MBA from Harvard Business School, and Ph.D. in Management of Technology from MIT. Her research investigates means for measuring and improving the process of new product development. Her latest research can be found in the book titled: *Serial Innovators: How Individuals in Large Organizations Create Breakthrough New Products*. Her 1993 article titled "Voice of the Customer" was awarded both the Frank M. Bass Dissertation Paper Award and the John D. C. Little Best Paper Award by INForms and has been named the 7th most important article published in Marketing Science in the last 25 years. She was the editor of the *Journal of Product Innovation Management* from 1998 to 2003. The PDMA named her as a Crawford Fellow in 2009, and she currently serves as the Vice President of Publications for the Association. She was on the Board of Directors of Navistar International, a $ 13 billion manufacturer of diesel engines and trucks from 1998 to 2009. Prof. Griffin is an avid quilter, hiker, and swimmer.

A BRIEF INTRODUCTION TO DESIGN THINKING[1]

Michael G. Luchs

Innovation and Design Studio, College of William & Mary

Introduction

Within the context of new product development (NPD) and innovation, design thinking has enjoyed significantly increased visibility and, for many, increased perceived importance over the last decade. For others, however, this term can be fraught with confusion, questions of relevancy and, for some, the perception of a fad. Within that context, the objectives of this chapter include the following: First, I briefly describe the concept of design thinking and its role within NPD and innovation. Next, I provide and describe a simple framework of design thinking, followed by a summary of some fundamental principles of the "mindset" of design thinking. Throughout, I identify linkages with the other chapters in this book. While this chapter provides an overview of design thinking as well as some context, the remaining chapters in this book provide significantly more detail and a wide variety of specific examples. Thus, this chapter concludes with a visual overview of the book to help guide you to the specific ideas, tools, and practices most applicable to the NPD and innovation problems and opportunities that you and your firm are facing today.

1.1 The Concept of Design Thinking and Its Role within NPD and Innovation

What is design thinking? At its core, design thinking can be construed as a creative problem-solving approach—or, more completely, as a *systematic and collaborative*

---

[1]This chapter was adapted from "Understanding Design Thinking: A Primer for New Product Development and Innovation Professionals." © 2014 College of William & Mary.

approach for identifying and creatively solving problems.[2] The term *design thinking* simply means that one is approaching problems, and their solutions, as a designer would. While this will be elaborated subsequently, an illustrative characteristic of the design thinking approach is that it is intentionally nonlinear. Designers, whether in the arts or industry, tend to explore and solve problems through iteration. They quickly generate possible solutions, develop simple prototypes, and then iterate on these initial solutions—informed by significant external feedback—toward a final solution. This is in contrast to a linear process, such as the traditional Stage-Gate™ new product development (NPD) process, in which prototyping is typically done toward the end of the process to reflect the culmination of the development phase and to explore manufacturability, rather than as a mechanism for gaining market feedback. A more thorough description of design thinking as a process and mindset follows, but first I address an important question for those involved with new product development and innovation: When is design thinking most applicable?

When to Apply Design Thinking

Generally speaking, design thinking is best applied in situations in which the problem, or opportunity, is not well defined, and/or a breakthrough idea or concept is needed, that is, an idea that has a significant and positive impact, such as creating a new market or enabling significant revenue growth. Design thinking methods have been used successfully in different ways within business including new venture creation, business model design, and process improvement. While our focus is on applying design thinking to the challenge and opportunity of new product development[3] and innovation, this book also includes several chapters that address other contexts, such as business model design (Chapters 18 and 19).

Within the context of NPD, design thinking is very well suited to use in markets that are quickly changing and when user needs are uncertain, such as the emerging market for wearable biometric devices. However, design thinking is equally applicable in more mature markets as a means to identify new, latent customer needs and/or in an effort to develop significant or radical innovations (Chapter 17). Whereas incremental innovations are also critically important to most companies, they typically are bounded by well-defined problems or established customer needs, such as improving gas engine fuel efficiency. In those situations, a more linear, Stage-Gate process is still appropriate. Nonetheless, even in these situations there may be specific elements of a design thinking approach—specific tools or techniques—that can improve a project's outcome.

For the right situations, however, a design thinking approach is more likely to lead to better solutions that address the most important customer needs, and do so more efficiently than traditional NPD approaches alone. One of the reasons for this is that design thinking helps to avoid the trap of investing too many resources too early in a

---

[2] Given a focus on the customer's perspective, I refer throughout to "solving problems," but in the context of NPD, it would clearly make sense to also think of problems as opportunities.
[3] For simplicity, I refer to products, where products can be physical goods and/or services.

project toward developing a specific, single solution. Rather than placing such a "big bet," design thinking encourages many "little bets" (Sims, 2013) about customer insights and possible solutions. Sims describes these little bets as "low risk actions taken to discover, develop, and test an idea." These little bets make it more likely that a project team will quickly converge on solution concepts with the highest potential market success. At some point, of course, specifications need to be well defined and the product needs to be developed and, ultimately, produced. In this sense, another way to think about design thinking is as a clarifying lens on the oft referred to "fuzzy front end" of NPD, whereby a project begins with an iterative, design thinking approach, followed by a traditional Stage-Gate process after enough has been learned about customer needs and possible solutions.

The Origins of Design Thinking

The methods and mindset of design thinking, although championed by progressive companies and design consultancies, draw from a wide field of disciplines including software development, engineering, anthropology, psychology, the arts, and business. Design thinking as it exists today has co-evolved across a variety of disciplines and industries. Over time—well over 50 years, and even longer depending on your perspective—the best and most generalizable methods and practices have emerged and converged in a quasi-Darwinian process of natural selection. These have been codified, integrated, documented, and championed by leading design firms (such as IDEO and frog) and academic institutions (such as Stanford's d.school, and the Rotman School of Management), and have increasingly been adopted by industry and popularized by the media under the shared moniker of design thinking.

While this co-evolution and vetting of design thinking has led to a robust set of methodologies, it has also contributed to some confusion given the proliferation of tools, methods, books, seminars, and, more recently, online training available. Rather than getting lost in the details from the start, a useful way to learn about design thinking methods is through the lens of an organizing framework. Even here, however, there are a variety of frameworks to choose from, each with its own nuances and biases. To the novice, this, too, can be daunting. Given the time to explore these, however, it becomes apparent that there actually is significant consistency across these frameworks. In a sense, each of these has been a prototype framework—building on the ideas and lessons of its predecessors. In that iterative spirit, I propose a framework for design thinking in the next section that is intended to reflect the shared elements of existing frameworks, with the objective of retaining the most important elements of design thinking and their distinctions, while simplifying their depiction and terminology. At the least, this framework introduces the major elements of design thinking as efficiently as possible and facilitates an exploration of the rich content contained within the other chapters of this book. Further, it will make it easier to quickly navigate other design thinking frameworks in use and, in so doing, enable an efficient exploration of the vast library of tools, techniques, and advice beyond these pages.

1.2 A Framework of Design Thinking

There are literally dozens, if not hundreds, of specific design thinking–related methods and tools available, and this book will explore many of these. Learning about just a few of these and understanding how they are used together is likely more valuable than trying to experiment with them without any context. The following framework is intended to provide that context, by organizing these methods and tools based on their role or purpose.

Design thinking, as a *systematic and collaborative approach for identifying and creatively solving problems*, includes two major phases: identifying problems and solving problems. Both of these phases are critical, but in practice most people and project teams within companies are more inclined to focus on the latter, that is, on solving problems. We are naturally creative beings, and given any problem—however ill-defined—most of us can generate a set of ideas. Unfortunately, these often will not be great ideas, that is, ideas that are both original and that solve the problems with the greatest potential. One of the most powerful features of design thinking is its emphasis on identifying the right problems to solve in the first place. This is, therefore, a key element of the following framework, as indicated by the two phases of design thinking depicted in Figure 1.1: Identify and Solve. Next, I describe the purpose of each of the modes within these two phases, followed by a discussion of the iterative nature of the process as a whole.

Discover

The purpose of the first mode of the design thinking framework (see Figure 1.2) is to Discover new customer insights. One of the challenges for many product development teams is that they are immersed in the world of products and, often, technologies. While that is clearly important expertise, it can limit their field of view and perspective; market information tends to get framed in terms of product specifications relevant to existing products. As a consequence, well-intended research, even when conducted with product users, is often unintentionally biased toward relatively minor modifications to

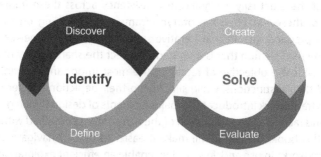

Figure 1.1: A framework for design thinking.

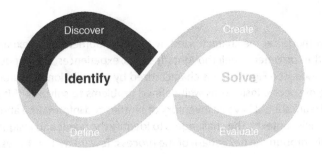

Figure 1.2: Discover mode.

existing products. Instead, a quest for breakthrough ideas often begins with an open exploration of customer[4] needs—especially latent, undiscovered needs that may be difficult to articulate—also referred to as customer insights.

So how does one identify customer insights that will inspire great ideas? While there are many specific methods, they generally are qualitative in nature and are intended to help the project team become immersed in their customers' context. This is typically described as a process focused on gaining empathy with customers, that is, developing an understanding of their context, experiences, and behaviors (Chapters 3, 4, and 7).

At some point during data collection, the project team needs to begin synthesizing the data that they have collected. This does not mean that their discovery work is complete. Indeed, the Discover mode is built on iteration between data collection and data synthesis, where data synthesis is the process of summarizing and deriving meaning from the data. Given the qualitative nature of the data (i.e., pictures, transcripts, audio recordings, etc.), the data synthesis process is very different than what is typically assumed with market research. Rather than relying on numerical data and statistics, the team needs to be able to translate qualitative data into specific customer insights. There is a variety of ways to do this, including coding transcripts, drafting personas and empathy maps of archetypical customers, and journey maps that describe the customer's current or ideal experience (Chapters 3 and 4).

Once again, while there are many different techniques available, an important principle of the Discover mode is to continually iterate between data collection and synthesis, that is, to attempt to synthesize insights throughout rather than wait until all data have been collected. This requires flexibility and patience but helps to ensure that the most appropriate methods are used as needed rather than rigidly prescribing exactly how the research will be conducted at the start of a given project. Once the team is confident that they have identified a set of significant customer insights to consider, then they are ready to proceed to the Define mode.

---

[4]For simplicity, I refer to customers throughout. These methods are not, however, limited to designing products for traditional customers. They are equally appropriate for any person or group that uses a product or service, or that is part of product or service creation and delivery, for example, a hospital nurse.

Define

The Discover mode can be characterized by the development of an expanded understanding of the customer—their thoughts, feelings, experiences, and needs. In contrast, the Define mode (see Figure 1.3) is characterized by a distillation of customer insights and framing of specific insights as well-defined problems to solve. At this point in the process, the team should have an inventory of synthesized information about their customers and their contexts. The challenge is to identify the needs and insights most worthy of pursuit through the next phase of the process. Toward that end, these needs and insights are often framed as discrete "problem statements" to use in the next phase as a basis for idea generation, the initial activity within the Create mode. These problem statements generally are short statements that describe the customer type, an unaddressed need, and the insight that explains why the identified need is especially worthy of addressing. For example:

- A busy parent of teenagers (customer type) …
- … needs a way to reconcile and integrate the dynamic schedules of all members of the family (the need) …
- … because the lack of reliable, up-to-date information about conflicting schedules is leading to missed activities and unnecessary stress (the insight that clearly explains why the need is worth addressing).

Next, the team needs to converge on a subset of these problem statements to address in the next mode: Create. Multivoting is one of the skills that is most useful at this point. While there are different ways to vote for ideas, or problem statements in this case, the intent is to take advantage of the evolving wisdom of the group that has collectively benefited from participation in the Discover mode (which depends on consistent team membership throughout the project).

Create

The purpose of the Create mode of design thinking (see Figure 1.4) is to develop a concept or set of concepts that can be shared with the target market for feedback and that, through iteration, can be improved upon. While customers can respond to an idea on its own, the best feedback will result from their engaging with a rough prototype of a

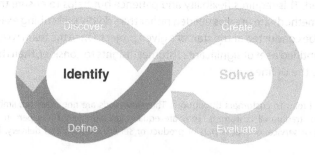

Figure 1.3: Define mode.

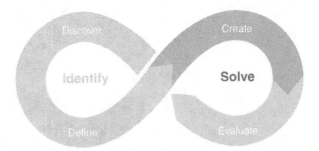

Figure 1.4: Create mode.

concept since a good prototype can provide an experience to respond to and another opportunity for designers to observe actual behaviors. Thus, the two primary activities of the Create mode are idea generation and prototyping. Although these will be described in sequence, in practice they are, once again, highly iterative in nature.

The first major activity within the Create mode is idea generation. There is a wide variety of tools and techniques available to do this (Chapters 5 and 6). Next, after grouping and refining ideas, the team can again use some form of multivoting to converge on the most promising ideas. At this point, it is appropriate to consider a broad set of criteria. A simple schema to consider might include (a) desirability (from the customer's perspective), (b) feasibility (the ability to deliver the product), and (c) viability (the ability for sustained business benefit, either financial or strategic). It is important, however, to remain focused on the identified customer insights and to avoid filtering ideas too much based on other criteria, since the idea is still nascent at this point and can be improved upon during the next activity, prototyping.

When practitioners of design thinking talk about prototypes, they are not referring to the camera-ready or fully functioning prototypes that appear in the popular press. Rather, they are referring to simple prototypes that provide a very basic experience of a product or feature of a product (Chapter 7). These are often referred to as "low-resolution prototypes" (d.school, 2014). These early-stage prototypes can be three-dimensional objects, a sequence of screen shots of a "software app" concept, or even a mocked-up service counter with actors as agents. One of the unique features of design thinking is that prototyping is used as another activity for exploring an idea—to accelerate and improve idea generation by considering different manifestations of the concept. Thus, a series of prototypes might be developed within the group before one or more are chosen to present to prospective customers for feedback.

Evaluate

The final mode of the design thinking framework is Evaluate, as shown in Figure 1.5. The purpose of this mode is to get feedback on concept prototypes, and the ideas and assumptions embedded within them. Within the design thinking framework, we typically assume that much of this feedback will be used to iterate and improve upon the concepts, especially in the first iteration of the four modes. In other words, this is not

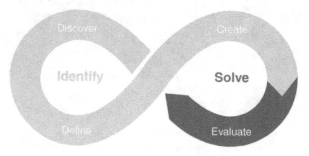

Figure 1.5: Evaluate mode.

the "final step." This will be elaborated on subsequently, but for now it is important to appreciate that the purpose of the feedback is initially as a mechanism to learn more rather than merely to validate.

There are typically two types of activities conducted with this mode. The first is to share prototypes with potential customers to gain feedback. To get the most valuable feedback, the prototype should be used to help simulate an experience for the user rather than serve as a prop for presentation. After the team has collected sufficient feedback, they proceed with a process of synthesizing the feedback. This activity is similar in spirit to the data synthesis completed during the Discover mode, with the obvious difference being that users now have a tangible solution concept to respond to. The objective is, however, quite similar: to gain further insight in addition to converging on the most promising solution or elements of a solution. Depending on the synthesis of the feedback, the team then decides where to go next in the design thinking framework. The ultimate objective, of course, is to move beyond concept prototyping to full development of the product or service. The assumption within design thinking, however, is that this is likely to occur only after multiple iterations of one or more of the modes of design thinking, to which we turn next.

1.3 Design Thinking as a Nonlinear Process

Thus far, I have presented the modes of design thinking as a linear sequence of activities. This is likely the easiest way to learn about these modes and, in practice, the first iteration of these modes will typically proceed as they have been described above: Identify (Discover and Define) and then Solve (Create and Evaluate). However, design thinking is not intended to be a linear process, nor would that be desirable in most situations. Instead, the design thinking approach is to create potential solutions as quickly as possible—knowing that our knowledge is incomplete and that these solutions will be incomplete and potentially flawed—and then use these initial solutions as a means of learning more, of developing more refined insights, and creating better solutions.

Thus, design thinking is best understood as an iterative approach to problem solving, rather than as a sequence of steps—hence the use of the term *mode* as opposed to *step*. The number of iterations depends on the project and is, to a certain degree, unknowable at the initiation of a project. This is a judgment call, based on the objectives and constraints of the project, as well as the perceived progress of the work. Deciding how to proceed on a given project is one of the key tasks of the team and its leader throughout the project. This includes deciding when to shift to a different mode and when, ultimately, to move beyond concept evaluation within the design thinking framework into a more traditional, linear product development process once the concept has been sufficiently described and evaluated.

At first glance, the lack of predefinition of which modes to use at a given point in the project, to what degree, and in what order may seem unnecessarily complex and at odds with the logic and efficiency of traditional Stage-Gate development processes. In the right situations, however, this approach provides important flexibility, increasing the odds of arriving at great solutions, while minimizing wasted time and effort. This requires a fundamental shift in mindset, a point to which we turn to next.

1.4 The Principles and the "Mindset" of Design Thinking

At this point, it is likely evident to the reader that design thinking is as much about a way of thinking and doing as it is about process. Process is clearly important, and there are specific, tested tools to consider within each mode, each with its own set of inputs, outputs, and well-defined activities. Beyond process, design thinking is also about mindset, where mindset can be thought of as an integrated set of beliefs and attitudes.

Several chapters in this book will address the mindset and principles of design thinking, as well as the implementation of design thinking in the firm (Chapters 8–14). However, I share below some common themes that can serve as an initial primer. Becoming familiar with these should enable a flexible approach to exploring the wide variety of topics addressed throughout the remaining 24 chapters of this book, collectively illustrated in Figure 1.6. In that spirit, some common principles of the design thinking mindset and philosophy include the following:

People-centric: A shift from a product and technology-centric orientation to a primary focus on the values, experiences, and needs of people; although products and technologies are clearly critical to ultimately addressing customer needs, they are viewed as enablers of solutions that follow from customers' needs.

Cross-disciplinary and collaborative: Using teams with a wide variety of backgrounds and training, and with team members that are open to the different perspectives and abilities of a diverse team. While team membership should be relatively consistent throughout the project, it may be wise to occasionally include participants external to the organization—such as customers, suppliers, and other subject matter experts—for select modes or activities.

Holistic and integrative: Although details are important, design thinkers are also able to see and consider relationships, interactions, and the connections between seemingly disparate ideas.

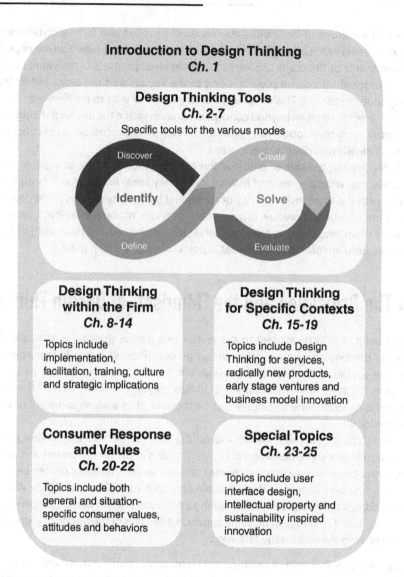

Figure 1.6: A brief guide to the chapters in this book.

Flexibility and comfort with ambiguity: Design thinking is best suited to addressing ambiguously defined problems and opportunities, and requires great flexibility with respect to both content and approach (e.g., through as-needed iteration of modes and phases).

Multimodal communication skills: A willingness to communicate and work in various modalities, including verbal, visual, and tactile. Design thinkers sketch and create prototypes, without being constrained by a perceived lack of ability or skill. And, last but not least ...

Growth mindset: A willingness to test ideas, concepts, and prototypes in an effort to learn, unhindered by a fear of failure.

References

d.school (2014). Prototype to test. Retrieved from https://dschool.stanford.edu/
Sims, P. (2013). *Little bets: How breakthrough ideas emerge from small discoveries.* New York: Simon & Schuster.

About the Author

DR. MICHAEL G. LUCHS is an Associate Professor and is the Founding Director of the Innovation and Design Studio at the College of William & Mary's Raymond A. Mason School of Business. He earned his PhD from the University of Texas at Austin in 2008. Prior to earning his PhD, Dr. Luchs worked for over a decade as a consultant and executive. As a principal with the consulting firm Pittiglio, Rabin, Todd & McGrath (now PwC), Dr. Luchs worked with a broad client base to improve their product development and marketing practices and performance. Clients included Fortune 500 companies as well as small and midsized businesses in the consumer packaged goods, consumer durables, computing and telecommunications equipment, and telecommunications services industries. In addition to his prior and ongoing consulting experience, Dr. Luchs worked in industry as the Senior VP of Marketing for Labtec Inc. (now Logitech), and as a Product Manager for Black & Decker Power Tools.

Part I
DESIGN THINKING TOOLS

2

INSPIRATIONAL DESIGN BRIEFING

Søren Petersen
Ingomar&ingomar-consulting

Jaewoo Joo
Kookmin University

Introduction

A *design brief* is a short document, usually 2 to 20 pages in length, that relays issues of "who, what, when, how, and why" to the design team (Petersen & Phillips, 2011). As a written explanation of the aims and objectives of a project, the design brief represents the desired outcome by relaying requests from management to design teams. A well-written design brief enables designers to understand their clients and to communicate with other designers in a team fluently, eventually helping them to develop concepts. As concept development reflects only 5 percent of development costs, yet influences 70 percent of the final product's cost (Andreasen & Hein, 2000), using a design brief to translate management criteria into measurable and actionable design concepts is critical.

Although the design brief plays an important role in concept development, there are few resources about how to write one. In general, the design brief is viewed as a competitive advantage and traditionally is guarded as a business secret. Research on writing a design brief is scant, and prescriptions for how to organize documents are heavily based on individual consultants' experiences. As such, most design briefs are the writer's interpretation of a request for proposals (RFP) or merely a reformulation of an existing business plan (Petersen, 2011).

Design Brief and Wikipedia

When asked to write a design brief, designers often consult Wikipedia. Wikipedia illustrates six elements of design briefs: company history, company profile, problem statement, goals, solution analysis, and synopsis. Unfortunately, these basic elements provided no insight into how to write a high-quality design brief.

The responsibility for writing a design brief is usually relegated to one department, and there is little or no cross-departmental collaboration. At the Industrial Design Society of America event in 2012, for example, design students and professional designers alike voiced their concerns about the design briefs they had seen. The design briefs written by engineering departments contained too much information and were overly restrictive, whereas the design briefs written by marketing departments contained too little information and did not inspire designers. Therefore, many designers read a design brief when a project is started and rarely revisit it afterward.

2.1 Nine Criteria of an Inspirational Design Brief

To begin, we consider how industrial designers work. Designers are inspired by a wide variety of sources, including nature, fashion, movies, automobiles, aviation, weapons, architecture, and cutting-edge technology. Although some sources may not apply to a specific project, they may help designers formulate a new concept at a later point. Along these lines, we define an inspirational design brief as not only a guide to follow but also a mind-set to help designers leverage constraints in ideation. More specifically, we examined a wide variety of applications submitted to worldwide design awards and identified nine common criteria. We categorized them into three groups—strategy, context, and performance—and introduced them as nine design quality criteria (DQC).

A. Strategy
   1. *Philosophy:* History, values, belief, vision, mission, and strategy of a company
   2. *Structure:* Domain, business model, and competitive advantage of a company
   3. *Innovation:* Area and type of innovation of a company
B. Context
   4. *Social/human:* Needs and activities about individual and/or group of consumers
   5. *Environment:* Requirements of and expectations for environmental concerns
   6. *Viability:* Expectations about economic performance
C. Performance
   7. *Process:* Budget and schedule of a project
   8. *Function:* Nature of deliverables including unique selling point
   9. *Expression:* Sensory styling and aesthetics of products

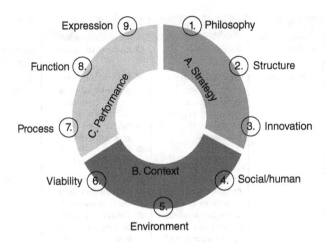

Figure 2.1: The nine criteria of an inspirational design brief.

More detailed explanations, questions to answers, and the conventional metrics of each criterion are provided in Figure 2.1 and Table 2.1.

Example of an Inspirational Design Brief in Product Design

Writing a design brief for an innovative product design project is an art as well as a science. Successful design brief writers elaborate their projects in detail using the DQC while keeping their final documents to a manageable length. Here, we introduce an example of an inspirational design brief for developing an innovative storage system project submitted to LEGO:

1. *Philosophy:* The name LEGO is an abbreviation of the two Danish words, *leg godt,* meaning "play well." The ultimate purpose of LEGO is to inspire and develop children to think creatively, reason systematically, and release their potential to share their own future—experiencing the endless human possibility. The LEGO toys have become a staple in the homes of creative families. The imagination of a child is what LEGO emphasizes.
2. *Structure:* The LEGO Group is owned by the founding family and its ownership is handled by KIRKBI, the investment company, and the LEGO Foundation. KIRKBI not only owns 75 percent of the LEGO Group but also owns 38 percent of the Merlin Entertainments Group who runs the LEGOLAND theme parks. The LEGO Foundations holds the remaining 25 percent of the Group. LEGO is one of the largest toy manufacturers in the world.
3. *Innovation:* In 2004, LEGO (a) listened to consumers, (b) utilized new technologies, and (c) refocused its business to successfully save it from a steady decline in sales. After listening to consumers, LEGO recognized that consumers consistently bought the sets having a story with a good character and an evil one, suggesting that good-bad conflict appeals. LEGO also continuously adjusted to new technologies to cut the development process from two years to one year.

Table 2.1: The Nine Criteria of an Inspirational Design Brief

Group	Criteria	Explanation	Questions to Answer	Conventional Metrics
A. Strategy	1. Philosophy	Design contributes by formulating, visualizing, and communicating the organization's philosophy	■ What is the history of the company as well as its values, beliefs, vision, mission, and strategic intent? ■ How is the brand communicated?	■ Achievement of strategic goal
	2. Structure	Design provides design-related knowledge to the Strength-Weakness-Opportunity-Threat (SWOT) portion of the Five-Forces analysis	■ In which business and category does the firm operate? ■ What is the firm's business model and how is it vertically and horizontally integrated? ■ What are its competitive advantages?	Not identified
	3. Innovation	Design co-creates innovative concepts, visualizes, and communicates innovation opportunities	■ What is the innovation area of the business (i.e., technology, finance, process, offering, or delivery)? ■ Is the innovation type breakthrough or incremental? ■ What is the organization's level of ambition?	■ Research and development budget ■ Number of patents, copyrights, and trademarks, cps ■ Percentage of revenues of new products
B. Context	4. Social/human	Design participates in user studies, tests conceptual ideas, and communicates findings	■ What are the users and other stakeholder's cultural connection, identity, needs, behavior, and activities?	■ Satisfaction (with product) ■ Satisfaction (with ease of use) ■ Employee satisfaction
	5. Environment	Design contributes to environment by exploring eco-friendly opportunities	■ What are the requirements to meet the environmental concerns?	Not identified
	6. Viability	Design provides design-related knowledge for the development of business models, including positioning, value creation, and cost reduction	■ What are the expectations regarding market share, earnings per share, and return on investment as related to the time horizon?	■ Revenue/sales ■ Market share ■ Net income/profit ■ Percentage of sales (new customers) ■ Percentage of sales (repeat customers)

Group	Criteria	Explanation	Questions to Answer	Conventional Metrics
Table 2.1: (*continued*)				
C. Performance	7. Process	Design co-creates the design brief, synthesizes concepts, refines them, and provides support in their subsequent development	■ What are the project's budget, schedule, and deliverables? ■ How are these aligned and coordinated with other projects?	■ Time to market ■ Number of design modifications ■ Cycle time with phase ■ Number of products completed
	8. Function	Design participates in integrating the provider and user aspects into functions and features		Not identified
	9. Expression	Design translates provider and user aspects into attributes, form, features, proportion, surface, and details; design creates a cohesive statement supported by a compelling story	■ What are the brand's attributes, design language, and design principles (i.e., proportion, surface, and details)?	Not identified

It designed products according to feedback and recognized failure early in the production cycle, solidifying its integrity. Finally, it stripped down from a wide variety of businesses including clothing, theme parks, and video games to a core brick business.

4. *Social/human:* Children assemble blocks randomly when they are young. As they grow, their projects become more complex, until they eventually incorporate stories as well as engineering and aesthetic components. People constantly push the boundaries of what is possible with LEGO with others, being adult LEGO fanatics. Therefore, kids and their parents are their main markets as LEGO bricks evolve with them.

5. *Environment:* LEGO bricks and storages are sold in boxes. We should consider reducing the size of the box to reduce the consumption of cardboard coming from sustainable forests.

6. *Viability:* In order to maximize the return on investment of the steadily growing LEGO, we should consider material choice, ease of disposal/recycling, safety standards (both American and European), and feasibility.

7. *Process:* We should present artworks using a given PowerPoint template with a maximum of 12 slides and 5 MB. We can submit a video to go along with the presentation: maximum length of 3 minutes; maximum size of 50 MB; and allowed file types are mp4, avi, flv, mpg, swf, and wmv.

8. *Function:* We should explore a different concept that can potentially replace the current Bricks & More storage boxes while keeping the following requirements. It needs to convince parents and gift givers of delivering great functionality and permanent storage in store and at home, suggesting that it survives a child's play life. It must also be feasible; the project must show how we produce and integrate it into LEGO's current product line.

9. *Expression:* We should clearly communicate the ideas of LEGO such as imagination, creativity, fun, and learning. Specific expression languages of the concept must follow; its form is geometric and static, edges are rounded, and its primary colors are bright.

Example of an Inspirational Design Brief in a Research Project

Our suggested nine DQC are versatile and can be applied to a very different type of project, such as when business decision makers approach a conventional challenge in a more innovative fashion. Traditionally, they made decisions by considering the analyses and suggestions made by internal researchers and external economists. However, these inputs often stem from a worldview based on outdated assumptions and fail to nudge decision makers to see an issue in a fresh perspective.

Take an example of sustainability. According to "A New Era of Sustainability" (Lacy, Cooper, Hayward, & Neuberger, 2010), a report released by the United Nations Global Compact and Accenture, " ... while the belief in the strategic importance of sustainability issues is widespread among CEOs, executives continue to struggle to approach them as part and parcel of [their] core business strategy." As a result, sustainability considerations often end up coming from random, ad hoc, or unrewarded contributions from passionate individuals, and seldom from strategically informed corporate policies. Although bottom-up processes are imperative for corporate culture to shift toward a more sustainable path, top-down initiatives are more influential in achieving significant change. Here, we introduced an example of inspirational design brief for proposing a research project submitted to corporate leaders. It aims to help them to reflect on the progress to date, the challenges ahead, and the impact of the journey toward a sustainable economy.

1. *Philosophy:* The underlying assumption of business is that growth is good. However, in the new market where the cost for food and energy increases, demographics change, and populations grow, the assumption that growth is good must be challenged. In order to explore new business models, new legal frameworks, and new economic systems that prosper in the contemporary market, the question we ask is, "How can we shift the current paradigm of corporate thought leadership into one that values innovative thinking for a sustainable future?"

2. *Structure:* The proposed project will consist of an autonomous team that makes decisions with the support of an expert advisory group. The outputs of the project will include an open research and thought leadership process, a collaborative content piece that looks at sustainable business practices, and a diverse community of co-authors.

3. *Innovation:* The proposed project will test open innovation techniques, such as crowdsourcing and crowd funding, in the context of corporate thought leadership research and development. The purpose of the project is to challenge and improve the current research paradigm of corporate thought leadership such that it invites more diverse thinking and problem-solving approaches.

4. *Social/human:* The proposed project aims to use social networks such as LinkedIn, Facebook, and Twitter as open research platforms from which we draw questions, ideas, and insights about sustainable business. We will tap into ongoing conversations, forums, and discussion boards from diverse communities of interest. The communities include design, science, technology, agriculture, health, education, and transportation.

5. *Environment:* Business leaders and scholars tackle the world's most pressing issues such as climate, poverty, inequality, and population using the existing forums such as United Nations Global Compact, World Economic Forum, and World Council for Sustainable Business. Sustainability needs to be more highlighted.

6. *Viability:* The proposed project will be independently funded through corporate foundations, government organizations, and academic institutions.

7. *Process:* We will establish best practices by conducting consulting projects with in-house and external teams, while soliciting ideas using open research platforms. We aim to not just provide solutions, but also to explore the development of dynamic capabilities and address challenges more deeply.

8. *Function:* The proposed project will provide an open research framework that will help to identify underlying assumptions and offer a new research approach for corporate thought leadership. It will convene diverse communities of interest, thereby acting as a catalyst for connection, collaboration, and innovation.

9. *Expression:* All project communications, internal as well as external, will reflect the values and intentions of the project. The values of integrity, community, and openness will be honored throughout the process, which will be reflected in the final deliverable.

2.2 Writing the Inspirational Design Brief

The optimal approach to writing an inspirational design brief is through co-creation. Studies show that the act of writing a design brief improves the quality of concepts by 20 percent on average and 25 percent for top-performing designers. Writing a design brief also changes research behavior; when novice designers invest time in writing a high-quality design brief, they conduct research for a longer period of time as well as identify more impactful opportunities for ideation. Moreover, writing a design brief collaboratively reduces team members' perception of ambiguity while increasing their willingness to take risks in the subsequent concept exploration phase (Petersen & Ryu, 2015). Therefore, joint development of a design brief and treating brief writing as an important phase has the potential to add value to a project, curb risk, and increase creativity. Co-creating an inspirational design brief consists of the following three steps, as illustrated in Figure 2.2.

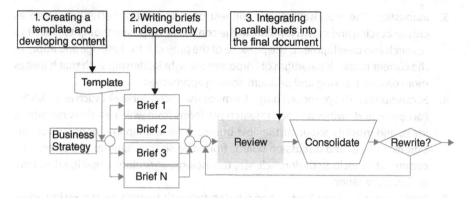

Figure 2.2: The three steps of co-creating an inspirational design brief.

1. Creating a Template and Developing Content

Prior to creating an inspirational design brief, team members on the project (e.g., designers, marketers, and engineers) usually have little or no systematic documented information about the previous projects including their design briefs and their outcomes. To remedy this, they are provided with the DQC as a generic framework to organize previous information under the nine criteria as well as a general guideline for good balance of the DQC content. This assists brief writers to consider the whole aspects of the project, increase their emotional investment, and mentally prepare to address each issue in the later phase.

2. Writing Briefs Independently

Each member writes a 500- to 1,000-word brief independently using the structure of the inspirational design brief template, aided by the content from previous projects. Doing so helps each member empathize with other members by formulating other functions' contributions clearly. Following the sequential process of the DQC, moving from philosophy to expression, supports the creation of the logical top-down architecture for the design brief. This facilitates building cohesive and comprehensive design requirements while assisting the individual members in seeing the project broadly as well as understanding the interdependencies between the criteria.

3. Integrating Parallel Briefs into the Final Document

Team members collaboratively review multiple briefs by considering the final performance of each design brief. When the final performance data is unavailable, they may rely on the quality or quantity of insights obtained from each design brief. Then, they consolidate multiple design briefs into a well-balanced and more effective design brief. As the team gains experiences, performance evaluations can be updated accordingly.

Time for Writing a Design Brief

Engineers at NASA's Jet Propulsion Laboratory (JPL) in Pasadena, California, recommend investing 25 percent of a budget on writing a design brief, while design consultancies generally spend in the neighborhood of 10 to 15 percent. Even if one does not design a space mission, dedicating significant resources (comparable to a 15 percent cost of managing the subsequent project) makes sound business sense.

2.3 Research Findings about Inspirational Design Briefs

Going over budget is a serious issue for product developers. They can avoid this issue by carefully examining the amount of the content allocated for the two design quality criteria, *process* (how to make a product) and *expression* (how a product looks and feels), in a design brief. Petersen, Steinert, and Beckman (2011) reviewed 81 briefs including 51 briefs from the projects performed at Stanford University and 30 briefs from the projects performed at several companies. Their collected briefs covered a wide variety of fields, including automotive, consumer products, health care, construction, and aviation. Projects ranged in complexity from shavers to earthmoving equipment, and in size from cell phones to aircraft interiors.

Interestingly, the authors discovered that the amount of the content for *process* is negatively correlated with the amount of the content for *expression* (see Figure 2.3). This suggests that the less information a brief contains regarding the outcome of the project (*expression*), the more information it requires to describe how the project runs (*process*). Indeed, one group of automotive product developers who distributed the amount of the content for the two criteria in a more balanced way ran into fewer

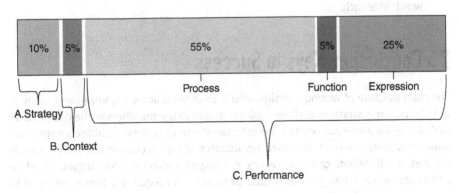

Figure 2.3: A suggested distribution of the DQC content in a design brief.

problems. However, the other groups of product developers who wrote too little about *expression* in their briefs lost control of their projects, leading to budget overruns and project failure.

2.4 Three Pitfalls to Avoid

We suggest that brief writers avoid the following three pitfalls when writing briefs:

1. *Content distribution:* First, brief writers often undervalue the importance of communicating strategy (*philosophy, structure, and innovation*) with their team members. It is an outdated belief that design is an afterthought and should not be integrated with the rest of the business. We suggest that the content regarding strategy should occupy at least 10 percent of the design brief. Second, some brief writers intentionally hide the complete information about the expected specifications (*function*) or the expected shape (*expression*) of the final product in order to encourage the blue-sky thinking of their team members. However, team members can only benefit from possessing the full information available. We suggest that the content about *function* occupies at least 5 percent and that the content about *expression* occupies at least 25 percent in the design brief.

2. *Balancing between **process** and **expression:*** As illustrated earlier, balancing the amount of the content between the two criteria determines the success of project. When brief writers include too much content about *process* (>55 percent), they may neglect the other criteria, potentially hurting the quality of the project outcome. When they include too little content about *expression* (<25 percent), they may ask team members to explore extensively, which results in a high risk of going over budget.

3. *Length:* Brief writers should benchmark the number of words used for their design briefs. The length indicates an aircraft interior, when in fact the product intended for development was a shower stall. Most effective briefs are usually 500 to 1,500 words in length.

2.5 Conclusion: Keys to Success

The main purpose of writing a design brief is to communicate organizational capabilities, the business strategy, and the business model to the members of the design team so that they are well equipped to synthesize novel, useful, and marketable concepts. Creative ideas come from well-informed individuals and teams. Leaving the design team in the dark is self-defeating; it only results in a negative effect on the budget, schedule, and outcome. Each design brief should be unique and requires a concerted effort to create. Recycling old briefs, with minor updates and modifications, does not lead to an innovative concept. Team members recognize "the same old briefs" on their desks and pay no attention to them.

In this chapter, we introduce the inspirational design brief as an answer for designers seeking to improve the current situation of misaligned business opportunities and design execution. We introduced its nine criteria called design quality criteria (DQC) and illustrated two examples, one for a product design project and the other for research project. Then, we illustrated the three steps of how to write design briefs in a co-creative fashion. We also provided research findings and clarified three pitfalls. We believe our proposed brief-writing method provides a unique opportunity for product developers in various industries to facilitate communication between their business managers and designers so that they can successfully leverage design in their new product development projects.

References

Andreasen, M. M., & Lars Hein. (2000). *Integrated product development*. Technical University of Denmark.

Lacy, P., Cooper, T., Hayward, R., & Neuberger, L. (2010). A new era of sustainability: UN global compact–Accenture CEO study 2010. (pp. 1–56). Accenture Institute for High Performance: Chicago, IL.

Petersen, S. (2011). *Profit from design*. South Pasadena, CA: Ingomar & Ingomar.

Petersen, S., & Ryu, B. H. (2015, April). Strategic comprehension driving risk-taking & design performance. 11th European Academy of Design Conference, Paris, France.

Petersen, S., & Phillips, P. (2011). Inspiring design—informed by metrics. *Design Management Review* 22(2), 62–71.

Petersen, S., Steinert, M., & Beckman, S. (2011, August). *Design driven portfolio management*. International Conference of Engineering Design, Copenhagen, Denmark.

About the Author

DR. SØREN PETERSEN is an international business consultant and design science researcher, author, and a regular contributor to *The Huffington Post* (The Creative Economy: www.huffingtonpost.com/soren-petersen). Throughout his 20-year career, he has worked with many top-tier international organizations, including Rambøll Group, BMW Group, Stanford University, Copenhagen Business School, and Hanyang University. He received his PhD from Stanford University in ME Design Research, MS ME from the Technical University of Denmark, and his BS in Transportation Design from Art Center College of Design. Over the past six years, he has published 24 scientific papers and over 150 articles on the Creative Economy, as well as authoring *Profit from Design*, a definitive book on design quantification. His areas of research include developing methods and metrics for bridging business and design. These include:

Design Driven Start-ups, Design & Business Model Experimentation, Design-Driven Portfolio Management, Gamification in Concept Design, and Crowdsourcing Design Research.

JAEWOO JOO is an Assistant Professor of Marketing at Kookmin University. He holds a PhD in Marketing from Rotman School of Management, University of Toronto. Jaewoo teaches and writes about design marketing and new product development through the lens of behavioral decision theory. He has served as a panelist for the *Business Week*'s World's Best Design Schools.

3

PERSONAS: POWERFUL TOOL FOR DESIGNERS

Robert Chen
LG Electronics

Jeanny Liu
University of La Verne

Introduction

During the past decade, personas in product design have received much attention from academicians and innovative companies (Blomquist & Arvola, 2002; Chapman & Milham, 2006; Faily & Flechais, 2011), an interest that is part of a positive trend toward building user-centric products. Personas provide designers with a user-centric reference tool that depicts an ideal user (Cooper, Reimann, & Cronin, 2014). This tool allows designers to maintain focus on the ideal user as they explore and develop solutions. This chapter explores personas as a practical tool for design. Many of the examples are software and technology oriented, but personas apply similarly to other product categories. We organize this chapter into several sections:

1. Defining personas—practical descriptions, underlying bases, and common types
2. Importance of personas—exploring the power of using personas during product development:
   a. During design
   b. During development
   c. As a communication tool
3. Creating personas—an overview of creating personas from ethnographic research
4. Illustrative application of personas—an example from three areas during product development
5. Limitations of personas—constraints and mitigation

In this chapter, we define designers broadly as cross-functional members of a team tasked with developing user-facing product solutions. We define user-facing solutions as product features with which users interact and experience. Personas are particularly useful for designers who work on user-facing solutions.

3.1 Defining Personas

Personas are a representation of ideal or prototypical end users, based on behaviors and motivations of real people (Cooper, Reimann, & Cronin, 2014). Personas represent clusters of users from research and are not derived from stereotypical assumptions (Cooper, Reimann, Cronin, & Noessel, 2014). Personas allow designers to relate to and empathize with users, and encourage them to view product problems from a user's perspective. Personas are created at the beginning of the design process. As representations of users, personas define both the target user and the problem for a design team. Minor iterations can be made to personas, but major revisions reset design to the beginning.

Two nonuser personas are often considered during design: the buyer persona (Scott, 2013) and the anti-persona (Cooper et al., 2014). In this context, we refer to buyer personas to signify those who make purchase decisions but do not necessarily use the product. For example, when an airline purchases a plane, pilots represent one user persona and passengers another. Pilots might be consulted during purchasing, but the buyer is usually a business decision maker at the airline—another persona. The buyer persona has disparate considerations for the purchase of an airplane. For example, a buyer considers financing, passenger capacity, maintenance costs, flying range, and fuel economy. Another example of differences between buyers and users can be found in children's products. Parents are buyers, concerned with child safety, purchase cost, and the ability to return a product. Modeling a buyer ensures that a product solves their concerns. Depending on the product, buyer and user personas can be the same person or different people.

Anti-personas illustrate actors who are not the intended users of a product (Cooper et al., 2014). Consumer product designers often create both user and anti-personas to differentiate targeted users from others. For example, a high-end, digital, single-lens reflex (SLR) camera targets expert users and photographers; the expert consumer is the user persona and the anti-persona is the casual consumer, focusing designers on designing a product for an expert. Labeling, memory storage, carrying cases, user manuals, and so on are designed for the expert consumer. Any infrequent or edge cases for the user persona that are common use cases for the anti-persona are ignored. Edge cases are experiences that influence some or all users, but occur infrequently (Cooper et al., 2014). One example of how anti-personas are used for a camera product asks, "What if the user does not understand the basics of operating a camera?" The issue is an edge case for the user persona but is a common-use case for the anti-persona. Since this applies to the anti-persona, design would ignore this issue.

3.2 The Importance of Personas

Personas during Design

Personas form the basis of problem definition for a designer; they define users and set parameters for design solutions, keeping designers from falling into a common design pitfall: designing for oneself (Cooper et al., 2014). Consciously or not, designers often infer and assume about users based on work experience and industry knowledge. Consequently, personas can be useful to avoid self-referencing, frame design problems from a user's perspective, and focus designers. Design teams often use brainstorming and storyboarding as tools for generating and exploring ideas. Brainstorming is the freeform generation of ideas, with minimal constraints or thought to feasibility. During ideation or concept phases, brainstorming facilitates conversation. Combined with appropriate personas, brainstorming allows designers to engage and express ideas for subsequent reflection. Storyboarding is a second example of when personas combine with another design tool during design. A storyboard is the visual telling of the story. Designers often storyboard ideas early during a concept phase to visualize either a problem or a solution, and sometimes both. The storyboard's protagonist is the persona, allowing a design team to form deeper empathy for users.

Personas during Development

During development, personas get an engineering team up to speed quickly. A clearly defined persona makes it easier for designers and engineers to achieve a common understanding about a user and the scope of a solution. It is critical that engineers understand the target persona so they can make the right decisions and trade-offs. For a flexible, iterative process, it is impossible and inadvisable to document every detail during design. To compensate, personas provide contextual understanding to an engineering team so it can interpret design documents.

Another benefit of personas is managing edge-case discussions between engineering and design (Cooper et al., 2014). A common design challenge is determining whether an edge case is important. Personas provide a reference point from which communication can be more efficient between designers and engineers. If an edge case is important to the persona, it should be part of the design. For example, the cockpit of a commercial airplane is designed for highly skilled and experienced pilots and crews. The cabin crew and passengers are not expected to be able to operate the controls in the cockpit. Consequently, a designer's persona is the pilot. The edge case of what happens when all capable pilots are unavailable is not a viable use case.

For a typical smartphone app, one edge case asks, "What happens if the user does not have an Internet connection?" The answer depends on personas defined for the app. Internet browsers on smartphones offer limited functionality without an Internet connection because designers determined that their personas understand how browsers behave without an Internet connection. Personas are useful to a development team so engineering can understand the scope of its work, and the quality assurance (QA) team will not waste time testing irrelevant edge cases.

Personas as a Communication Tool

Personas are useful when it comes to communicating with other business functions such as marketing, management, and sales. Personas provide a clear definition of a target market and assist a marketing team with aligning a product from inception through promotion. Buyer personas provide sales and marketing a method of collaborating with the design team. Pitching a product concept to executive managers in a corporation or potential investors (e.g., in a start-up company) involves communicating abstract, contextual information. Personas help decision makers understand a problem from a user's perspective and provide a context for evaluating the product concept. Therefore, personas are useful for obtaining corporate support or financial investment for a start-up.

3.3 Creating Personas

When creating personas, the first step is to identify and select a group of users to research. Choosing users who belong to an appropriate market segment is key to yielding useful insights, often requiring a product manager to possess intimate knowledge of a market and various market segments in the industry. In practice, product managers often rely on secondary research and internal records or conduct a small-scale study to define various personas.

The next step is to collect data. Ideally, personas are created by clustering or consolidating real-life people and experiences from primary research that includes ethnographic studies and user interviews (Cooper et al., 2014). Ethnographic research is the deep, qualitative study of users in the context of their environment when using a product. There are various methods for collecting data during an ethnographic study. We often conduct user interviews, conduct observational studies, and (if possible) use video recordings of users using a product and photos of their environment. Interviews uncover user problems and their underlying causes. Interviews help designers understand user motivation and a user's state of mind while using a product. However, user responses alone are unreliable since users are often unaware of their own needs (Rosenthal & Capper, 2006). Mixed methods may explore user needs more fully. Observational studies and video recordings capture users performing tasks, techniques that are effective when conducting efficiency studies. Using these methods, ethnographic research is a reliable source for uncovering behavioral responses and user problems. When capturing a user with video, audio, or photos, researchers must always ask permission from the user before recording and guard the user's privacy.

The third step is consolidating data from the studies and grouping insights based on common user problems. Often, this is done with a broader design team so all designers have the opportunity to learn directly from the researchers. This also offers the advantage of building personas with the designers so they can internalize user models. Researchers typically look for patterns in responses and organize them into clusters, which are then grouped based on common user problems. Researchers sometimes find that users from multiple market segments share similar problems.

Finally, the team examines the notes and merges various clusters to create a series of personas. The team looks for a dominant profile or common demographics within the cluster. The profile becomes the basis for a persona as long as it does not focus on a single, real person. The team also looks for attributes of its subjects that are impacted by the user problems, and build these same attributes into the persona. For example, a busy, active lifestyle might be an important attribute in the cluster of test subjects. The persona built from this cluster must have this same trait. Although we discuss user personas, the same reasoning extends to nonuser personas, which must also be based on data. Various clusters of problems coalesce into personas, and prioritization of these personas determines which represent personas and anti-personas. Buyer personas can be different people from users, and separate ethnographic interviews might be needed to study them.

3.4 Illustrative Application of Personas

This example is based on a software product, though application of personas is the same for other types of products. Although it is common for a design team to work with multiple personas, we demonstrate only two—one user and one anti-persona—for the purpose of expediency.

Product Manager at ACME

The example begins with Anne, an experienced product manager at ACME Tech, a technology company that makes a variety of productivity software for consumers across devices: personal computers, smartphones, and tablets. ACME's business model is to provide products free for users to download, and include advertising from third parties in the products. ACME receives the bulk of its revenue from advertising. Anne works on the company's mobile-apps team, which makes apps for smartphones and tablets. According to ACME's marketing team, there is an opportunity in the marketplace for a better-productivity mobile app. The marketing team sends her an analyst's report that suggests all mobile productivity apps in the market are disappointing and used rarely. Anne is excited and wants to seize this opportunity to launch a new app for smartphone users, with the goal of adding new customers and expanding ACME's customer base.

Stage 1: Creating personas. With her goals firmly in her mind, Anne turns to a colleague on the user research team, and commissions an ethnographic study. Her colleague recruits eight highly productive users in Texas for the study. The researcher recommends that Anne and any interested members of the design team help her with the interviews. Anne thinks this is a great idea and convinces a few designers and engineers to participate as interviewers, observing in pairs during the series of ethnographic studies. After completing the studies, Anne and her cross-functional team review the data under the guidance of the researcher.

After an initial review of results, Anne organizes a working session to cluster information from the interviews and derive personas. All the designers and engineers who participated in the interviews attend Anne's working session. Two dominant profiles emerge: a tech-savvy, self-employed persona and a tech-aware, corporate persona. From the original market-segmentation data, many of the segmentation characteristics were deemed not useful while clustering subjects based on user problems. The dominant attributes that mattered were *willingness to adopt new technology* and *corporate versus noncorporate work background*. Anne found some indication that users who looked for new technology were most dissatisfied with existing productivity solutions. Aware that there were only eight interviews conducted during the study, Anne commissions a survey to validate results. The survey results validate the initial findings. At this point, Anne is ready to build personas. Mindful of how personas can be misused, she decides to name her personas Wilma (the tech-savvy, self-employed) and Fred (the tech-aware, corporate). These personas will be used as a point of reference throughout the project (Figures 3.1 and 3.2).

Stage 2: Method of inquiry based on personas. Anne wants to frame the user problem for the designers so she uses the personas to represent the target users. Anne organizes a series of brainstorming sessions with the cross-functional design team to discover ideas for exploration. Although the design team is familiar with user research, Anne presents her personas and tapes a printout of each to a whiteboard in the conference room used for the brainstorming session. During the session, all participants can be reminded of the intended users of their ideation brainstorm. Anne uses the personas to set the context and ensure that participants focus on generating ideas that will benefit the target user. Personas play a strong role during the brainstorm by getting designers in the mind-set of thinking of personas before generating ideas. This keeps brainstorming focused on users.

Stage 3: Communicating with engineering. Although it is early during the design, Anne and the lead designer review results of the brainstorming session with the lead engineer on her cross-functional team to get his technical feedback. They use the personas to summarize their research and frame ideas generated from the brainstorming session. An issue the lead engineer raises is that corporate users encounter integration issues with corporate security and e-mail. He estimates that solving these issues alone will take up the majority of engineering resources.

Stage 4: Prioritizing personas. Based on the technical feedback, Anne reassesses the personas and decides that Wilma is more important; Fred is demoted to an anti-user persona but is not ignored. Anne knows from the survey data that if she excludes corporate users, she will be leaving out a large pool of potential customers. She plans to go to market focused on the Wilma persona and incorporates the needs of the Fred persona in subsequent versions of the product.

Stage 5. Concept storyboarding. Based on prioritization of the Wilma persona, Anne and her lead designer return to the design team and reset the team's focus. They review brainstorming ideas, and three stand out as being interesting for Wilma.

Figure 3.1: Anne's persona #1, Fred, tech-aware, corporate.

Name: Fred Dallas

Age: 42

Martial Status: Married

Lives: Dallas, TX

Occupation: Fred is a midlevel manager at a business services company. He manages a team of about 30 people and does some sales and brings in customers. He specializes in the quarry industry, so his team and customers are spread across the United States. Dallas airport is very convenient for Fred since he flies frequently to visit his team and customers across the country.

Family Life: Fred grew up and went to college in Dallas. He and his family live in Plano, Texas, an affluent Dallas suburb.

Technology: Fred grew up with technology, but he focuses on the same products he uses daily, many of which are work related. Fred lives on e-mail; he uses it to keep up with his team and customers. He gets the standard smartphone issued by his company's information technology (IT) department and relies on IT to ensure that his e-mail, calendar, and productivity apps work.

Problems: Fred spends half his time on the road, so when he is in town, he spends as much time out of his office as possible, working from home and spending time with family. His wife complains that he spends too much time glued to his smartphone. Fred wishes he could find a better way to stay on top of all of the to-do items from work and family.

Figure 3.2: Anne's persona #2, Wilma, tech-savvy, self-employed.

Name: Wilma Houston

Age: 33

Martial Status: Married

Lives: Houston, TX

Occupation: Wilma works as an independent wedding planner, the perfect job for her because she is a romantic at heart, is highly social, and has a passion for photography.

Family Life: Wilma grew up on the East Coast, and after going to college in Texas, she settled in Houston. Wilma is married and has a two-year-old baby girl. Wilma's husband is an engineer in the energy industry.

Technology: Wilma uses social media as a primary method of finding clients, tracking friends, and blogging about her business and photography. She uses the latest smartphone because she finds it easy to use and has all of her data and contacts at her fingers. Since she owns her own business, Wilma uses all of the latest cloud services for e-mail, contacts, scheduling, and social media.

Problems: Wilma prides herself on being organized. She is able to juggle her busy home life and her business, but she is a victim of her own success; it is getting harder to stay on top of all her vendors, subcontractors, clients, and schedules.

Figures 3.3 through 3.6 comprise a simple story board that Anne uses to describe a use case of how her persona (Wilma) uses the software product.

Figure 3.3: Anne's storyboard reminder action: Scene 1 (Wilma as a phone call with a client, and they agree to schedule an event in the future).

Figure 3.4: Anne's storyboard reminder action: Scene 2 (Based on the phone call, Wilma creates an action of her smartphone using the new application).

Figure 3.5: Anne's storyboard reminder action: Scene 3 (The action appears automatically on Wilma's calendar).

Figure 3.6: Anne's storyboard reminder action: Scene 4 (The action e-mails client automatically about the appointment so the client can confirm).

To understand each product concept better so the team can decide, each concept is storyboarded. Using Wilma as the protagonist, Anne and her design team sketch the three concepts with a storyboard for each. Figures 3.3–3.6 are a storyboard for one of the product concepts: the *reminder action*. The storyboards describe how Wilma will use and benefit from the concept. After each concept is storyboarded, they evaluate all concepts together, choosing the *reminder action* as the focus for their app.

Stage 6: Interpret and communicate design proposals to stakeholders. Anne uses the personas to communicate and pitch the design team's concept to ACME Tech's product committee for approval. During the presentation, there was insufficient time to delve into the details with the committee on the ethnographic user research. Anne uses the personas to get the committee to empathize with the target personas. Anne takes the committee through the problem scenarios using the personas. She uses Wilma as the voice when describing how dissatisfied users are with existing productivity apps. Anne's intent is to get the committee to view potential opportunities and challenges from the target user's perspective. Her presentation goes well, and the committee approves the project.

Stage 7: Working with engineering. With the approval to build the app, Anne is able to add software engineers to her cross-functional team. As part of the induction meeting for the new engineers, Anne introduces them to the basic ethnographic user research that she and the cross-functional team used. Anne then introduces her two personas and gets the engineers to understand Wilma as the user persona and Fred as the anti-persona. She gives every member of the cross-functional team a printed profile of Wilma to post at their desk so they are reminded that they are creating an app for a specific type of user—Wilma.

3.5 Summary

Anne and her cross-functional design team originally created personas after reviewing the ethnographic user research, and she clustered results in a way that made sense for her productivity app idea. She validated findings using a survey, allowing her to create personas that summarized the user research findings. Her personas helped her place a human face on the user research information. Anne created two personas (Figures 3.1 and 3.2) but eventually prioritized Wilma over Fred during design. Throughout development, Anne maintained her persona to ensure that all stakeholders focus on the right user model.

Limitations of Using Personas

Although based on data, personas are dependent on subjective decisions—which market segment to study and which user problems to cluster. Product managers and designers create personas at the start of design to frame a problem for the design team, and, consequently, conducting research with real users is the best way to gain insights into user problems; the purpose is to uncover insights the researchers did not know before studying the user. These insights catalyze product innovation. The danger is creating personas too quickly based on existing knowledge held by team members. One study suggests that some teams struggle to relate to personas (Blomquist & Arvola, 2002). Their difficulties consist of poor communication among team members and a sense of distrust for a primary persona. Their persona was based on presuppositions of system administrators and lacked empirical evidence.

A persona is a model, not a substitute for product testing. User testing with real users must validate design solutions derived from personas. To test that personas are valid, begin with original market segmentation data to recruit participants to test the solutions, and observe whether participants respond to the new product solution. In the case of designing a digital camera, if a single expert user persona represents both professional and expert-amateur photographers, user testing must assess this early during design of prototypes. If both expert and amateur photographers struggle with the prototypes, this might signal that the persona was too broad to represent both. Consider a new set of personas that separate the users and revisit the design solutions. Over time as user testing validates both personas and designs, and as the solutions become specific and more detailed, testing recruitment shifts to using persona profiles instead of market-segmentation information.

Prioritizing personas during design is an extremely important and subjective task. Omitting user personas reduces the usability of the product for those users, and omitting buyer personas creates obstacles to purchase and adoption. However, having too many personas dilutes the value of the product. Documenting the goals of a product at the beginning of the definition phase provides a framework for prioritizing personas.

3.6 Conclusion

The purpose of personas is to provide design and development teams with a representation of a target user so everyone shares a focus to build a user-centric product. Personas emerge at the beginning of design because they are part of the problem definition for designers. Multiple personas emerge often, including users, buyers, and anti-personas, and these personas need to be prioritized for a design team. Prioritized personas are powerful tools during design, development, and communication to business stakeholders.

References

Blomquist, Å., & Arvola, M. (2002, October). *Personas in action: Ethnography in an interaction design team*. Paper presented at the Second Nordic Conference on Human-Computer Interaction, Aarhus, Denmark.

Chapman, C. N., & Milham, R. P. (2006, October). The personas' new clothes: Methodological and practical arguments against a popular method. *Proceedings of the Human Factors and Ergonomics Society Annual Meeting*, 50(5), 634–636.

Cooper, A., Reimann, R., & Cronin, D. (2014). *About face 3: The essentials of interaction design*. Indianapolis, IN: Wiley.

Cooper, A., Reimann, R., Cronin, D., & Noessel, C. (2014). *About face: The essentials of interaction design*. Indianapolis, IN: Wiley.

Faily, S., & Flechais, I. (2011, May). Persona cases: A technique for grounding personas. *Proceedings of the SIGCHI Conference on Human Factors in Computing Systems* (pp. 2267–2270). Vancouver, BC, Canada.

Rosenthal, S. R., & Capper, M. (2006). Ethnographies in the front end: Designing for enhanced customer experiences. *Journal of Product Innovation Management*, 23(3), 215–237.

Scott, D. M. (2013). *The new rules of marketing & PR: How to use social media, online video, mobile applications, blogs, news releases, and viral marketing to reach buyers directly*. Hoboken, NJ: Wiley.

About the Authors

ROBERT CHEN is a Product Manager at LG Electronics Silicon Valley Lab. He earned his bachelor's degree from the University of British Columbia, Vancouver, and an MBA from Indiana University, Bloomington. He is passionate about working with smart people to create innovative and inspiring consumer-software products. His interests include product innovation, design thinking, and creative management.

JEANNY LIU is an Associate Professor at the University of La Verne's College of Business and Public Administration. She earned her PhD in Marketing from the University of Turin, Italy, and an MBA from California State Polytechnic Pomona. Her research interests include marketing strategy, branding, design thinking, and teaching pedagogy. Her research has been published in the *Journal of Marketing, Journal of Organizational Psychology,* and *Journal of Education for Business.* She has received multiple research awards, including the Young Scholar Award by the University of La Verne Academy, and the Drs. Joy and Jack McElwee Excellence in Research Award. Correspondence concerning this chapter should be addressed to Jeanny Liu, Department of Marketing and Law, University of La Verne, La Verne, CA 91750. E-mail: jeanny.liu@laverne.edu

4

CUSTOMER EXPERIENCE MAPPING: THE SPRINGBOARD TO INNOVATIVE SOLUTIONS

Jonathan Bohlmann
North Carolina State University

John McCreery
North Carolina State University

Introduction

The increasing complexities of the competitive marketplace make innovation ever more challenging for companies. Differentiation and innovation beyond the incremental become more difficult as customers become better informed and more demanding and as competitors move more quickly within compressed product life cycles. To meet these challenges, business scholars and practitioners have increasingly called for more focus on the total customer experience, in contrast to more traditional approaches of feature-based product development and innovation. Christensen, Anthony, Berstell, & Nitterhouse (2007), for example, advocate a "job to be done" perspective, whereby product and service development is related to customer motivations (what problem is being addressed?) and the benefits the customer extracts through the product/service experience. Prahalad and Rangaswamy (2003) discuss "next practices" that lay out an experience-based view of product/service design for enhanced innovation.

Consistent with the experience perspective (and likely predating its more recent emphasis in the business press), a deep, empathic, human-centered approach is the critical first step in any design effort. As articulated by Tim Brown (2008), knowledge of "human behavior, needs, and preferences" is what helps "capture unexpected insights and product innovation" that will be more desired by consumers. Brown further claims, consistent with many design and innovation consultancies we have encountered, that most successful innovations "are inspired by a deep understanding of consumers' lives."

The total customer experience is therefore comprehensive or holistic, since consumers derive value from a product or service through their usage experiences within a particular context (e.g., the "job to be done" perspective). The total experience is what customers evaluate, leading to satisfaction, loyalty, and word-of-mouth behavior—all goals for a company creating a product or service. The problem is that traditional new product development (NPD) processes and marketing research often uncover a list of perceived needs or product attributes, but many important aspects of the customer experience, which may help generate new insights, fall through the cracks. In other words, outcomes of an NPD process will be rather different when the focus is on designing a better experience versus merely designing an improved product (e.g., Brown, 2009).

A primary method to understand the total customer experience and integrate it with the NPD innovation process is experience mapping. Sometimes also called journey mapping or an experience blueprint, experience mapping is a method to help understand, synthesize, and form insights about the total customer experience. The goal is to create an experience-based springboard for product design and innovation. Experience mapping is part of many design thinking toolboxes (e.g., Fraser, 2012; Kumar, 2013) and is directly linked to other methods in the design process such as personas, ideation, and stakeholder value exchange. We discuss the essential elements of experience mapping in three parts:

1. Understanding the total customer experience as inputs to the experience map.
2. Making the experience map.
3. Utilizing the experience map as a springboard to developing innovative solutions.

We do not attempt to create a "one size fits all" technique for making the experience map. Instead, we discuss key considerations about inputs and process steps pertinent to any experience mapping endeavor. Importantly, we describe how the experience map can be effectively utilized to envision and design innovative solutions for users.

Whose Experience, and with What?

Throughout this chapter, we generally use the terms *customer* and *user* interchangeably. The user/consumer of a product or service may not be the purchaser or immediate customer, but ultimately the experience the firm maps for improved innovation design is that of the user (whether an individual consumer or a firm's employee), who determines the ultimate value being derived from the product. We also use *product* to refer to a company's entire product/service/brand offering that the user experiences in a usage context. The experience mapping process is useful in both product and service domains.

We utilize a running example of a patient who requires physical therapy services. Imagine someone suffering from an injury or ailment that requires extensive and closely supervised physical therapy for recovery. The physician (MD), a specialized orthopedic surgeon, prescribes a therapy regimen so the patient can reduce pain and regain normal physical function. The physician refers the patient to a physical therapist (PT) to devise and implement a specific treatment plan, which requires both home exercise and regular therapy sessions. The problem is that both the MD and the PT practice in a major city, many miles from the patient's rural home. Fortunately, a small clinic with a resident physical therapy assistant (PTA) is located relatively close to the patient. While the PTA can administer basic elements of the needed physical therapy, the patient still requires regular contact with the PT and the physician to assess the patient's progress, adjust the therapy regimen, and receive more specialized care as needed. This example will illustrate the experience map and how it can be used to design a new and improved physical therapy solution.

As a preview, a somewhat condensed experience map for the physical therapy is shown in the "as is" map (Figure 4.1). The "as is" denotes that the map corresponds to a user experience as it currently exists, before the project team has envisioned a new concept. The map depicts a flow of experience stages that include the patient's initial consultation with the MD in the city hospital, the consultation with the PT supervising the therapy plan, sessions with the PTA at the local clinic, and in-home exercises the patient performs. Periodic visits with the MD and PT at the city hospital assess patient progress. Note that several stages repeat, especially the PTA session and at-home exercising. The map also highlights important information flows and perspectives of the patient.

The Physical Therapy Project

The physical therapy example is derived from a sponsored student project in the Product Innovation Lab, a project-based graduate course for engineers, industrial designers, and MBAs at North Carolina State University. The example and the actual project differ in several details to allow for a concise exposition. In the physical therapy example and throughout the chapter, we refer to the "team," representing the design or NPD team tasked by a company to devise an innovative new product solution.

4.1 Inputs to the Experience Map

The key input to any experience map is, simply, a deep understanding of user experiences with a given product or service in a usage context. Gathering information to serve as inputs in creating the map entails several key considerations. We focus on four main issues:

1. What types of users should be researched to understand their experiences?
2. What methods are commonly used to research user experiences?
3. What are the touch points and key elements of the product or service that define the user's total experience?
4. How can the user research be synthesized to glean important insights?

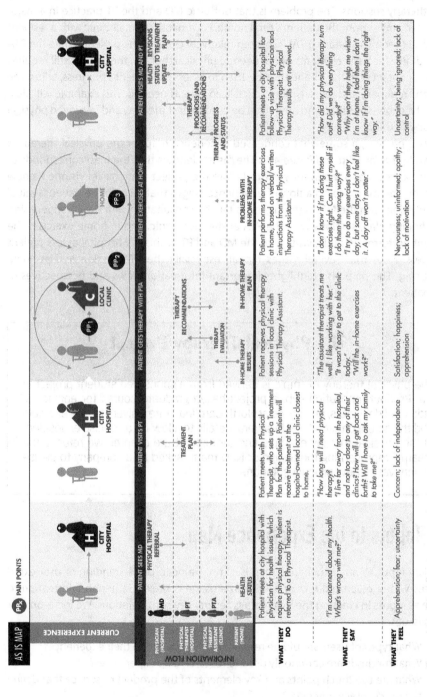

Figure 4.1: "As is" map.

Types of Users and Their Experiences

For any product or service, a variety of user experiences (good and bad) will exist, so it is important to capture experiences for different user types and usage contexts. One or more prototypical experiences can be devised to capture the range of user experiences and how they might differ along important dimensions. The key, however, is for the team to test any assumptions about the user experience, learn firsthand about user experiences, and comprehend user experiences in a deep and holistic fashion so that opportunities for improvement can be identified and explored.

The goal is not to arrive at statistical precision in order to explain/predict user experiences. Instead, the team is primarily looking for insights that help them envision and devise better solutions. "Extreme" users (e.g., novices or experts in using a product) form an important part of any user research endeavor, since they can often better reveal behaviors or needs that typical users leave unarticulated or mask as they devise ways to work around product shortcomings. Researching the experiences of nonusers can also reveal insights about product alternatives and disadvantages that the innovation team could potentially address in a product improvement. Each type of user segment or prototypical user will typically have its own experience map, reflecting significant differences among user types in their experiences.

Research on a variety of user types should be complemented with understanding various usage contexts. The user experience can differ based on social aspects of the experience (e.g., traveling alone or with the family), personal situational factors (e.g., being fatigued or in a hurry), among other factors that constitute the user's experience space (Prahalad and Rangaswamy, 2003). Observing users in context is an important activity for the team if a deep, empathic understanding of the user experience is to be achieved (Brown, 2009; Christensen et al., 2007). The variety of usage contexts can be depicted within an experience map (e.g., noting different branches the experience might take depending on a specific situation) or shown in separate maps, each of which depicts a different usage scenario.

To illustrate with the physical therapy example, different types of rural patients could include patients with acute (recovering from knee surgery) or chronic (spinal injury) conditions, older patients with slower expected recoveries, or situations where the need for physical therapy is part of a broader, complex issue (Alzheimer's). Important usage contexts that could impact the user experience might include whether the patient lives alone or with other family members who could assist with home exercises or travel. Defining users and contexts is an iterative process—as the innovation team learns more about customer experiences, it better understands key issues that drive better or worse outcomes for users. Of course, the user research is conducted in the context of the design problem (rural physical therapy), but the project team can give itself freedom to explore adjacent opportunity spaces. For example, what factors besides proximity to the city suggest a need for better in-home therapy? The team's research should challenge assumptions that might limit the team's ability to innovate for a wider set of users facing similar problems.

Methods to Understand the Total Customer Experience

Generally, a combination of methods—both qualitative and quantitative—is used to comprehensively study the experience, which extends beyond the product usage itself. User research should capture elements that show motivations for usage, need arousal, and what happens after a usage situation ends to better understand any consequences from the customer's experience with the product/service.

Observation techniques are at the heart of research efforts to gain knowledge about user experiences (Brown, 2008, 2009). Leonard and Rayport (1997), for example, discuss the "empathic design" process based on observation and ethnographic techniques. It is important to observe users in a natural usage context and capture observational data through images, audio, video, and field notes for later analysis by the team.

Even though it often provides rich and insightful information about the user experience, observation is not a magic bullet. Lead user analysis and voice-of-the-customer interviews, for example, can complement observations to reveal attitudes and opinions the customer has about the experience (e.g., Christensen et al., 2007). Reflective user journals and photo elicitation can also convey stories about the user's experience, triggering insights about what is driving positive or negative experiences (e.g., Fraser, 2012). Regardless of the method, the experience context should be noted and understood by the team as it collects the data. The number of users to research is typically determined by the team's assessment of a sufficient variety of experiences that can generate insights, with the research continuing as long as new information is being learned (and budget and time allow). The team may also have access to existing data within the company, such as customer service logs, salesperson reports, or point-of-sale data, all of which can provide essential detail to certain aspects of the experience. The combination of methods utilized should capture the behavioral (what users do), emotional (how users feel), and cognitive (what users report thinking through articulation) elements of the experience (Brown, 2009).

In the physical therapy example, observational research would be conducted through home visits to understand the patient's home environment and exercise routine, and observing therapy sessions with the PTA. Interviews of the patient, PTA, PT, and MD would reveal attitudes about the therapy regimen, areas of uncertainty or concern, and opinions about what may be helping or hindering desired therapy outcomes. A patient journal could document the type and frequency of home exercises (how well does it conform to the therapy plan?), as well as contemporaneous reflections on achievements and frustrations. In the "as is" map, we give several examples of the patient's perspective of the experience in what they do, say, and feel.

Identifying Touch Points and Key Elements of the Experience

With the variety of customer experience data in hand, the team can begin to fill in additional details about what is driving each part of the experience. Who participates in each phase of the experience (e.g., salesperson, third-party agent)? What activities by the firm and other entities define and create each part of the experience? The deep customer experience research, by definition, takes the customer's perspective. The purpose here is to fill in the background perspective of the company and other players/stakeholders that are involved in making the experience happen.

The firm's perspective in the customer experience is often portrayed as a series of touch points that detail each occasion whereby the customer interacts with the product, service, and brand components of the firm. Obviously, whenever a touch point occurs, this will necessarily be reflected as part of the customer experience. In this sense, specific interactions captured by a touch-point analysis will overlap with an experience map. However, a touch-point perspective is often incomplete because it tends to focus on customer interactions or transactions with the company, without understanding the more comprehensive customer experience. Rawson, Duncan, & Jones (2013) discuss the pitfalls of a firm trying to optimize each touch point without considering the overall customer experience.

In the physical therapy example, one important factor that defines the user's experience is the information exchanged among various individuals about the patient's progress (see the "as is" map). For example, the MD and PT communicate to devise the best therapy plan for the patient. The patient does not directly experience these communications, but the plan is conveyed to the PTA, who then administers in-clinic sessions that the patient experiences. Based on the therapy plan, the PTA also directs the patient to conduct specific in-home exercises with a certain frequency. The patient may or may not adhere to the home regimen and may incompletely, or inaccurately, communicate with the PTA about what is being done at home. The noise in such information exchange makes it difficult for the PTA to adjust and administer exercises that might be more effective for the patient, and the PT and physician consequently have difficulty knowing whether the planned therapy is being as effective as it could be for the patient. Note that some, but not all, of these information exchanges would constitute a touch point. The goal in researching drivers of the experience (details of which may be hidden from the user) is to not only understand what the customer experience is, but also to gather valuable information about how/why the experience unfolds as it does.

Synthesizing for Insights

Once the experience information is gathered, it is time to organize the raw data into a usable form. The goal is to identify key insights about the user experience that can be translated into opportunities for innovation or product improvements. Sophisticated analytical techniques for qualitative data can sometimes be used to identify patterns or categories of emergent themes. However, the team should not let analytical sophistication substitute for their involvement in synthesizing and understanding the user experience to gain insights.

Teams will engage in a series of sorting, mapping, and clustering exercises to organize the information in a usable form. Fraser (2012) and Kumar (2013), for example, summarize and depict numerous techniques that can be used. The goal of these techniques is to help the team begin to abstract from the raw data and synthesize meaningful insights. Brown (2009, p. 70) calls this a "fundamentally creative act," related to divergent thinking, that can identify new opportunities for enhancing the user's experience. The series of activities involving data analysis and synthesis of insights is naturally an iterative process meant to generate inspiration for the team to generate new solution ideas. Numerous methods exist to aid the team in the synthesis process, but it takes practice and skill for the team to arrive at actionable insights about innovation opportunities (Brown, 2008; Dyer, Gregersen, & Christensen, 2009).

4.2 The Experience Mapping Process

Once the inputs are gathered, the specifics of creating and using an experience map will, in part, be a function of the project's context and who is involved in the effort. The process of creating an experience map will sometimes reveal gaps in the team's knowledge about users and require additional research and iteration. Nonetheless, there are some common activities that the team will need to do if the experience map is to be a useful tool for innovating new solutions. Broadly defined, these activities are:

1. Develop and utilize one or more personas that represent relevant type(s) of users.
2. Create a map that captures the user's journey through the experience.
3. Use the experience map and its rich set of inputs to identify critical user pain points.

Utilizing Personas

A user persona is a composite character that encapsulates data gathered and synthesized from the user research (Fraser, 2012). Although it may be feasible or even advisable to draw a separate experience map for individual users the team has researched, the team at some point will create prototypical maps that correspond to specific user types and usage contexts deemed most relevant. These maps will depict the typical experiences of the user personas that can reflect key insights from the user research. Although seemingly analogous to customer segmentation, personas are driven by the experience data the team synthesizes into an insightful and somewhat prototypical understanding of user experiences.

Rarely will a single persona capture all the research and user insights developed by the team. There is certainly a trade-off between too many personas that would be cumbersome for the team to address with a single new solution and too few personas that might omit promising opportunities that address user needs. A rule-of-thumb might be between 3 and 10 personas to capture and frame the range of experiences and usage contexts (Kumar, 2013).

The physical therapy example could involve several personas. One might be a relatively young patient recovering from a sport injury, where a proven physical therapy regimen should lead to full recovery. Another could entail an older patient suffering from a chronic condition where physical therapy is designed to maintain certain functions as well and as long as possible, given a changing patient condition as time progresses. Each of these two personas would capture distinct user types (acute versus chronic) and usage contexts (short-term and well-defined regimen versus a longer-term flexible regimen) to cover a range of user experiences.

Creating the Map

It is critical to have a cross-disciplinary, diverse team working collaboratively to create the experience map. To begin, the team should create a timeline on which to construct the experience map. Users navigate experiences through time, and the team's perspective should match the user in this regard. The team will then work to populate the timeline with steps, or stages, of the user experience. There is no hard-and-fast number of steps that will always work best. The trade-off is one of detail and accuracy versus usability. Making an experience map too detailed with too many steps can bog the team down in minutiae, when their job is to consider the user journey in just enough detail to allow for insights and eventual improvements. Conversely, too few steps can result in a lack of insights into the user journey, which can limit the range and quality of innovative solutions that can ultimately come from the experience mapping effort. If the team wants to consider more user experience detail as they move forward in the innovation process, they can always return to the experience map and hierarchically drill down into specific steps as needed.

Once the steps are connected to the timeline, the experience map can be fleshed out by considering the surrounding environment within which the user journey takes place. During the journey, the user may interact with other people, information, physical objects, supporting services, and so on. Depending on the specifics of the experience, these considerations may drive a deeper understanding of why the current user experience is less than ideal and how it might possibly be improved.

The output of the mapping effort should be a clear, visual, accessible map and associated narrative that engages team participants as they develop it, and fosters interaction with others once it is completed. Once an initial snapshot of the experience map is developed, the team should share the map with others who may be able to provide useful feedback. The first version of the experience map is rarely a perfect and complete representation in all aspects, and iteration should be expected and even welcomed.

The experience map for our physical therapy example was previewed earlier in the "as is" map figure. For simplicity, details of the user type or persona are not shown, but the map corresponds to what might be experienced by an older patient with chronic mobility difficulties. Each of the five steps or stages depicted could be drawn in more detail to reflect the user's more specific activities in each stage.

Two key elements discussed as inputs to the experience map have been highlighted in the example. First, each stage of the experience shows important findings from the user research in terms of what users do, say, and feel for that part of the overall experience. These reflect the behavioral, cognitive, and emotional data gathered in the research. In drawing an experience map, the team will often select quotes or data that reflect critical insights about the user. Second, the map shows important information flows that are key elements of defining the user experience. Some of these flows are touch points, such as when the PTA communicates an in-home therapy plan with the patient, but other flows are not directly part of the user experience, such as a therapy evaluation the PTA sends to the PT.

Identifying Pain Points

Armed with a well-defined user experience map, augmented with important "do-say-feel" elements and key factors that help the team understand and explain how/why the experience unfolds, the team can summarize important insights. This is frequently done by identifying important pain points that users experience, reflecting gaps in the experience that a new solution should potentially address. Pain points generally reflect specific aspects of the user experience that result in reduced value or benefits to the user (or opportunities for increased value), reflecting stated or latent needs that are relatively important to solve.

Pain points can exist at different levels of granularity. A single step in the experience map can be the source of pain for the user, or a pain point may come from a group of related or connected steps that are a portion of the entire journey through the experience. The entire journey should also be considered as a whole, to examine whether user needs are ultimately satisfied.

When identifying pain points, the variety of inputs to the experience map matter. Note that pain points should not be limited to the user's physical activities in an experience (e.g., the patient has difficulty traveling to the clinic for sessions). More complete and even promising insights about the user experience will also arise in other areas. This is part of the motivation for ensuring that experience map inputs reflect not only physical behavior (what they do), but also user emotions (what they feel) and how users articulate their attitudes about the experience (what they say).

One popular method to help broaden the scope of user insights that reflect opportunities for improvement is the SPICE framework (Fraser, 2012). SPICE is an acronym for the social, physical, identity, communication, and emotional components of user needs and experiences. In our physical therapy example, many of the user concerns are not just physical. The experience map also reflects social (family's role in travel), identity (lack of self-worth due to thinking "something is wrong with me"), communication (information on doing home exercises correctly), and emotional (lack of motivation) aspects. Often, innovative solutions with real benefits to the user arise from addressing the nonphysical part of the experience.

In the physical therapy "as is" map example, three different pain points have been identified by the team. We will discuss and utilize the pain points to illustrate how a well-crafted experience map can help springboard the team to devise innovative solutions.

4.3 The Experience Map as a Springboard to Innovative Solutions

Once the "as is" experience map (or several maps) is created and shared, it is time to use it as the springboard for identifying innovative solutions. The initial challenge in doing so is to use the experience map and associated pain points as inputs and seek out opportunities for changing the user experience for the better. This can be done by reframing the situation so that it can be reconceptualized in ways that benefit the user. As opportunities are recognized, the team can modify the existing experience map or create an entirely new one to capture the changes that would be necessary to provide an enhanced experience to the user. This new map is then used as the basis for developing, testing, refining, and possibly implementing the solution. In other words, the experience map serves as a method to aid the three main "spaces" in the design process (Brown, 2008): inspiration to define or reframe the innovation opportunity, ideation to generate and develop new solutions, and progress toward implementation that includes testing and prototype-based experimentation of critical issues the team needs to resolve about their new solution concept.

Reframing the Opportunity

Translating pain points into potential opportunities arises from the team's insights and by changing their perspective about the situation in different ways. This can include thinking about the experience from other stakeholders' perspectives, thinking more broadly about what is within the bounds of feasibility for a new user experience, changing perspective by questioning assumptions and standard ways of operating, and calling for order-of-magnitude improvements in performance metrics that matter to the user. This is expansive, divergent thinking.

An important method for opening up possibilities is to ask questions that often begin with "Why?" or "What if?" and relate to discovery skills such as associating and deep questioning (Dyer et al., 2009). A common phrase to begin reframing an opportunity is "How might we … ?" (HMW), which is part of a series of innovative questioning to frame opportunities (Berger, 2014). HMW works best with design challenges that are ambitious, yet also achievable. Often, the original frame or scope of the team's project will change based on what was discovered from the user experience research.

For the physical therapy experience, we can frame specific opportunities for new solutions that correspond to each of the identified pain points (marked as PPx in the "as is" map). In creating the "as is" experience map, the team noted that patients often have trouble traveling to the clinic (PP1), sometimes causing missed appointments or even increased physical pain from travel. Also, the team noted considerable doubt about whether in-home exercises are being done effectively (PP3), which can compromise coordination of an effective therapy plan (PP2). At some point, the team may use information about the market context to prioritize which pain point reflects the most compelling opportunity from a business perspective (e.g., the highest profit potential). However, the team often lacks such data early in the development process, and the goal is to ideate possible solutions that can then be examined in more detail from the user, technical, and business perspectives. Opportunity statements frame the team's activities in devising new ideas that can address the pain points and improve the user's experience, as depicted in Table 4.1.

Conceptualizing a New Solution to Enhance User Value

Using the "as is" experience map as a foundation, the team now has a good sense of what the key user pain points are and has identified associated opportunities to reduce

Table 4.1: Pain points for physical therapy from the "as is" experience map

	Pain Points	Opportunities
PP1	Patient must travel to the local clinic to receive physical therapy from the physical therapy assistant.	*How might we* decrease the travel burden for patients who have difficulty getting to the local clinic?
PP2	The city hospital care team is not well connected to the patient during delivery of the physical therapy treatment plan.	*How might we* create more communication and coordination across the whole care team—physician, physical therapist, and physical therapy assistant?
PP3	The patient performs physical therapy exercises at home without direct medical supervision or guidance.	*How might we* establish better interaction between the patient and the physical therapy assistant when the patient is doing therapy exercises at home?

or eliminate those pain points. Ideation for new solutions can now be conducted by brainstorming and other effective techniques such as storyboarding, role-playing, storytelling, analogous thinking, and rough-cut prototyping.

Given the identified opportunities for the physical therapy example, consider a new concept developed by the team called "Tele-PT" (see Figure 4.2). This concept would allow the patient to video-connect with the PTA while doing home exercises. In-home sessions could be recorded for later viewing by the PTA, or the PTA could watch live during the exercises and offer real-time advice on how the patient could more effectively complete the exercises. Not only would the PTA now be able to help the patient with in-home exercising, but the need for the patient to have therapy sessions at the clinic would be reduced due to increased effectiveness of home exercise. The Tele-PT concept has sophisticated video and 3D imaging technology so the PTA can see important details about the home exercises.

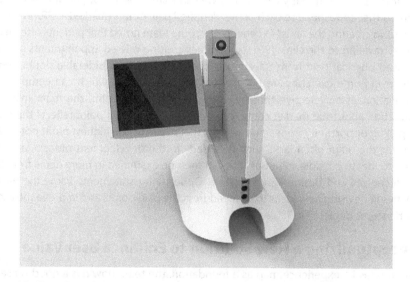

Figure 4.2: Tele-PT concept.

Note how the new concept addresses the three main pain points and opportunities identified by the team. With greater confidence that in-home exercises can be effective, the PTA can reduce the number of clinic sessions, reducing patient travel. By being able to directly observe the patient's home exercise regimen (either real-time or via recorded session), the PTA can give better direction to the patient about effective exercise techniques. The increased interaction and more accurate assessment of the in-home regimen help the PTA communicate and coordinate the overall therapy plan with the PT and the physician.

With a specific concept in mind, the team can now reimagine what the user experience (again, for one or more personas) might look like. The team would draw a "to be" experience map that reflects how the user experience would be re-designed for increased benefits to the user. In the "to be" map (Figure 4.3), a major portion of the experience map is highlighted. Since the new concept is meant be used in the home and address pain points related to in-home therapy, the at-home stage of the original "as is" map is now configured as a series of experience stages with the Tele-PT concept. As illustrated, the team focuses on stages that would be required for the patient to get the most benefit from Tele-PT, including delivery and installation, user training so Tele-PT usage can be most effective, actual Tele-PT usage during home exercises with PTA interaction, and the need to make any required servicing as efficient as possible. An actual reenvisioned experience map would contain more detail than what is illustrated, and several maps may be drawn to reflect different personas and usage situations.

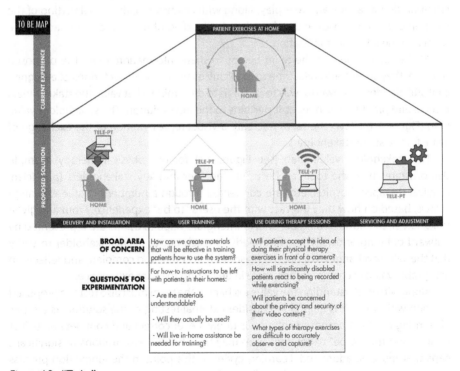

Figure 4.3: "To be" map.

A "to be" experience map (see Figure 4.3) helps the team accomplish several things. First, drawing a revised experience map as new concepts are being developed helps the team stay grounded in the goal of improving the user's experience. Second, by considering how a new concept potentially changes the user experience, the team will discover important issues that the evolving new concept design may lack. For example, the team may come to realize that the Tele-PT should be usable for home exercises in various postures (sitting, standing, lying down), with direct implications for the product design. More generally, by mapping the new experience, the team can identify specific gaps or areas of concern with the new solution that should be explored and tested.

Testing and Refining the New Solution

To address any areas of concern and determine whether a proposed solution is plausible and valuable, the team must take the proposed solution to users and other critical stakeholders for testing, refinement, and evaluation. This need not be a highly formal or structured process, at least early on, nor should it be a one-time effort. Instead, user and stakeholder engagement should be done as soon as possible and iteratively as the solution takes shape.

Before engaging users, however, the team should communicate with others internally to clarify and examine the overall viability of the solution. The experience map serves as the team's user-based understanding of their value-enhancing solution, which can be shared and tested in more detail through the use of sensory techniques such as storyboards, narratives, and role-play. Along with clarification, this initial vetting of the solution builds consensus and a shared understanding of the new concept across the team and the internal organization.

At the same time that the team begins to share information about the proposed solution, they should also take time to explicitly evaluate the expectations of all significant stakeholders. A powerful way to do this is to determine what value the stakeholders receive and provide to others, as it pertains to the new solution. The stakeholder value map (Figure 4.4) allows the team to create a visual representation of the exchange of value for the set of stakeholders.

The stakeholder value map (see Figure 4.4) for our physical therapy example demonstrates how the tool can be used. The user and key stakeholders (physician, physical therapist, hospital, etc.) are connected through a number of value exchange arrows, based on how they interact with the user's "to be" experience. From any given stakeholder, the flows of tangible and/or intangible value to others are represented by outward-pointing arrows. The team needs to reach out to each stakeholder to verify that the value exchange as defined is reasonably accurate and complete and reflects an acceptable value exchange for their role in delivering the user experience.

Along with understanding the value exchange, the team must also test the proposed solution with users. It is helpful to consider external testing of the solution as a series of learning cycles. A proposed solution is rarely if ever correct and complete in its first version, and the "to be" experience map may have hidden assumptions or significant gaps that must be addressed. Learning cycles at this point in the innovation process consist of rapid experiments and good use of prototypes. The speed and ease of rapid

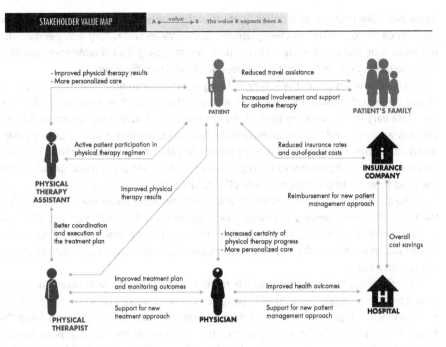

Figure 4.4: Stakeholder value map.

experiments has a beneficial side effect: it fosters excitement and brings energy to the team, and keeps potential solutions alive while they get refined and improved.

The "to be" experience map notes two areas of concern to the team if the Tele-PT concept is to effectively deliver an improved user experience. User training and user acceptance issues raise a number of specific questions around which the team could devise experiments to refine the concept. For example, the team could role-play a series of exercises in different postures (standing, lying down, etc.) to test how well the Tele-PT device captures the range of exercise motions likely to be performed by patients. The team could also test training material with several types of users to see how well patients understand the Tele-PT features and usage. By conducting a variety of experiments, the team can quickly learn about any shortcomings in the new concept, take corrective action, and move the concept closer to successfully implementing a more valuable user experience.

4.4 Conclusion

This chapter offers a high-level look at customer experience mapping. While we reviewed a number of concepts, tools, and methods related to the topic, readers interested in more detail can find many more sources of relevant information in our cited references and in the growing list of books and articles about user experience and innovation. Our example of a patient's physical therapy experience demonstrated how experience maps can be effectively utilized to add value and satisfy user needs. Since the focus is on the user's perspective, experience mapping is a valuable component of the innovation process for any product/service/brand combination.

Each innovation project will vary in budget, time constraints, staffing, and a host of other practical constraints. Specific challenges of any experience mapping effort will therefore vary from project to project. An organization should nonetheless try to consistently apply several aspects of experience mapping. The first is to stay user-focused, recognizing that understanding at least something about users is better than knowing nothing. It is also imperative to keep the team involved, even if outside organizations are used to assist in the user research efforts. Since the project team ultimately must craft the innovative solution, it is important that knowledge about users resides within the team. Of course, good user research techniques should be utilized, even if budget or time constraints do not allow the team to do everything it would want. Finally, consistent with most design thinking techniques, the team should give itself room to learn, refine, and iterate.

We offer a few final suggestions. First, the best way to build skills in experience mapping is to learn by doing. Try it! Should you and your team make a concerted effort to use it, experience mapping will almost certainly result in improvements. Second, keep the focus on the user throughout the project. There is a time in the development process to give detailed attention to issues such as financial viability or technical feasibility, but the deep user insights gained from experience mapping should continuously be at the forefront. Finally, engage your product and service development teams about experience mapping, and increase the chances that your development efforts will truly deliver more valuable user experiences.

References

Berger, W. (2014). *A more beautiful question: The power of inquiry to spark breakthrough ideas*. New York, NY: Bloomsbury USA.

Brown, T. (2008). Design thinking. *Harvard Business Review*, 86(6), 84–92.

Brown, T. (2009). *Change by design*. New York, NY: HarperCollins.

Christensen, C., Anthony, S., Berstell, G., & Nitterhouse, D. (2007). Finding the right job for your product. *Sloan Management Review*, 48(3), 38–47.

Dyer, J., Gregersen, H., & Christensen, C. (2009). The innovator's DNA. *Harvard Business Review*, 87(12), 60–67.

Fraser, H. (2012). *Design works*. Toronto, Ontario, Canada: University of Toronto Press.

Kumar, V. (2013). *101 design methods*. Hoboken, NJ: Wiley.

Leonard, D., & Rayport, J. (1997). Spark innovation through empathic design. *Harvard Business Review*, 75(6), 102–113.

Prahalad, C. K., & Rangaswamy, V. (2003). The new frontier of experience innovation. *Sloan Management Review*, 44(4), 12–18.

Rawson, A., Duncan, E., & Jones, C. (2013). The truth about customer experience. *Harvard Business Review*, 91(9), 90–98.

About the Authors

DR. JONATHAN BOHLMANN is Professor of Marketing at the Poole College of Management, North Carolina State University (NCSU). His research and teaching deal extensively with innovation, new product development, and product strategy. He has published in numerous leading journals, including *Journal of Product Innovation Management*, *Journal of Marketing*, and *Marketing Science*, among others. Prof. Bohlmann is coordinator of a new multidisciplinary "Innovation and Design" faculty excellence program at NCSU, partnering the Colleges of Management and Design in new innovation initiatives. He received his PhD at MIT's Sloan School of Management and was formerly an R&D and design engineer in the aerospace industry.

DR. JOHN MCCREERY is Associate Professor of Innovation and Operations at the Poole College of Management, North Carolina State University. He is a faculty lead for the Product Innovation Lab, a company-sponsored, project-based course for graduate-level engineers, industrial designers, and MBAs. This course recently was recognized by *Forbes* as one of the top 10 most innovative business school courses in the United States. His research and teaching focus on product and service innovation, project management and leadership, and operational excellence. Prior to joining academia, Prof. McCreery worked as a biomedical engineer, a systems consultant, and chief operating officer for a medical device firm. He received his PhD in Management at The Ohio State University.

Acknowledgment

Engin Kapkin created the illustrations in this chapter. Mr. Kapkin received an MS in Industrial Design from Anadolu University in Turkey. He currently teaches design classes at North Carolina State University, where he pursues his PhD in Design as a Fulbright grantee. He is active in several design studios and has worked in design at Ford and IDEO.

5

DESIGN THINKING TO BRIDGE RESEARCH AND CONCEPT DESIGN

Lauren Weigel

Empire Level—Division of Milwaukee Tool

Introduction

This chapter includes the following objectives: It starts by outlining why people involved in new product development have challenges when coming up with new ideas. It then explains why there is a need for a systematic method, or approach, to connect the people responsible for coming up with new product solutions to the user. It goes on to describe a method based on design thinking principles, which can be used to help bridge user research findings to concept generation and concludes by explaining how this method can be applied in industry.

5.1 Challenges in Idea Generation

Coming up with new product ideas and innovations is not an easy task. The process of coming up with new ideas can sometimes feel challenging for a number of reasons. Sometimes the team coming up with a new product idea has been working on that particular product line for a long period of time. They may have years of experience working on one product, and for that reason they may consider themselves experts in that category. In this case, their experience with what the product *can* and *can't* do can actually create a barrier to their creativity when attempting to reinvent or even refresh a product. They may be very good at improving a product's performance or optimizing its technology, but they may struggle when it comes to effectively evaluating the relevance of the product to a user. However, when deciding to enter a new category, the team faces different obstacles in ideation. The team may have limited exposure to the product category that they are entering and they may not be familiar with users of that particular product. This lack of familiarity can limit their ability to create a competitive product with innovation that is meaningful to the end user. Ideation can also be challenging simply because it can be hard to come up with a new ideas. Even if we have substantial research and clear findings on the end user, the transition from research findings to concept generation is challenging. Often, the solutions that come from the concept generation phase lack a meaningful connection back into the user research.

5.2 The Need for a Systematic Method to Connect to the User

A deep understanding of the user and his or her experience can help us develop more meaningful solutions. Unfortunately, it is at times hard for people tasked with designing a new product or system to understand the user. Even experienced industrial designers, engineers, marketing professionals and other new product development (NPD) team members struggle with understanding the needs of users. This is challenging to team members for several reasons.

The first reason is that team members may simply lack a comprehensive understanding of their users. Often, this is a result of a lack of ethnographic research on their end users' needs. The team may not have conducted ethnographic research and they may lack real insight into what their end users' experiences are. They may have relied exclusively on quantitative data to formulate a profile of their end user. While quantitative data is important in constructing a user profile and can provide directional information, it doesn't truly expose any rich end-user experience information. Sometimes even when a team has conducted ethnographic research, the research that they have conducted may lack depth and may not be truly representative of their users' experience.

One of the other reasons that team members face obstacles when understanding their end user is that it is natural for the people on the teams to think of themselves as the intended user, when in fact they are not. This is evident in a person's quick reaction to denounce or eliminate an idea during a brainstorming session because they do not particularly care for it. Their response to declare the idea good or bad is instinctive because they are evaluating the idea from their personal perspective and experience. Their personal judgment of an idea being good or bad indicates that they are thinking of themselves as the end user. This is a false assumption on their behalf because their personal preferences, demographic information, needs, pain points, and the problems and frustrations they have with a product and/or within a system, may be very different from the intended users'. Getting team members to separate themselves and their personal judgments from what is important to the end user can be challenging.

These challenges in getting team members to truly understand the user inhibit their ability to come up with meaningful and impactful ideas and innovations that users care about. This can result in stagnant product innovations that may be unsuccessful because they do not address the user's true needs and therefore may lack real value. Because it is challenging for people tasked with developing new products, systems, and services to understand and identify with end users, there is need for a systematic method that helps them make that connection. The connection that they make has to go beyond reading and reviewing data in order for it to be effective. It has to be an active method where the team members can deeply understand users' experiences and pain points. The method has to be systematic in order for it to effectively bridge the collection of research findings into the generation of concepts. A systematic approach also allows the method to be repeated and applied consistently over multiple projects.

5.3 The Visualize, Empathize, and Ideate Method

The Visualize, Empathize, and Ideate method was specifically designed to help bridge the gap between research, product ideation, and conceptualization. Often, during the development of a new product, there is a lot of velocity in the research phase. Many research insights are collected, but the research findings don't always make their way into new ideas and innovations. Sometimes people "stall" in taking the research findings and turning them into a product idea. This method, shown in Figure 5.1, helps people take ethnographic research findings, extrapolate key insights, and form new product innovations and ideas that respond to the research.

This method was inspired by a combination of methods used in industry during the NPD process, specifically for envisioning new products and for reimagining stagnant products and product categories. While working in new product development and participating in and conducting cross-functional brainstorming sessions, I found a few key methods that helped to drive new ideas in static product lines. After transitioning from industry to academia, I attempted to extract and streamline some of the things that worked from these sessions into one method that could be executed within a three-hour studio time frame that mimicked the industry environment. The method needed to take findings discovered through ethnographic research and provide a means for the students to deeply understand them so they could come up with user inspired product solutions during the concept generation phase. It was important that the method had a process and a pace that would actively engage the students in a collective realization of what the research findings meant and would allow them to identify and arrive at their own separate approach to a product solution.

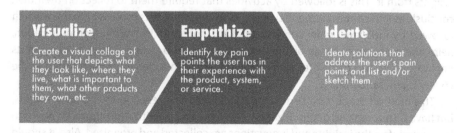

Visualize
Create a visual collage of the user that depicts what they look like, where they live, what is important to them, what other products they own, etc.

Empathize
Identify key pain points the user has in their experience with the product, system, or service.

Ideate
Ideate solutions that address the user's pain points and list and/or sketch them.

Figure 5.1: The Visualize, Empathize, and Ideate method.

Like many design thinking methods, this method has its own strengths and weaknesses that should be considered prior to implementation. One of its strengths is that it can be completed in a short period of time relative to the number of ideas that it produces. It also allows a group of people to establish an in-depth understanding of the end user, including the users' pain points, and then generate a large quantity of ideas that directly respond to these pain points. This helps guide a team to come up with ideas and innovations that are inspired by the end user, rather than inspired by the team's personal interests. This method also helps a team take the post-research mountain of user insights and synthesize those observations into concepts that can be built on during the development phase by breaking the information down into more digestible components.

There are several things that should be considered before employing this method. This method should be used between the research and concept generation phases. It requires minimal setup, but its effectiveness is contingent on good ethnographic research findings that are vital in serving as the foundation for user-inspired innovations. In the first part of the method, the findings from the research have to be distilled into key insights that are shared with the team so that the team has a comprehensive understanding of the user that is based on research and not on their own personal perspective. This part of the process is essential to its success. Without a thorough briefing on the user, the participants will not be effective—their participation may be limited by their own unsubstantiated views of the user and the user's experiences. The method takes several hours to conduct and relies heavily on active engagement of the participants. This active engagement works only when there is minimal distraction to the participants (i.e., e-mail checking, texting, phone calls, etc.). In the business environment, this is most effective when the team can go offsite so that the participants can avoid interruptions and have the chance to focus. This method also depends on a strong facilitator who is capable of getting the team to work together and to share their ideas.

The method follows the three key steps shown in Figure 5.2. The first step is getting the participants to have a deeper understanding of the user. The second step is getting the participants to identify what the user's pain points are. In the final step, the participants use the information generated in the previous two steps and ideate solutions that are specific to the user. These activities are summarized in three key steps: visualizing, empathizing, and ideating. This method also employs alternating periods of action followed by subsequent reflection. The team is required to reflect on the ethnographic research they conducted together and draw their own personal insights from it. This is followed by activities that require them to reflect on their own conclusions and communicate their insights to the team. This process of team research followed by individual reflection and thinking, and then interactive communication, helps the team make sense of the research and gain a deeper understanding of its implications. It also helps the team move forward with a clearer product vision in the concept generation phase.

This method produces a large quantity of ideas that respond to end-user insights, but have no real ranking or prioritization. A ranking and prioritization approach needs to be applied after the insights and innovations are collected and organized. Also, it should not be assumed that the ideas that are a result of this method are immediately usable.

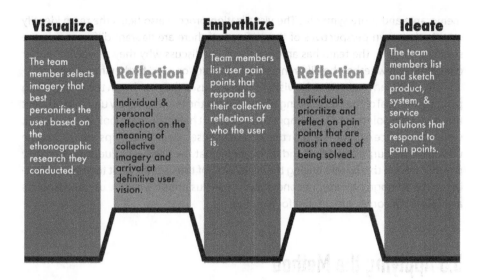

Figure 5.2: Using the Visualize, Empathize, and Ideate method.

The ideas are in their earliest phase and usually just a starting point that will need to be built on in order for them to become a usable concept.

5.4 The Importance of Visualizing and Empathizing before Ideating

The order of this process is intentional and significant to the outcome. The goal of this method is to specifically get *user-inspired* ideas that will generate innovations, not ideas that are a result of personal preference or the preference of the team or organization. In order to accomplish this, visualization of the user must be completed first. The team members need to "see" the user before they can establish an exhaustive list of what the user's pain points are. The way that the visualization process is applied can be entertaining for participants and is relatively easy for them to do making it a good starting point to get the group engaged. Team members are asked to reflect on the ethnographic research and to visualize the end user and then assemble a visual map of the user. The visual map includes images cut from magazines that show things about the users' lives, including what products they may own, the type of house they may live in, things that are important to them, and so on. The process of creating a visual map is important because it allows the team to build consensus by coming to a collective definition of who the user is. Individuals start by selecting images that they think best represent the user. Once everyone has placed their images together in a map, the team can collectively discuss who the user is, what the user values, and why. This allows the team to arrive at a unified vision of the user. The imagery also creates a more substantial impact on the team, as opposed to reading reports on the users' profile. The imagery is more memorable and more symbolic. The visualization process also helps the team identify gaps in their own perspectives of the end user. If there are flagrant differences in the selected imagery, the team has an opportunity to discuss why they perceive the user differently and whether the differences are important or need further investigation.

Once the team has a cohesive vision of the end user, they can begin the empathizing process. The goal of the empathizing phase is to get the team to deeply understand what the user's pain points are. This is important because it helps the team look at the product, system, or service experience from the view of the user. Doing this helps the team come up with more purposeful ideas and innovations that the user will value and ultimately pay for. They do this by referencing the visual map of the user they put together, reflecting on the ethnographic insights, and drawing conclusions on what the users pain points are. This list becomes the catalyst for ideating.

5.5 Applying the Method

The following section outlines how this method was applied in an industrial design studio with third and fourth year undergraduate students. While this example focuses on a classroom application, this method can easily be adapted to industry. For example, what was accomplished in one three-hour session could be accomplished in an offsite workshop or over several shorter meetings. The studio was composed of 12 students. It was an industry collaborative studio, where students worked with a company to individually develop a new product solution. The company tasked the students with designing a product in a category in which the company was not yet competing. The students were challenged to bring user-centered innovation to their design solution.

After the project was kicked off with the client, the studio was divided into three teams to conduct ethnographic research. The three ethnographic research teams included user, technical, and market research teams (Figure 5.3). The three research teams represented the cross-functional teams that make up the NPD process in industry. The user research team represented the roles that industrial designers, product managers, and/or marketing research team members have in industry. The technical team was tasked with uncovering technical findings similar to what engineering and/or research and development would contribute. The market research team focused on looking at market trends, competitive analysis, and benchmarking, mimicking the product management and marketing functions in industry. The goal of the research was for students to understand what users' pain points were with the existing solutions in that product category. Each team was responsible for compiling insights from their research. The ethnographic research that was collected during this phase became the foundation for the Visualize, Empathize, and Ideate method.

The technical research team included four students. The team started by conducting performance testing on competitive products. They set up tests that mimicked how users would interface with the product. They also tested the product to failure to fully understand its capabilities. Then they disassembled them. During this product tear-down, they timed themselves to determine how long it would take to access maintenance and service components. This team also interviewed service and maintenance professionals to identify what were common issues with existing products. They also researched new technologies that were emerging in this category and related categories that could be employed in their design solution.

The market research team was composed of four students. The team started by putting together a competitive comparison chart that showed product specifications, features, price, and the like for each of the major competitors' models. They highlighted strengths and weaknesses in competitors' product line-ups. They also outlined major competitors' strengths and weaknesses in both brand perception and product perception. They also put together market opportunity maps that highlighted potential areas in the market for differentiation and entry. They went on in-store visits at retailers and interviewed store associates about existing products that were sold in that particular store. They also analyzed online customer reviews of existing products to further understand customer perception at the brand and product level.

The user research team also consisted of four students. The students started by putting together a quantitative user behavior survey that they launched digitally. They then began identifying users in residential, commercial, and industrial markets. After identifying users, they visited each of them and conducted ethnographic observations and interviews. The team conducted product interceptions, where they took existing products into public places and interviewed people about their experience with that particular type of product. During these product interceptions, they had people try to start and use the product. The team videotaped and timed each person to get a better understanding of how long it took them to figure out how to start the product and to identify where the users' frustrations were when starting the product. The team also put together user personas and profiles and they mapped users' sequence of use when interacting with the product.

The teams were not only assigned with providing raw research data (images, videos, transcripts, etc.), but they were also tasked with providing meaningful and actionable key insights. To discover these key insights, the teams had to ask themselves, "What does this data mean? And how can it inform a new solution?" These key insights helped them clearly articulate not only what they had done but why they had done it and what was important about it.

After the ethnographic research was collected, organized, and analyzed, the Visualize, Empathize, and Ideate method was implemented. The studio was set up in advance to facilitate an effective environment for design thinking. One wall was divided into three sections, each section representing a market segment (residential/home users, professional users, and do-it-yourself/light professional users). The market and user research teams identified these three segments in this product category after conducting their research. Each section of the wall was labeled with the target user segment name. The output from the research was hung up around the room making it visible and accessible to all of the student participants. These research outputs included: product opportunity maps, brand profile charts, sequence-of-use charts, exploded views of existing products, product comparison charts, and so on. Work areas were put together in the center of the room. Groups of desks were arranged as designated work areas for the students to sit at while participating in this phase of the design thinking

Figure 5.3: Technical, market, and user research teams.

method. The work areas included a diverse selection of magazines so that the students had many images to select from, sticky notes, markers, tape, and scissors.

Visualizing the User

The first part of this method required students to visualize the user for each of the three segments. Creating a visual map of the target users helped the students see who they were designing for and gave them a visual priority of what was important to the user. The visual maps are different than personas. Visual maps uses imagery to show what users value and their experience with the product versus a persona where much of the content about the user is captured through written descriptions. Students were asked to reflect on the ethnographic research they had conducted and individually select images that they felt answered: "What does a user in this segment look like? Where do they live? What other products do they own? What things do they value and/or care about?" Then they taped the images that they selected under the respective user segment. Students included images that answered the questions above, and any additional images that they felt represented the user. The compilation of imagery created a visual "collage" of each of the user groups for which the students were tasked with designing (Figure 5.4).

Once the visual "collages" of the users were complete, students were asked to describe to the rest of the class some of the images they selected and explain the image's significance to that particular segment. During this phase, students identified similarities in the imagery that was selected and established a common view for each user segment.

Empathizing with the User

Once a strong visual of the user was established, the empathizing phase began. Students were asked to identify the pain points that each of the user segments had with existing

Figure 5.4: Creating visual maps.

products and product solutions. The empathizing phase began with students imagining themselves as the user segment they created with imagery. They were then asked, "What challenges would you have with the existing product solutions if you were this user?" and "What challenges did you see the user have with this product when you observed them during the ethnographic research?" As the students responded, a pain points list was created next to each of the visual collages that represented the user groups.

Ideating

After pain points were identified in each of the user segments, students were divided back into their ethnographic research teams (market, user, and technical). The students were divided this way because they had built rapport with each other and had developed a positive and productive team synergy that they leveraged to come up with new ideas. Each team was given a different colored stack of sticky notes and markers. The teams were asked to come up with solutions, in the form of product ideas, that would solve these user pain points by writing down their ideas or drawing sketches of their ideas on the sticky notes. As they came up with ideas, they placed their sticky notes on the visual map of the user segment that the idea was most applicable to (Figure 5.5).

After the ideating phase of the method was complete, students were tasked with finding inspiration for design solutions from existing products in other product categories. Looking for inspiration in other product categories helped them build on their ideas from the ideating session and think of new ideas altogether. Students were asked to bring in five images of existing products, in different product categories, that

Figure 5.5: Ideating.

demonstrated functional solutions that could be applied to the ideas that the team came up with during the ideating phase. The purpose of this exercise was to help them find examples of products with functions that they could draw inspiration from during the concepting phase, the phase where they started transforming ideas into product solutions. Each student presented his or her five existing products to the rest of the class and identified their significant function and how their function might be applied in a product solution. As each student presented their five products, the rest of the class was given sticky notes to write and draw new ideas on as they were listening. After each student finished presenting a product and its function, the rest of the students placed their new ideas on and around the product image.

5.6 Conclusion

This method can be used to help NPD teams make an effective transition from research insight collection and analysis to concept generation. Although this method was used in a classroom environment with undergraduate industrial design students, it has relevance outside of this application and could be used with NPD teams in industry. It can help teams synthesize a large quantity of information, thoroughly and deeply comprehend it, and then act on it by generating a large quantity of new ideas in a relatively short period of time. When used in the classroom, the students were able to come up with over 200 ideas across three user segments within several hours.

The three different teams that the students were initially divided into (technical, user, and market) were designed to mimic industry. In industry, the technical team represents engineering, manufacturing, research and development. The user team may include industrial designers, anthropologists, interaction designers, psychologists, and so on. And the market team could include market researchers, analysts, product managers, and the like. In most cases in industry, these teams are fully integrated and are working side by side throughout the product development process. There are different ways to approach this method, with pros and cons to each approach.

Additionally, and most importantly, through active participation and synthesis of the research data, participants are able to understand all aspects of the research findings, and examine them at a level deeper than through a summarized report or presentation. Engaging all of the cross-functional participants tasked with bringing a new product to market in one participatory exercise where they are challenged to visualize the users, empathize with them, and then ideate by coming up with solutions helps them embed the findings into their product solutions.

About the Author

LAUREN WEIGEL is a Product Manager at Empire Level, a division of Milwaukee Electric Tool in Wisconsin. Prior to working at Empire Level, she worked at Generac. She started there as an industrial designer and later joined the engine-powered tools team as a product manager for the portable generator category. She has taught industrial design at Auburn University in addition to teaching at the Art Institute of Wisconsin in Milwaukee. She earned a Master of Industrial Design and a Bachelor of Science in Environmental Design with a focus in industrial design at Auburn University.

6

BOOSTING CREATIVITY IN IDEA GENERATION USING DESIGN HEURISTICS

Colleen M. Seifert
University of Michigan

Richard Gonzalez
University of Michigan

Seda Yilmaz
Iowa State University

Shanna Daly
University of Michigan

Introduction

When facing a design problem, designers across disciplines often fall into familiar patterns, and have difficulty producing creative designs. Where do new design ideas come from? This chapter presents a new tool to help with idea generation called *Design Heuristics*. These heuristics capture the cognitive "shortcuts" that designers know to help them produce many candidate designs with interesting variations to choose among. Through empirical studies of industrial and engineering designers working on a variety of consumer products, a total of 77 Design Heuristics for use in new product development were identified. This empirical evidence supports the value of the Design Heuristics tool in generating new designs across disciplines.

Next, we describe how to use Design Heuristics for idea generation in design contexts. Design Heuristics help designers by suggesting specific ways to develop new concepts, and to modify and extend existing concepts. Each heuristic offers new possibilities for introducing variation. These easy-to-use guidelines are described through examples of designers generating new concept ideas. Design Heuristics can be applied to any type of product design, and examples of their use in award-winning consumer products are provided. Studies of designers using this Design Heuristics tool have shown its effectiveness in generating more, more varied, and more creative designs. As a design tool based on evidence from the practice of designers, and empirically tested for effectiveness, Design Heuristics are a helpful tool for designers in any area interested in uncovering new ideas.

6.1 Where Do New Design Ideas Come From?

In the earliest phases of the design process, design thinking typically focuses on identifying user needs. Using these insights, designers begin to identify possible solutions. The best opportunity to identify creative solutions depends on considering many *different* ideas. But designers attempting to generate new ideas often fall into a trap: while the first idea or two may come easily, it is often difficult to generate more and different ideas. This is simply a result of the way we think: what comes to mind may be the most obvious, familiar ideas based on existing designs. As a result, designers are often "fixated" on their first ideas (Purcell & Gero, 1996). In the early concept generation phase, a goal is to create as many *different* ideas as possible. If many alternative concepts are generated, then the best of these potential designs can be selected for further examination. So how can designers generate many different concepts to consider? How can designers best explore possible concepts and consider many different ideas during the early stages of product design?

By studying how designers create a variety of concepts, it may be possible to learn about methods and strategies they find useful. Our goal was to examine *how designers think* when generating ideas and to identify the strategies evident in their thinking about concepts. By systematically comparing their steps in creating new concepts, we hoped to uncover underlying cognitive strategies. Close observation of design thinking in the earliest stages of idea generation may provide some evidence of how successful designers create new concepts, and provide guidelines for other designers to use when generating new ideas.

6.2 A Tool to Assist with Idea Generation: Design Heuristics

To answer this question, our research group set out to empirically study the ways that designers generated concepts. Consider this example from a project described to us by an industrial designer: The task was to create a set of novel desktop accessories that could be manufactured and given to clients to promote an office products company. For inspiration, the designer looked through a magazine and came across a flower vase

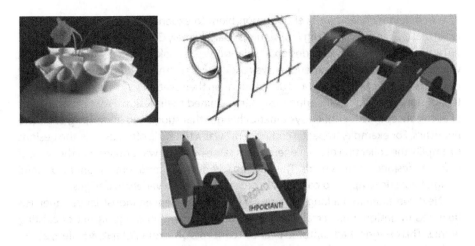

Figure 6.1: The first image shows a "scroll"-like embellishment on a vase. The designer exaggerates the scroll in her drawing (center), and then changes the shape to open the ends. She then "flips" the center element on its axis, creating the desk accessory shown in the bottom image.

that made use of circles with overlapping edges. By expanding on this form, she created a drawing of circular shapes with one "long end" hanging from each circle, leading to a "J"-shaped object (see Figure 6.1). Then, to add interest to the form, she "flipped" the larger, middle piece to go in opposition to the others. This "flip" to change the orientation of one piece created an office accessory (bottom image) that is striking in its creativity. In this example, the designer described following a strategy of refining the form by "flipping" an element.

How can we characterize the cognitive processes involved in this creative design? We propose the notion of Design Heuristics as the implicit strategies designers use to explore variations in concept elements (Daly, Yilmaz, Christian, Seifert, & Gonzalez, 2012a). By generating more and different concepts, designers introduce variations in their designs, and can then determine which concepts are worth pursuing. For experienced designers, cognitive strategies based on their past experiences may simply "come to mind" during design and are therefore implicit. As researchers, we set out to identify the use of heuristics by designers while creating concepts. Our hope was that the resulting set of heuristics would serve as a useful tool set to share with other designers across disciplines.

6.3 How Design Heuristics Were Identified: The Evidence Base

In our empirical studies, we set out to sample design thinking in idea generation in two domains: industrial design and engineering design. By studying new idea generation across these two disciplines, we hoped to uncover potential strategies in design creation that will be helpful to designers in any discipline as they develop new products.

We began by giving a novel design problem to experienced designers from both industrial (12) and engineering (36) settings (Daly, Yilmaz, Christian, Seifert, & Gonzalez, 2012b). Their problem was to design a "solar-powered cooking device" for families that was both portable and inexpensive. We asked them to "think aloud" during the short design task and to write labels and descriptions on their sketches. An experienced industrial designer and an engineering designer examined their design. From close analysis of each concept, we identified systematic changes that suggested underlying cognitive heuristics. For example, "repeat a component" was a heuristic often used in the designs to amplify the collection of solar energy. This same heuristic was apparent in the work of different designers across fields. This suggests "repeat a component" might be a useful design heuristic to apply to concepts in order to introduce variety in designs.

Next, we examined a long-term project by an experienced industrial designer. His goal was to design a universally accessible bathroom within the footprint of existing homes. This designer had captured over 200 concepts on scrolls (Yilmaz, & Seifert, 2011), and this serial record of designs allowed observation of the many ways this designer intentionally varied the concepts he generated. Over 34 separate Design Heuristics were observed in this case study, supporting their usefulness in creating diverse concepts for a single product. Another study analyzed over 400 consumer products identified as innovative in award competitions, and the same underlying Design Heuristics were observed and reported by independent coders (Yilmaz & Seifert, 2010).

Across design activities and across disciplines, with existing and new products, the use of Design Heuristics was strikingly evident. They included concepts from designers working on familiar consumer products, both durable and nondurable goods, interior designs, and technology-based products. While the results may also be applicable to the design of services and industrial needs, our emphasis in the studies was on consumer product designs. Through this major review of a large body of concepts generated by professional designers, we identified a final set of 77 different Design Heuristics. While the evidence of Design Thinking was observed from individual designers, we expect Design Heuristics to be just as evident in groups working together (as we describe in Section 6.4).

6.4 77 Design Heuristics for Idea Generation

Our next goal was to take these systematic observations of evidence about Design Thinking and turn them around to use as guidelines to help other designers. The Design Heuristics were formulated as a tool that would help designers apply each observed heuristic within new design problems. Each heuristic was named and described, and published on a 4-by-6-inch card along with an abstract image (see Figure 6.2). The back of each card depicted two existing consumer products that showed how the heuristic was used in other contexts. For every heuristic, one of the illustrations shown was a design for a chair. Designers continue to create interesting variations of this product; for example, one industrial designer is conducting a project to design 1,001 new chair concepts.[1]

---

[1] http://1001-chair-sketches.blogspot.com

UTILIZE OPPOSITE SURFACE

76

Create a distinction between exterior and interior, front and back, or bottom and top. Make use of both surfaces for complimentary or different functions. This can increase efficiency in the use of surfaces and materials, or facilitate a new way to achieve a function.

UTILIZE OPPOSITE SURFACE

76

FARALLON CHAIR
fuseproject
The back side of this chair has a pocket for storage.

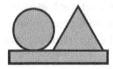

980 TATOU
Annika Luber
The laces wrap around the bottom of this shoe and connect with the sole.

Figure 6.2: Information on each heuristic is depicted on two sides of a card, serving as a tool for designers to use while working on new concepts.
© Design Heuristics, LLC.

In the Design Heuristic shown in Figure 6.2, the suggestion is to consider using an "opposite surface" to add new functions for the product. For example, a shelf is designed to hold objects on top of it, but it can also serve a purpose with its opposite side, such as hanging hooks. This heuristic provides direction by drawing the designer's attention to unused space on the product so that they can consider whether it can be employed as part of their design. In the product example, shoes traditionally tied with strings on the top surface make use of the bottom surface to provide additional tension to tighten the shoe's fit.

The 77 Design Heuristics cover a wide range of possible variations for concepts (see appendix) (Yilmaz, Daly, Seifert, & Gonzalez, 2014). Some of the heuristics address ways to change the *form* of the product, such as changing its geometry; twisting, rolling, or nesting; or stacking, telescoping, or folding to conserve space. Simple changes in shape resulting from these heuristics can add interesting diversity to the look of the resulting concepts. Other heuristics address changes to *function,* such as using multiple components for one function, using a common base to hold components, redesign joints, and adjusting function through movement. Each Design Heuristic serves as a prompt

USE PACKAGING AS FUNCTIONAL COMPONENT   73

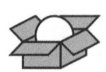

Instead of disposable packaging, incorporate packaging within the product to perform a supportive function. This can reduce waste and provide a storage or organizational option.

USE PACKAGING AS FUNCTIONAL COMPONENT   73

WHEELED CUBE
Heinz Julen
This chair can be folded into a wooden box with wheels when not in use, protecting interior cushions.

FLIPBOX PENCIL CASE
Faber Castell
This set of colored pencils comes inside a package that also serves as a stand during use.

Figure 6.3: An example card depicting the nature of a specific Design Heuristic. This heuristic suggests using packaging as part of the product, and it is illustrated in two consumer products.
© Design Heuristics, LLC.

to encourage designers to both introduce intentional variation and to consider more efficient functional qualities within their designs.

A critical theme within the Design Heuristics is to develop concepts based on *user* needs. For example, "adjust functions for specific users" suggests altering the concept to accommodate differences among users, such as height or age. Other heuristics focused on user needs include suggestions to incorporate user input, provide sensory feedback, change surface properties (to guide users), and allow the user to customize, reconfigure, reorient, and assemble the product. The needs and role of the user are central to product development, and, consequently, the heuristics observed in expert designs often addressed user needs.

Another design concern reflected in the observed Design Heuristics is *sustainability*. This is represented by heuristics such as reduce material, use recyclable materials, use packaging as a functional component, repurpose packaging, use human-generated power, and make products recyclable. Bringing sustainability issues into conceptual design is important to both users and manufacturers, and these and other heuristics help to drive changes in concepts toward this goal. For example, the "use packaging as functional component" heuristic raises the notion of planning packaging that becomes part of the product. In the product example depicted in Figure 6.3, a packaging case holding a set of colored pencils is adjustable so it can serve as a stand for them during use.

6.5 How to Use *Design Heuristics* to Generate Design Concepts

To begin, select a problem statement that you would like to address, for example, "design a chair." Consider the card "bend" (see Figure 6.4), and its description, along with the product examples. Now, think about a standard chair design, and then think of a couple of ways you could apply "bend" to that chair to come up with a new concept. Take a few minutes to sketch out each of your ideas. Try to go in a different direction with each of your concepts by applying "bend" to different parts of the chair, or considering different materials.

When people are given this task, we find they are able to generate a wide variety of chair designs using just this single Design Heuristic. For example, consider these three designs by participants in our studies (Figure 6.5). In the first example, the material is one folded sheet ("of metal, wood, or plastic") bent to form the legs, seat, arms, and back of the chair. In the second example, the design uses a continuous surface that can be rolled up to form a seat or a lounge chair. In the third, a round tube is bent into a bench seat and contoured for more comfortable sitting.

The Design Heuristic aids the designer by suggesting a "prompt" or direction for the design, adding a more specific intent to the creation of a concept. However, there is still plenty of latitude for the designer in that they can choose different parts of the concept, materials, angles, forms, and even functions to alter with the use of "bend."

For example, if you are focusing on user needs (Chapter 1) and want to explore concepts related to it, you can choose heuristics such as "adjusting based on demographics,"

BEND

16

Form an angular or rounded curve by bending a continuous material, and assign different functions to its surfaces. This can reduce material, improve product uniformity, and create additional functions.

BEND

16

OVERLAP TRAY
Offi
Using a single continuous material, this bent tray and the negative spaces it creates serve a variety of functions.

LAPTOP2 CHAIR
Christian Flindt
This chair for two people is made from one continuous fiberglass form. It is bent to make use of both sides, which are coated in different materials.

Figure 6.4: The "bend" heuristic adds changes to surfaces to introduce contours.
© Design Heuristics, LLC.

or think about how to "incorporate user input" to customize seating. However, you can consider any of the 77 Design Heuristics to allow a playful exploration of possible designs. Trying out different heuristics in any order may lead you to surprising ideas. Each heuristic can be applied to any problem, and often, more complex designs may be created by repeatedly applying the heuristic, or by applying another heuristic to the same concept. Through this method, a chain of concepts can be generated where more variations are introduced by adding more heuristic prompts to your thinking. For example, in Figure 6.6, a participant created a concept that combines the "bend" and "synthesize functions" heuristics in a single design.

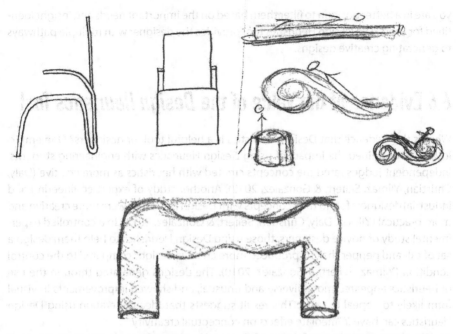

Figure 6.5: Three different chair designs by study participants where the heuristic "bend" is observed.

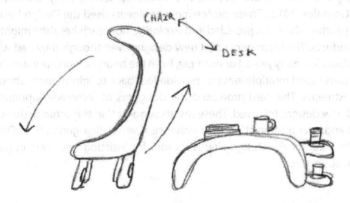

Figure 6.6: A participant's design of a chair can be used as a coffee table when placed facedown, combining use of the "bend," "synthesize function," and "convert for a second function" Design Heuristics.

The advantages of the Design Heuristics tool include its ease of use, with simple prompts to encourage designers to think in a given direction, and the option to change directions easily by adding another heuristic. As a result, an endless variety of concepts can be generated. And by increasing the number of different concepts generated, there is a larger set of potential designs that may meet the needs you are considering for the user context. In other words, by creating more, more diverse, and more creative designs, you are in a better position to filter them based on the important needs and insight identified for users. The Design Heuristics tool provides the designer with multiple pathways to generating creative designs.

6.6 Evidence of the Value of the *Design Heuristics* Tool

What is the evidence that Design Heuristics is a helpful tool for designers? One empirical study examined the impact of using Design Heuristics with engineering students. Independent judges rated the concepts created with heuristics as more creative (Daly, Christian, Yilmaz, Seifert, & Gonzalez, 2012). Another study of expert engineering and industrial designers found their concepts with Design Heuristics were more creative and more practical (Yilmaz, Daly, Christian, Seifert, & Gonzalez, 2012). In a controlled experimental study of novice designers, those using Design Heuristics to help them design a set of salt and pepper shakers produced more creative designs compared to the control condition (Yilmaz, Seifert, & Gonzalez, 2010). The designs generated through the use of heuristics appeared more diverse and unusual, and showed improvements in visual form likely to appeal to users. This result suggests that idea generation using Design Heuristics can have immediate effects on conceptual creativity.

We also tested the usefulness of Design Heuristics for a design team working together as a group. This involved a very experienced group of engineering designers redesigning a commercial product line in a workshop setting (Daly, Christian, Yilmaz, Seifert, & Gonzalez, 2012). These professional engineers used the Design Heuristics by discussing each of the cards provided and exploring how each heuristic might apply to specific products. The team generated new designs even though they had worked on these products for many years. For example, from the heuristic "incorporate user input," the team considered multiple ways to provide feedback to inform users about how to make adjustments. This card prompted the designers to reconsider options for user input, and new designs followed. These results suggest that the combination of Design Heuristics and group interaction may enhance diverse idea generation. Our studies confirm the value of the Design Heuristics tool for assisting designers in generating more creative concepts.

6.7 Conclusion

Designers across disciplines face substantial challenges in generating creative concepts. Thinking of many diverse concepts may be very helpful in selecting the most promising designs based on user needs and context. But it is often difficult to think of designs that are different from the ones already generated. To address this problem, our research has investigated how expert designers introduce variations within their concepts. The resulting tool, 77 Design Heuristics, captures each of the observed design strategies observed across disciplines and illustrates their use in existing products. Studies of both expert and novice designers using Design Heuristics have verified their utility in generating creative designs. This chapter provides information about the set of 77 Design Heuristics, and describes how to use them in design problems.[2] Design Heuristics are a useful tool for designers in any domain who want to maximize the diversity of the concepts they generate in order to create their best designs.

6.8 Appendix

77 Design Heuristics Extracted from Designers' Concepts

#	Design Heuristic	Definition
1	Add Levels	Identify different levels of the product functions and add a series of gradual changes to facilitate gradual transitions of uses.
2	Add motion	Apply motion as part of the product's function. Consider how this can decrease the need for user activity or act as a playful attribute.
3	Add natural features	Explore relationships between the product and nature to improve function or aesthetics.
4	Add to existing product	Add an existing item to the product's functions. Consider physical attachment, creating a system, or defining relationships to products.
5	Adjust function through movement	Allow users to adjust function through moving the product or parts. Consider different motions (e.g., rotating, sliding, rolling) and controls.
6	Adjust functions for specific users	Design functions around a user population based on age, gender, education, and diverse abilities; allow each user to adjust functions.
7	Align components around center	Arrange extra components around a main function. Consider arrangement or configuration around a circular design element.
8	Allow user to assemble	Make the user part of the process by having them assemble if too large for packaging or if adds to user understanding of function.
9	Allow user to customize	Involve the user by giving them customization options. Consider how this provides the user with a sense of ownership and awareness.
10	Allow user to rearrange	Allow the user to change the configuration of components for adjustable functions by simple attachments or alignments of components.
11	Allow user to reorient	Allow user to flip the whole product or parts vertically or horizontally to perform different functions.
12	Animate	Give lifelike qualities to the product by replicating human or animal features, gestural forms, and emotions.
13	Apply existing mechanism in new way	Consider how function is accomplished in other products and determine how they can be applied to your product when adapted to its new use.
14	Attach independent functional components	Identify different parts or systems with distinct functions and combine them by assigning form to each, and add a connection between parts.
15	Attach product to user	Make the user part of the function by attaching the product to a body part, such as user's head, finger, or feet, and redefine product use.
16	Bend	Form an angular or rounded curve by bending a continuous material in order to assign different functions on the bent surfaces.
17	Build user community	Consider how two or more users can work together to operate the product, or how one user's operation affects another.
18	Change direction of access	Use different ways of approaching the product, such as from the side instead of the front, to create more flexible solutions.
19	Change flexibility	Change material properties with different or modified material; Consider durability, collapsibility, function, and adjustability.

(continued)

[2]http://www.designheuristics.com

#	Design Heuristic	Definition
20	Change geometry	Use a simpler geometric form to achieve the same functions. Changing from familiar forms redefines user interaction with the product.
21	Change product lifetime	Consider the assumed lifetime of a product or its parts and alter the number of times it can be used.
22	Change surface properties	Highlight areas where the user interfaces with the product by using different colors, textures, materials and forms.
23	Compartmentalize	Divide the product into distinct compartments or add a compartment.
24	Contextualize	Envision the detail of how and where the product will be used and fit the product to this context.
25	Convert 2D to 3D object	Create a three-dimensional object by manipulating two-dimensional materials through bends, twists, creases, or joints.
26	Convert for second function	Design the product or its components with multiple stable states, where each state defines a separate function.
27	Cover or wrap	Overlay a cover, form a shell, or wrap the surface of the product and its parts with another material to customize, add function, and protect.
28	Create service	Develop a service by defining interactions between the user and a service provider.
29	Create system	Identify the core processes and define a multistage system that synthesizes those processes to achieve an overall goal.
30	Divide continuous surface	Divide single, continuous parts or surfaces into two or more elements or functions that can then be repeated and reconfigured.
31	Elevate or lower	Raise or lower the entire product or its parts to provide adjustability in use by allowing ergonomic solutions or suggesting additional functions.
32	Expand or collapse	Design the product to get larger or smaller to adjust or change function. Consider fluids, inflatables, flexible materials, and complex joints.
33	Expose interior	Show the inner components of the product by removing the outer surface or making it transparent for user perception and understanding.
34	Extend surface	Widen or expand the functioning surfaces of the product to enhance, adjust, or add new functions.
35	Flatten	Compress the product until flat with flexible materials or joints. Consider the effects on portability, structure, and storage.
36	Fold	Create relative motion between product parts or surfaces by hinging, bending, or creasing to improve packaging and storage.
37	Hollow out	Remove parts from the product for better fit to other products, functions, or the user's body.
38	Impose hierarchy on functions	Present functions in a set order to assist product use. Make the steps for reaching each function clear by controlling access to functions.
39	Incorporate environment	Use the living or artificial environment as part of the product by designing around it rather than distinguishing from it.
40	Incorporate user input	Identify product functions that are adjustable and allow users to make changes through an interface. Integrate in a cohesive, intuitive way.
41	Layer	Build the product through a series of layers of similar or different materials to provide various functions and interest.
42	Make components attachable or detachable	Make individual parts attachable or detachable for additional flexibility, ease of use, carrying, or repair/replacement.
43	Make multifunctional	Identify a secondary complimentary function for the product and create a new form to accomplish both functions.
44	Make product recyclable	Replace disposable components with reusable ones or vice versa. Modify the design according to the capabilities of the new material.

#	Design Heuristic	Definition
45	Merge surfaces	Join the surfaces of two or more components with complementary functions.
46	Mimic natural mechanisms	Imitate naturally occurring processes, mechanisms or systems.
47	Mirror or Array	Reflect or repeat elements about a central axis or point of symmetry to distribute force, reduce manufacturing cost, and improve aesthetics.
48	Nest	Fit one object within another. Design the inner form of the containing object to match the outer form of the contained object.
49	Offer optional components	Provide additional components that can change or adjust function, purchased separately or included, and where they are stored.
50	Provide sensory feedback	Return perceptual information (i.e., tactile, audio, visual) to the user, reducing errors, confirming actions, and informing of product function.
51	Reconfigure	Define relationships between functional components and change their configuration; attachments or alignments of components.
52	Redefine joints	Identify the ways product parts are connected and modify by removing, covering or changing the orientation of joints.
53	Reduce material	Remove material from the product by eliminating unnecessary components or shaving structural elements to make more efficient.
54	Reorient	Design the product to perform different functions based on orientation. Consider flipping the whole product or its parts vertically or horizontally.
55	Repeat	Copy components or an entire product to enhance function, allow for multiple simultaneous functions, distribute load, and decrease costs.
56	Repurpose packaging	Convert leftover packaging after the product is removed. Consider turning the packaging into a game, decoration, or other useful product.
57	Roll	Revolve a part or the entire product around a center point or a supporting surface by adding flexible materials.
58	Rotate	Move components of the product about a pivot point or axis, or allow the user to move components to adjust or change function.
59	Scale up or down	Change any of the physical dimensions of the product or its parts. Consider how changes in size and proportions can affect function.
60	Separate functions	Define functional components of the product and separate them into individual forms.
61	Simplify	Remove unnecessary complexity from the product to reduce costs and waste, or make the product more intuitive.
62	Slide components	Move one component smoothly along a surface in order to open and close surfaces, rearrange components, or adjust size of the product.
63	Stack	Stack individual components or make the entire product stackable to save space, protect the inner component, or create visual effects.
64	Substitute way of achieving functions	Replace an existing component to accomplish or enhance the same function. Consider different materials or forms to achieve the function.
65	Synthesize functions	Combine two or more functions by joining them to form a new device. Consider how the two functions can complement each other.
66	Telescope	Identify long components and split them into sections that can slide into each other. This can help to reduce product size when not in use.
67	Twist	Turn simple geometric forms in opposite directions, single or multiple times, to create a playful, iconic product; provides a larger surface area.
68	Unify	Cluster elements according to intuitive relationships such as similarity, dependence, proximity, to unify them for visual consistency.

(continued)

#	Design Heuristic	Definition
69	Use common base to hold components	Aligning modules on the same base or railing system to reduce the number of parts needed, allow users to rearrange, and make compact.
70	Use continuous material	Find ways to create connections between parts, and apply one continuous material to them to reduce parts, joints, and complexity.
71	Use different energy source	Replace expected energy source and redesign accordingly. Possibilities include chemical, geothermal, hydroelectric, solar and wind.
72	Use human-generated power	Make the user act as the power source for both primary and secondary functions, and the synthesis of multiple energy sources.
73	Use multiple components for one function	Identify the core function of the product and use multiple components to achieve the same function, with components specialized in tasks.
74	Use packaging as functional component	Embed packaging within the product, create a shell or cover for a component or entire product using the package, and uncover for use.
75	Use recycled or recyclable materials	Explore the use of recycled or recyclable materials within the product. Consider how structure and context will change.
76	Utilize inner space	Hollow out the inner volume of the product or its parts, and use the space for placement of another component.
77	Utilize opposite surface	Create a distinction between exterior and interior, front and back, or bottom and top for complimentary or different functions.
78	Visually distinguish functions	Create visual relationships among product functions by changing individual design elements.

References

Daly, S. R., Christian, J., Yilmaz, S., Seifert, C. M., & Gonzalez, R. (2012). Assessing design heuristics in idea generation within an introductory engineering design course. *International Journal of Engineering Education (IJEE)*, 28(2), 463–473.

Daly, S. R., Yilmaz, S., Christian, J. L., Seifert, C. M., & Gonzalez, R. (2012a). Uncovering design strategies: A collection of 77 examples helps students tap their own creativity. *ASEE Prism Magazine–Journal of Engineering Education Selects (JEE Selects)*, 22(4), 41–41.

Daly, S. R., Yilmaz, S., Christian, J. L., Seifert, C. M., & Gonzalez, R. (2012b). Design heuristics in engineering concept generation. *Journal of Engineering Education*, 101(4), 601–629.

Purcell, A. T., & Gero, J. S. (1996). Design and other types of fixation. *Design Studies*, 17, 363–383.

Yilmaz, S., Christian, J. L., Daly, S. R., Seifert, C. M., & Gonzalez, R. (2013). Can experienced designers learn from new tools? A case study of idea generation in a professional engineering team. *International Journal of Design Creativity and Innovation*, 1(2), 82–96.

Yilmaz, S., Daly, S. R., Christian, J. L., Seifert, C. M., & Gonzalez, R. (2012, May 21–24). How do design heuristics affect outcomes? In M. M. Andreasen, H. Birkhofer, S. J. Culley, U. Lindemann, and D. Marjanovic (Eds.), *Proceedings of 12th International Design Conference (DESIGN)* (pp. 1195–1204). Dubrovnik, Croatia.

Yilmaz, S., Daly, S. R., Seifert, C. M., & Gonzalez, R. (2014, June 16–18). Design heuristics as a tool to improve innovation. *Proceedings of the Annual Conference of American Society of Engineering Education (ASEE)*, Indianapolis, IN: American Society for Engineering Education.

Yilmaz, S., & Seifert, C. M. (2010). Cognitive heuristics in design ideation. *Proceedings of 11th International Design Conference, DESIGN 2010* (pp. 1007–1016), Dubrovnik, Croatia.

Yilmaz, S., & Seifert, C. M. (2011). Creativity through design heuristics: A case study of expert product design. *Design Studies, 32*, 384–415.

Yilmaz, S., Seifert, C. M., & Gonzalez, R. (2010). Cognitive heuristics in design: Instructional strategies to increase creativity in idea generation. *Artificial Intelligence for Engineering Design, Analysis, and Manufacturing, Special Issue on Design Pedagogy: Representations and Processes, 24*, 335–355.

About the Author

COLLEEN M. SEIFERT is an Arthur F. Thurnau Professor in Psychology at the University of Michigan and holds a PhD in Psychology from Yale University.

RICHARD GONZALEZ is a Professor of Psychology, Statistics, and Marketing at the University of Michigan, and Director of the Research Center for Group Dynamics, Institute for Social Research. He holds a PhD in Psychology from Stanford University.

SEDA YILMAZ is an Assistant Professor of Industrial Design at Iowa State University. She holds an MFA in industrial design and a PhD in Design Science from the University of Michigan.

SHANNA R. DALY is an Assistant Research Scientist and Adjunct Assistant Professor in the College of Engineering at the University of Michigan. She has a PhD in Engineering Education from Purdue University.

A tool developed by their cross-disciplinary research team, the *77 Design Heuristics for Inspiring Ideas,* is available from Design Heuristics, LLC, www.designheuristics.com

7

THE KEY ROLES OF STORIES AND PROTOTYPES IN DESIGN THINKING[1]

Mark Zeh

Introduction

Stories and prototypes play essential roles within a design thinking process. They are the "glue" that binds the process together. Together, they contain both the problem to be solved and a hypothesis about how to solve it.

Stories and prototypes serve as a means of communication between customers and product developers, enabling the mapping of rational and emotional customer needs to concepts and ideas. This chapter contains a description of the roles of stories and prototypes within a design thinking product development process and discussion on how to create and use them. These are illustrated with an example from industry.

7.1 A Design Thinking Product Development Framework

Since the product development processes in every organization differ by things such as number of phases, important milestones, and so on, they must integrate design thinking into their processes in differing ways (Brown, 2008, 2009; Martin, 2009). In this section, a general product development process will be used to describe how stories and prototypes are created and evolve throughout the development cycle.

The product development process diagrammed in Figure 7.1 has been divided up into three general phases of work: Identify User Needs and Find the Value Proposition; Build, Test, Iterate, and Refine; and Validate and Communicate Broadly.

[1] Acknowledgments: Thanks to Dr. Sven Schimpf of Stuttgart, Germany, for helping with the initial structure, draft, and research. Thanks to Pelham Norville PhD, ECE, of Framingham, Massachusetts, for critical editorial reviews and feedback.

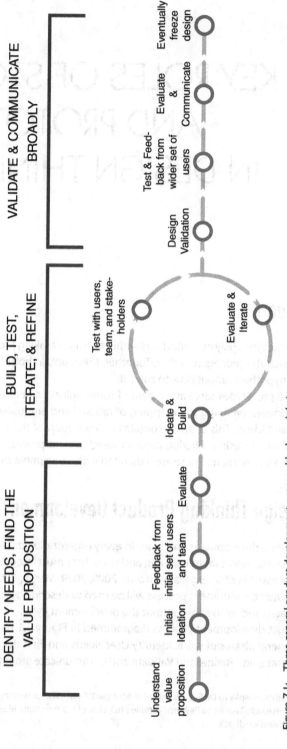

Figure 7.1: Three general product development process blocks and their components.

The purpose of the work in the first phase is to create an understanding of user needs and test the first hypotheses of the development team. These early story fragments are usually focused on describing the need. They start out as statements, like: " ... every man knows how hard it is to pick out what he's going to wear to a special event with his partner, like a dinner party or concert. It would be great if we could find a way to help with this decision."

This story fragment already contains testable elements: How many men have this need? What is the context that leads up to this problem? What are the emotional and practical aspects of making this decision? What kinds of practical problems do men who have this problem face? After these questions are answered through user testing, a more complete story, supported with prototypes, should be built to allow potential users to interact with the story.

The second phase, Build, Test, Iterate, and Refine, is used for development of the stories and supporting prototypes. This is done through cycles of testing with users, evaluating feedback from users, and cycles of iteration. The result of this phase of work should be a set of stories and prototypes that can be used to describe the user needs and problems, along with concepts that resolve them.

The final phase, Validate and Communicate Broadly, is used to validate the concepts developed in the first two phases. At this point, the stories and prototypes are refined into use cases, product architectures, and product descriptions. These are validated through focus groups and quantitative user testing. Additionally, they are communicated broadly through the product development organization, by employing personas, scenarios of use, and preliminary product specifications. All of these terms will be defined in the coming sections as they are employed.

7.2 What Is a Story?

Stories are the basis of human communication for abstract concepts. The foundational element of storytelling is the creation of a narrative, upon which a story can be built: establishment of a plot, a point of view from which the story is related, players in the story, settings in which things take place, and so on.

Stories are used to reinforce cultural values and to help us visualize a situation or scenario that is a departure from our personal or cultural experiences. They are also used to teach, reinforce memories, or serve as a means of validating cultural values. They help us visualize future states, inspire creativity, or see things from the perspectives of other people. A good story transforms a collection of facts and experiences into shared concepts and meaning.

In a design thinking product development process, stories allow concepts to be visualized and experienced before they have been designed and developed. Initially, the development team builds the stories and then shares them with the other stakeholders in the product development process. Stakeholders may include end users and potential partners.

The function of stories within the product development process is to create shared definitions of the types of problems to be solved, the contexts in which the problems occur, and the types of solutions that could resolve the problems. Stories allow quick communication within the complete product development team, its intended customers, and its extended stakeholder chain.

The user's or customer's point of view is the basis of the story narrative. Stories told from the viewpoint of the end user of the product are the foundation of business-to-consumer concepts. Business-to-business stories require creation of many variations of a particular story, each from the narrative points of view of the various customers within a value chain.

A good product development story informs its audience about the functional activities and interactions among people, products, and systems. It also reveals the emotional and rational needs of the people in it. Understanding these things allows the audience of the story to feel empathy for the people within it and develop a "feel" for how credible interactions within the described context could work.

Since the purpose of stories used within product development is to quickly communicate an idea and build shared meaning, they should be constructed using some basic principles:

1. **They should be short.** It should be possible to understand them within a few minutes. Presentation format must be thought through.
2. **They should start by introducing a context.** Where does the problem occur? Who has the problem? Who is involved in the experience or solution?
3. **They should describe the problem, as experienced by a representative person, or sets of people.** Composite personas should be built up, using the characteristics of customer types of interest.
4. **They should be limited to a time period in which an end-to-end experience of the user problem occurs.** How does it start and how is it resolved? Some of the basic story forms are:
   a. Scenarios built around a use case.
   b. A "day in the life" of a user.
   c. Product journeys.
5. **They should be supported with sensorial information: sketches, photos, renderings, prototypes, and example products.** These should show:
   a. Where the story takes place.
   b. What the people in the story look like.
   c. What potential problem solutions could look like or work like, or how the problem is presently being resolved.

Learning efficiency increases dramatically with the use of photographs, cartoons, sketches, and video, rather than words. The human brain processes semantically complex information more quickly than it does words (Hockley, 2008). Hence, visually rich communication tells a story faster and with more subtlety than does a long text.

Some common formats for telling stories are:

a. Spoken storytelling.
b. Acting them out (in person and with video).
c. Diagramming and storyboarding.
d. Written text.

Story fragments are best worked out through use of spoken storytelling, acting out, and diagramming. Story fragments arise as the team tries to understand the problems that need to be solved. They originate from a wide variety of sources, including user research and marketing knowledge. Fragments are almost always oral in nature, allowing them to be rapidly exchanged and iterated. Often, they are declared in forms like: "… the people we saw use text for almost all communications now, but can't do this well while walking. None of them felt comfortable using speech commands in public either … " This fragment can be instantly acted out, tested in real situations, and rapidly expanded into a full scenario of "texting while walking." After this scenario has been developed, the story may be used to generate solution hypotheses. These may be included in the story and similarly acted out, tested in real situations, and expanded further through iteration.

As an example of how to rapidly develop a story to help the product development team advance their understanding of a user problem, let's use the previously introduced example of the man trying to select clothing to wear on a concert date. Print format restricts this to using methods c and d from above:

> Ed is a 50-year-old engineer living in Berlin. He has been standing in front of his closet for a while, trying to figure out what to wear tonight. In an hour he's going to pick up his girlfriend Elise, the owner of a chain of luxury hotels, to go see La Traviata at the city opera.

> Afterward, they'll go for a drink at a popular new bar, then go to dinner at a cool new restaurant, where they'll meet some friends. Ed asks himself nervously: "What to wear so that I'm not over- or underdressed, Elise is pleased, and I'm comfortable?!?"

This story describes one of the scenarios where an example target customer is experiencing a problem that he doesn't feel confident to resolve alone. The story is five sentences long and is supported by photographs, as shown in Figure 7.2, illustrating the key

Figure 7.2: A storyboard collage. Ed, Elise, the city opera, a popular local bar, the cool new restaurant. Courtesy of Sabine Muth.

elements in the story. One of these photos was created by a team member acting out the situation, others were assembled from stocks of images. The characters are described in very thin detail, but enough so that the reader can visualize them.

This story can already be tested for resonance with customers and used as the basis of a brainstorming session for potential solutions. It takes only moments to understand the situation, the characters, the contexts, and the problems, but it is not written at such a high resolution that it appears definite. It serves the functions of communicating the problem, starting discussions, and brainstorming solutions.

"Ed" and "Elise" are personas—composite characters created to represent important types of customers. They have functional, emotional, and personal characteristics shared by people of selected typologies. Personas bring focus to a story, forcing you to tell it through the actions and views of people relevant to your business (Mulder & Yaar, 2007). Besides the functionally focused criteria of the "job" (Christensen, Anthony, Berstell, & Nitterhouse, 2007) that the customer types are trying to accomplish, personas should be used to convey nonfunctional attributes of customer needs: What makes something desirable for this customer type? What are they trying to achieve on emotional and functional levels that presently can't be done?

Creating personas at the beginning of the product development process enables the team to pick the right types of users for the testing and refinement steps, and allows the team to frame the questions to be answered through prototyping, user testing, technical explorations, and business evaluations.

7.3 What Is a Prototype?

In a design thinking process, prototypes are created to answer a set of questions, test assumptions, and demonstrate how something works, or could work. A prototype could be as complicated as the first fully operational build of a new submarine design, or as simple as a first model for an idea for a hair dryer grip, made from a soda can and modeling clay.

The level of resolution and complexity of prototypes developed at the start of a product development process should be much lower in resolution, finish, and function, than those in prototypes used for testing and validation before production start (Benyon, Turner, & Turner, 2005). Whatever types of prototypes are used, their purpose is to communicate and allow interaction with an experience, without the major investment of creating a real, fully functioning version. Simulation has the added benefit that iterations can be made very quickly and easily.

The term *prototype* connotes something of substance to most people, but prototypes do not need to be physical objects—simulation is a powerful prototyping tool. This can include video showing how an as-of-yet undesigned product and service could work, or a digital animation demonstrating how a software interface could look and function. Prototypes can also be created using one of the many new, easy-to program microcomputers, such as a Raspberry Pi or an Arduino.

A key principle in design thinking is to learn as much as you can as early as you can in a design and development process. The maxim *"Fail early to succeed sooner,"* often attributed to David Kelly of Stanford University and IDEO, summarizes this thinking. This maxim paraphrases a much-older, similar saying by Helmut von Moltke, the famous military strategist, who observed "No battle plan survives contact with the enemy." In other words, it doesn't matter how well the team plans, or how much experience they have, or how smart they are, all concepts contain some inherent, unknown flaws in their assumptions and execution, so it's best to find out what they are as quickly as possible and correct them. In addition to imparting the benefit of creating more desirable products, this leads to lower development costs and faster product development cycles.

Figures 7.3 and 7.4 show examples of quick prototypes built to test early ideas about a system for use by the driver of large construction equipment. Figure 7.3 shows a simple, quick-to-build, electronics breadboard prototype. This prototype features a bounceless switch, using a momentary-on pushbutton. The switch toggles between on and off states, when the button is pushed. Rather than hooking the switch up to the entire system for which it is intended to be used, an LED shows whether the switch is on or off.

Figure 7.4 shows a prototype that can be used to test placement and modulation of the switch with a potential user in the actual control area. With this type of prototype, it is possible to quickly gain a wide variety of initial user feedback, including whether the entire physical control architecture is valid, or whether another type of control modality, such as a switch on a panel, would present better solution architectures.

Figure 7.3: Electronics breadboard prototype.

Figure 7.4: A placement and modulation prototype.

It would be possible to connect the prototype in Figure 7.4 up to the breadboard circuit shown in Figure 7.3, but doing so would hinder the purpose of the prototypes as a tool for co-development with users. The Figure 7.4 prototype is deliberately constructed to allow other stakeholders, including selected potential customers, to interact with it and modify its form. They can cut into it, tape things onto it, position it in various places within the control cockpit, and so on, without risking damage to any of its other functions. After stakeholders have modified the prototype, a next set of prototypes might combine the breadboard electronics into some of the forms and volumes developed using this prototype.

Table 7.1 summarizes the purposes, testing locations, audience for the tests, and level of effort required to build each prototype.

Prototypes serve two important purposes in product development: They are tools to learn and to communicate. Prototypes make a concept tangible and allow it to be shared and developed with people who are not engineers or designers.

However, anyone who has ever visited a product development organization and seen old prototypes lying around knows that unless they are produced at a very finished level, prototypes only make sense within a product story, where their purpose is to demonstrate how key parts of that story could happen. Separated from their stories, they become orphans: useless objects, the purposes of which usually can be recognized only by the people who created them.

Table 7.1: Summary of the Initial Goals of Building Each Prototype

	Switch Breadboard (Fig. 7.3)	Cardboard box prototype (Fig. 7.4)
Purpose of building prototype:	■ Test reliability. ■ Test how much tolerance to input modulation the switch needs, to make it predictable for the user. ■ Verify power requirements. ■ Communicate switch behavior to other project stakeholders.	■ Understand whether this is the right type of actuator for the system being controlled. ■ Create a starting point for co-design of form, placement, and switch feel with users and other project stakeholders.
Test with:	Development team and selected users.	Development team and selected users.
Test where:	In the lab, workshop, and development build area.	In the development build area and in-context with users in the machines they are driving. Users try it in the cab of the machine, testing various physical placements and modifying the geometry.
Time invested to build:	2 hours	20 minutes
Materials cost to build:	<$10.00	<$2.00

7.4 Putting It Together—Combining Stories and Prototypes

As already described, the first phase of the process shown in Figure 7.1 begins with the construction of story fragments and scenarios, created to describe the user within a specific context. These first scenarios have the primary purposes of the describing the user, the context, and the problem. They may also include first hypotheses about how the user could solve the problems outlined in the story.

In the middle of the first phase, ideation usually begins. The team forms hypotheses about how users' needs and problems could be resolved by asking "What if" questions. Using the previously introduced example of Ed and his difficulties in finding something to wear, it is reasonable to ask "How can we recommend something for him if we don't know what he has in his closet?"

An ideation session could then begin, with the purpose of finding ways to determine and maintain an inventory of Ed's closet. Using the properties of his persona (he's a software engineer), it seems reasonable to assume that he is open to a solution based on technology.

Proposed solution spaces should include all problem areas, but might include a scanner app on his smartphone (But would he remember to use it when he's bought something new? Or when he's decided to remove something from his closet? What about the things he already owns?), or a scanner in his closet (Where would it be located? How would it be powered?), or maybe it's a service (Someone takes an initial inventory and is notified of every new clothing purchase?).

All of these solutions can be incorporated into the first scenario and tested for resonance with customers. Evaluation of customer feedback should narrow the range of potential solutions and also pose more questions. Are any of the proposed solution spaces perceived to be better than the status quo by customers? What combination of solutions will create value for them?

This initial testing ends when the team has identified a needs set and problem that customers believe would be valuable to solve. When this happens, it is time to enter the second phase of work.

During the Build, Test, Iterate, and Refine phase, user needs, story fragments, hypotheses, and early prototypes from the first phase are transformed into a complete product story (Davidow, 1986). A complete product story describes an end-to-end customer experience: how the customer is attracted to the product, what the first moment of "meeting" the product is like, the transactional experiences with the product, the experiences of setting it up and interacting with it, and the experience of what happens when the product becomes obsolete.

This phase begins with the team completing the product stories for each selected customer type, based on the outcomes of the first phase of work. This exercise enables the development team to frame the questions required as inputs for further ideation activities.

At this point in the process, first concept prototypes usually are created. A principle of design thinking is that you should "build to learn" (Kelley, 2001) That is, you should start without knowing most of the answers, designing the things you build in a way that allows you to test hypotheses or discover how things could work. This is accomplished through the process of building things that frame out a first hypothesis, then through getting feedback on the prototypes as quickly as possible. This allows the solution to emerge from the process of building, as well as from user testing and feedback.

Continuing with the example story of Ed and his closet: Three hypotheses for a solution to the question of how Ed could maintain an inventory of his closet were proposed. In this phase, rapid prototypes of the various types of solutions could be built and tested with potential users in their closets, in order to get feedback on the concept and answers to the questions raised.

The app idea could be tested by using one of the many app prototyping tools to build a quick simulation, showing interactions and screen flows. The scanner idea could be tested by placing a simple foam prototype inside users' closets and asking them to act out the scenario. The service idea could be acted out by users with an app prototype.

In all cases, the prototypes should be simple "architecture": placeholder representations of an undefined product. All of the prototypes should be deliberately "undesigned." Colors would be neutral, elements basic, and any service or digital elements should be focused on fundamental interactions. The role of the prototypes should be to allow the team and users to interact with the functions of the elements when acting out the scenario of use. The goals should be to gain feedback on whether the interactions would be credible within the scenario, how they could be improved, and what other types of problems the solutions might create.

Prototype building and evaluation rapidly expose false assumptions and flaws in the story. They also allow better feedback, involving cognition from the haptic and visual portions of the brain (Latour, 1986).

After the first set of stories and prototypes has gone through a cycle of testing and feedback from stakeholders and users, their feedback is evaluated: What did we learn? What additional problems did we discover? What can be combined to build a more complete product story or prototypes that are more refined? What types of prototypes need to be created to better illustrate how certain parts of the experience could work? How does the story fit with the company strategy and business model? Basically, what worked, what didn't work, and what needs to be changed or improved?

Following the evaluation step, there is generally another ideation cycle. These iterative ideation cycles must be carefully planned since they usually overlap with one another. In the first rounds of this phase, it is normal to learn things that invalidate early assumptions. This means that the learnings from the previous ideation, build, and test steps need to be prioritized: it is a waste of resources to iterate details of a part of a story or prototype, if something about the value of the overall concept has been called into question.

Returning to the example about Ed, the team may have included the scanner idea into the story, then used a packing tape dispenser as a rapid prototype. This would enable test subjects to act out the scenario in some detail. They would need to place the "scanner" somewhere in the home, where the user could locate it when it was needed. They would also need to determine how users would actually use the scanner. Would they take items out of the closet, scan them, and then hang them back up, if they weren't determined to be suitable? They would also have to determine how the scanner would be powered and connected to data. This could be prototyped with cardboard boxes and extension cords.

The team may find that the scanner is easy to hold and that the connection problems are minor, but they may also discover that the scanning process and getting feedback is considered onerous to the user. In this case, there wouldn't be any point in developing the details of the scanner until questions about the overall interactions were answered. In the end, the team may go with one of the other ideas, in order to provide better inter-action and feedback.

The team goes through many cycles of the Ideate, Build, Test, Evaluate, and Refine stage (Figure 7.1), until a complete product story emerges. This story should describe a complete experience for the target users, meeting their functional and emotional needs. The complete story is communicated through a variety of types of assets, including storyboards, videos, simulations, or a written text. Communication is supported with sets of refined prototypes. These should demonstrate how the product works, how customers interact with it, how it might look and feel, and how it could be built.

At this stage, the resolution of the prototypes that support the stories will vary by organization and product development process. In many organizations, there are separate "looks-like" and "works-like" prototypes at the end of this stage. The work of the next stage is to integrate these aspects.

After building a complete product story, it must be validated and communicated before it can be transformed into an implementable product definition. During this final phase of work, the product story and prototypes are communicated to a wider variety of stakeholders than in the previous phases. In this phase, quantitative user testing, partner presentations, presentations to government and regulatory agencies, and so on take place. These are used to validate the utility, desirability, and viability of the concept.

Different types of communication materials are prepared for each type of audience. The point of view of the product story needs to change, depending on the interests and needs of each type of stakeholder in the value chain. For example, a distributor will want to hear about the user story so that they can understand the business appeal, but they will also be very interested in the operational aspects of a new product. Government officials and agencies will be less interested in the customer appeal of a product, but they will want to understand how the product is used and where it sits within a broader social and legal framework.

Generally, the design of the product is frozen during this last phase of work. This means that its appearance and the technology that enables it are fixed, so that it can be designed for production and deployment. At this point, the main points of the business model making it viable have been framed out and approved by internal and external stakeholders.

An Example from Industry

Orbit Baby: Using Rapid, Rough Prototypes, Spoken Stories, and Acting to Gain Key Insights at the Beginning of a Product Design Process

The following example tells the story of how Orbit Baby employed prototyping and story development to create one of their successful products for babies.

Orbit Baby began as a product startup in Silicon Valley, a place better known for its digital companies. Then-president and co-founder Joseph Hei (now chief design officer at Ergobaby) explained:

> "Bryan White, my former business partner, and I had noticed that the baby product area seemed like a market which still had quite a few user problems and hadn't seen any real innovation in a while, so it seemed like a good area to start a company in."

Mr. Hei and Mr. White worked quickly, turning initial ideas into rough proto-types, so that they could gain feedback from parents.

> "Like most people, we started with a hunch—a notion of how and why car seats and strollers could be better. We thought about some scenarios of how people would load and unload their baby from a baby seat, then quickly hacked together some really rough prototypes of a rotational interface idea we had for a car seat, using existing products as the starting point," reported Mr. Hei.

Mr. Hei and Mr. White moved rapidly to gain feedback from potential users of their ideas, using their rough prototypes to help communicate the customer experience they were envisioning.

> "We immediately tried to get them in front of parents that we could recruit, just to get some initial reactions to the concepts. The prototypes were what allowed us to walk them through what we thought the story of their day might

be," related Mr. Hei. "What was interesting is that we got some of it wrong in our heads, getting the story right was more important than anything else."

Using the process of Build, Test, Iterate, and Refine, Mr. Hei and Mr. White were able to quickly identify unexpected errors in their assumptions and then make corrections to their overall concept. Their early work on the design of the handle of their infant car seat is a good example of how they used rough, rapid prototyping and storytelling to get the product experience and feature set right at the very beginning of the product design process.

> "When we were thinking about how to design a better infant car seat, one of the things we focused on was carrying it," related Mr. Hei, "We implemented a suitcase-style handle, so that you could more easily carry the seat by your side—we were picturing people interacting with the car seat and thinking 'oh, this would be the most comfortable way to carry it for long distances.' But, when we actually put it in front of moms, we were surprised to learn that they didn't want the handle to work that way at all. They wanted to carry the baby seat on a bent elbow, like a basket or purse." (See Figure 7.5).

Figure 7.5: An early prototype of the elbow-carry handle.
© Orbit Baby, Inc. Photo courtesy of Joseph Hei.

This unexpected result allowed the Orbit Baby design team to gain more empathy for their customers and understand their needs much better.

> "Part of it was that we are men, so we made some assumptions about how people would like to carry something, based on ourselves," explained Mr. Hei. "But the other part was that we got the story wrong—we were assuming that it

would be a longer-distance carry—longer in duration, but what the parents we interviewed told us is that what they were really looking for was a comfortable way to transfer the seat from their car to a stroller. In the end, they were not interested in optimization for longer-distance carrying: they thought the handle we designed and the scenario we described were cool, but our concepts didn't have anything to do with their daily experiences."

Based on that learning, the team abandoned their early assumption and re-designed the handle, so that it became easy to grab it with two hands. The final version of the product, with the elbow carrying handle, is shown in Figure 7.6.

Figure 7.6: The elbow-carry handle.
© Orbit Baby, Inc. Photo courtesy of Joseph Hei.

Interview conducted June 3, 2014.
Link: www.orbitbaby.com/

7.5 Employing Stories and Prototypes in Your Process

There are a few key points to remember when creating and developing stories during product development work:

1. **Communicate as efficiently as possible.** Involve as many senses as possible. Don't use words when you can use a picture or sketch. Act things out and make quick videos. Support pictures and videos with prototypes.

2. **Keep in mind where you are in the product development process.** Develop a basic understanding of the problem first, follow up with hypotheses of solutions, then actual solutions. Keep developing your stories to reflect current states of knowledge and hypotheses.

3. **Don't build a prototype until you know what you want to learn from it.** How will it be used? What do you plan to learn from building it?

4. **Don't try to learn everything with one prototype.** Build many rapid prototypes to test subcomponents of concepts. Wait to combine functions until after they have been tested and iterated separately. Appearance and function should not be combined until late in a product development process.

5. **Build scenarios to explore use cases, rather than trying to boil all the use cases down into one big story.** People and organizations rarely use one product or service to solve all of their problems.

Some common pitfalls into which companies fall, when trying to apply storytelling and prototyping methods:

1. **Being too much in love with themselves.** Remove yourself, your company, and your products from your stories. Describe products and services in generic terms. Be confident enough to call all of your present value propositions, business models, and understandings of customer behavior into question.

2. **Relying too much on present successes and understandings of past customer behavior.** Don't worry about "cannibalizing" your existing business. If you don't reinvent it, someone else will. Customer behavior is not static. Brand loyalty must continually be re-earned.

3. **Trying to do too much in one story.** Focus on a use case and succinctly depict how it works.

4. **Hanging on to unsupported concepts and use cases.** If some part of the story or a function of a prototype did not resonate with customers, it requires change, even if it was one of your most clever ideas, or was politically popular.

5. **Trying to polish things too early.** In phase one and in the early cycles of iteration of phase two, it should be possible to iterate stories and prototypes many times a day. Avoid data- or production-heavy methods of storytelling and prototyping. If the stories and prototypes must be sent out to a contractor for iteration, either the wrong tools are in use or the working level of resolution is too fine.

Build a plan to learn:

1. **Don't overthink the first story fragments and prototypes.** Plan to develop them through rounds of customer and stakeholder feedback.

2. **Use parallel paths.** Test several possible scenarios for the same problem. Combine the parts that work; drop the parts that don't.

3. **List out the complete value chain and test your stories and prototypes with all stakeholders in it.** Learn what is important to them.

4. **As stories and prototypes gain polish, shape the stories to reflect the point of view of the stakeholders being interviewed.** Make the stories relevant to them and elicit their feedback.

7.6 Conclusion

Prototyping and stories are inextricably intertwined with one another—a prototype can communicate an experience in a way that words never could. Also, a good story is needed to make the relevance of any prototype evident to anyone. This is especially the case for new-to-the-world products or services that meet needs in new ways. Stories and prototypes enable communication of a future vision in ways that allow customers to also visualize and interact with it.

Using stories and prototypes to communicate with stakeholders and users helps product development teams build a narrative about ideas and their usefulness. These tools give expression to customer needs, how people behave, and how they could interact with a new product. Stories facilitate the formation of hypotheses about viable solutions and help frame problem questions. These are all necessary to get any kind of useful output from a creative activity.

Any organization looking to incorporate design thinking into their product development processes must pay careful attention to how they integrate stories and prototypes into their processes. Overly complex and expensive prototypes cannot fill in the shortcomings of a poorly articulated or inadequate story. Better products can be built more quickly, by focusing on better stories, supported by prototypes of the appropriate resolution.

References

Benyon, D., Turner, P., & Turner, S. (2005). *Designing interactive systems: People, activities, contexts, technologies* (pp. 253–260). Essex, England: Pearson Education Limited

Brown, T. (2008). Design thinking. *Harvard Business Review*, 86(6), 84–92.

Brown, T. (2009). *Change by design*. New York, NY: HarperCollins.

Christensen, C. M., Anthony, S. D., Berstell, G., & Nitterhouse, D. (2007). Finding the right job for your product. *MIT Sloan Management Review*, 48(3), 38–47.

Davidow, W. H. (1986). *Marketing high technology*. New York, NY: Free Press.

Hockley, W. W. (2008). The picture superiority effect in associative recognition. *Memory and Cognition*, 36(7), 1351–1359.

Kelley, T. (2001). Prototyping is the shorthand of design. *Design Management Journal*, 12(3), 36–37.

Latour, B. (1986). *Visualization and Cognition: Thinking with eyes and hands*, Knowledge and Society Studies in the sociology of culture past and present, Jai Press, Greenwich, CT (Vol. 6, pp. 1–40).

Martin, R. (2009). *The design of business: Why design thinking is the next competitive advantage*. Boston, MA: Harvard Business School Publishing.

Mulder, S., & Yaar, Z. (2007). *The user is always right: A practical guide to creating and using personas for the web* (p. 22). Berkeley, CA: New Riders.

About the Author

MARK ZEH is a Design and Innovation Consultant based in Munich, Germany. His formal work with design thinking began with IDEO in Palo Alto, California, in 2000. Following his seven-year career with IDEO, he has worked with global firms, including Bose Corporation, Steelcase Inc., and the Commonwealth Bank of Australia, employing storytelling and prototyping processes to identify and create new products. He has lectured on aspects of design thinking at Stanford University, the Catholic University Eichstätt-Ingolstadt, and the Women's Forum Global Meeting. Additionally, he leads a design thinking–based entrepreneurship master's program at the Munich Business School in Munich, Germany.

Part II
DESIGN THINKING WITHIN THE FIRM

8

INTEGRATING DESIGN INTO THE FUZZY FRONT END OF THE INNOVATION PROCESS

Giulia Calabretta
Delft University of Technology

Gerda Gemser
RMIT University

Introduction

Managers recognize the importance of the *fuzzy front end* (FFE) for successful innovation (Reid & De Brentani, 2004). During the FFE, the innovation team identifies and selects interesting innovation opportunities, generates and selects ideas addressing these opportunities, and integrates the most promising ideas into product or service concepts for further development (Koen, Bertels, & Kleinschmidt, 2014).

A well-managed FFE will result in better innovation outcomes. However, the FFE remains a challenging step in the innovation process, due to its intrinsic uncertainty and the need to make choices based on incomplete information. Design professionals are particularly helpful for dealing with FFE challenges. Design professionals combine a sense of commercial purpose with a positive attitude toward change, uncertainty, and intuitive choices. Indeed, companies increasingly recognize the important role design professionals can play in the FFE and use them not solely for executing new product/service concepts resulting from the FFE but also for co-creating solutions during the FFE.

In this chapter, we provide guidelines on how design professionals and their practices and tools can help companies overcoming FFE key challenges. These guidelines (and the related examples) are derived from an analysis of prior literature and case studies on innovation projects in which design professionals were involved in the entire FFE, either as external design consultants or as internal design employees.

In the remaining paragraphs, we first discuss three FFE key challenges: defining the innovation problem(s), reducing uncertainty by managing information appropriately, and getting and maintaining commitment from key stakeholders. Then we discuss which and how design professionals' practices and tools can help overcome these FFE challenges. We conclude with advice on how to optimize collaboration with designers in the FFE.

8.1 Challenges in the FFE

FFE activities confront business practitioners with three key challenges for which design professionals' practices and tools are particularly helpful.

Problem Definition

At the beginning of the FFE, defining the innovation problem properly—for example, in terms of the target market, the needs to be addressed, the innovation objectives—can enable firms to identify and select valuable opportunities, which can steer idea generation and concept development toward unique solutions. However, innovation problems are generally complex, ill structured, and highly demanding in knowledge breadth and depth. When confronted with such problems, managers often only identify the most obvious symptoms or those to which they are most sensitive (e.g., current offerings' sales). The resulting problem definition might be too simple or too narrow and lead to new concepts that, for instance, do not follow a portfolio strategy or address only short-term market needs.

Information Management

All FFE activities require significant information management. Given the unpredictability of innovation outcomes, managers tend to reduce FFE perceived uncertainty by collecting as much diversified information as possible (e.g., market intelligence, technological knowledge, financial information). However, given the limitations of human information processing capabilities, simply accumulating information does not necessarily reduce uncertainty. To increase the odds of successful innovation, information should be selectively retrieved, meaningfully organized, and effectively communicated.

Stakeholder Management

Most innovation projects involve different stakeholders to access a broad spectrum of expertise and resources. While this approach is needed to address the complexity of innovation problems, stakeholders' individual objectives and interests (e.g., different departments, institutions/companies, career goals) might differ and even conflict.

The resulting frictions may lead to suboptimal problem definition, deviation from initial objectives, and even stakeholders' resistance to engage in the innovation project.

In the following paragraphs, we illustrate how specific design practices—that is, designers' way of working—and tools can support business practitioners in addressing the above-mentioned FFE challenges, thus building a case for a more prominent role for designers in the FFE. These design practices and tools for FFE are summarized in Figure 8.1. Any design professional involved in FFE activities generally adopts the design practices in Figure 8.1, which represent the true added value of designers. Conversely, the design tools can vary depending on designers' preferences or the specific context. Thus, our list of tools is not exhaustive but based on our field research and experience.

8.2 Design Practices and Tools for Assisting in Problem Definition

Design Practices for Problem Definition

As shown in Figure 8.1, design professionals use reframing and holistic thinking to support innovation managers in making sense of ill-defined challenges and overcoming biased and narrow problem formulations.

Reframing refers to design professionals' practice of stating a problematic situation in new, different, and interesting ways, thus paving the way for more creative solutions (Paton & Dorst, 2011). During the FFE, business practitioners generally define the innovation problem on the basis of their expert knowledge of the company, their information on current sales and market needs, and their past experience. For instance, a typical problem definition for innovation projects aimed at reinvigorating sales would be developing new offerings based on salespeople's suggestions, competitors' offerings, and incremental improvements of current products' technical features. This problem definition might be biased and short-term oriented, leading to new products/services with limited market impact and financial returns. When involved in the early stages of FFE, designers in general try to get to the "problem behind the problem" by means of reframing. Reframing is based on deconstructing a problem into its building blocks (e.g., subproblems and influencing factors) in order to highlight relevant aspects disregarded in the initial problem definition. By generating a different, sometimes broader perspective on an innovation challenge, reframing facilitates opportunity identification and points idea generation in more appropriate directions. Thus, design professionals supporting companies in innovation projects for addressing decreasing sales would take a broader array of influencing factors into account (e.g., lack of a deep understanding of user needs, lack of a clear vision for the industry/company future, lack of a distinctive brand image, narrow view on technological evolution) and reframe the problem definition into, for instance, developing a distinctive style for different user segments and a strong brand identity to incorporate into current and future offerings.

Designers' ability to look at a broader range of influencing factors for reframing the innovation challenge derives from their holistic thinking practice. *Holistic thinking*

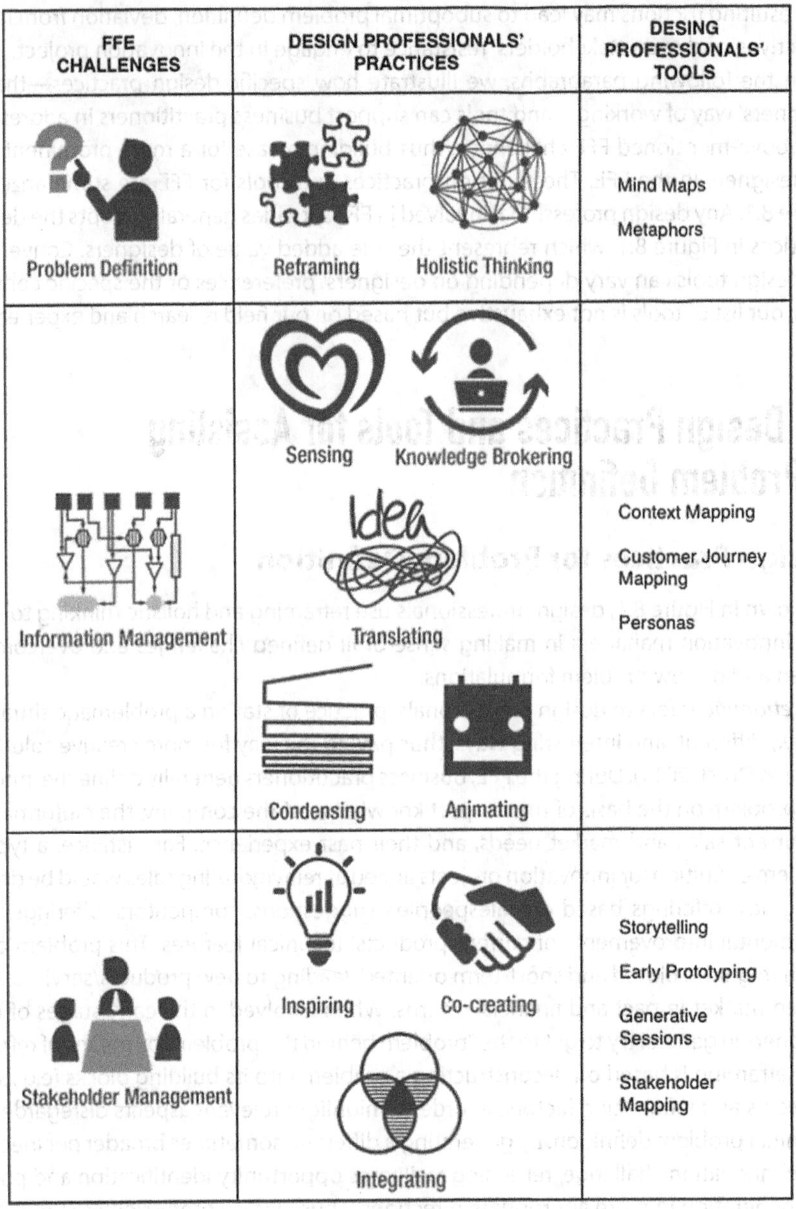

FFE CHALLENGES	DESIGN PROFESSIONALS' PRACTICES	DESING PROFESSIONALS' TOOLS
Problem Definition	Reframing — Holistic Thinking	Mind Maps Metaphors
Information Management	Sensing — Knowledge Brokering Translating	Context Mapping Customer Journey Mapping Personas
Stakeholder Management	Condensing — Animating Inspiring — Co-creating Integrating	Storytelling Early Prototyping Generative Sessions Stakeholder Mapping

Figure 8.1: Design professionals' practices and tools for FFE.

involves taking a comprehensive perspective on a problem, recognizing patterns, and making connections on the basis of (experience-based) intuition rather than a thorough analytical process. Through holistic thinking, design professionals can point innovation managers toward relevant, disregarded cues and connections (i.e., influencing factors) for effectively reframing their FFE problem definition.

Design Tools for Problem Definition

Design professionals can use several design tools to support their reframing and holistic thinking practices, including mind maps and metaphors.

Mind maps are diagrams for visually representing and connecting all the information regarding a certain theme (Buzan, 1996). In problem definition, designers use mind maps for identifying the issues influencing an innovation challenge and illustrating how such issues relate to each other. The resulting map offers a thorough overview that facilitates holistic associations for reframing an innovation challenge. Figure 8.2 provides an example of a mind map made by a design professional to deconstruct the innovation challenge of developing a new website for a public transportation service.

The process of developing a mind map should be loose and unstructured to stimulate holistic thinking in identifying relevant factors affecting FFE problems. However, design professionals usually follow the key steps below:

1. Write the name or description of the innovation problem (or subproblem) in the middle of a paper or any other drawing area.
2. Brainstorm on the major elements/factors/drivers of the innovation challenge, placing the thoughts on departing branches. Design professionals facilitate this step by keeping people engaged in the brainstorming and by maintaining the organic structure of the mind map.

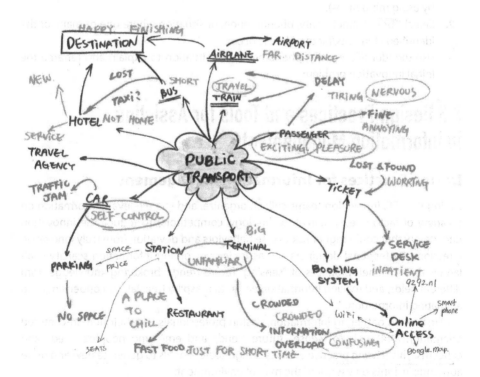

Figure 8.2: An example of a mind map.

3. Identify and emphasize connections (e.g., by using colors, shapes, and connectors).
4. Reflect (individually and collectively) on the resulting mind map to trigger new connections and problem reframing.

Metaphors allow for interpreting and illustrating a phenomenon through comparing it with something else (Hey, Linsey, Agogino, & Wood, 2008). In the FFE, design professionals use metaphors for better understanding the innovation context (e.g., the market, the users, the opportunities) and for opening up the problem space initially conceived by the business practitioner. By using compelling figurative expressions, designers' metaphors encourage managers to defer judgment, release the biases with which they approach innovation projects, and develop a deeper understanding of the innovation challenge. Combining metaphors with visual stimuli—as, for instance, in a mood board—makes this tool particularly effective. Mood boards favor a more open discussion, as the images use a metaphorical language, and do not immediately lead to innovation solutions. When creating mood boards, design professionals use images of different objects to convey the essence of a specific user group and to stimulate the client to adopt a future-oriented user perspective when generating new product concepts.

Design professionals usually undertake the following steps to identify and effectively use metaphors for FFE problem definition:

1. Define the key elements/factors/drivers of the innovation problem (for instance, by using mind maps).
2. Search for a distinct entity, phenomenon, or situation where one or many of the identified elements/factors/drivers occur.
3. Use the identified entity, phenomenon, or situation to explain and reframe the initial innovation problem.

8.3 Design Practices and Tools for Assisting in Information Management

Design Practices for Information Management

During the FFE, innovation teams collect, organize, and share relevant information on a variety of factors (e.g., market, technology, competitors) to reduce the innovation uncertainty that could trigger risk-averse behaviors and deviation from truly innovative directions. Professional designers can assist in overcoming information scarcity challenges through their practices of "sensing" future trends, brokering knowledge from different fields, and making information easier to grasp by translating, condensing, and animating information.

First, when gathering information, design professionals use their human-centered orientation and tools to "sense" future trends and emerging people's needs and concerns. This *sensing* practice enables design professionals to generate new and more authentic insights on users and the market environment.

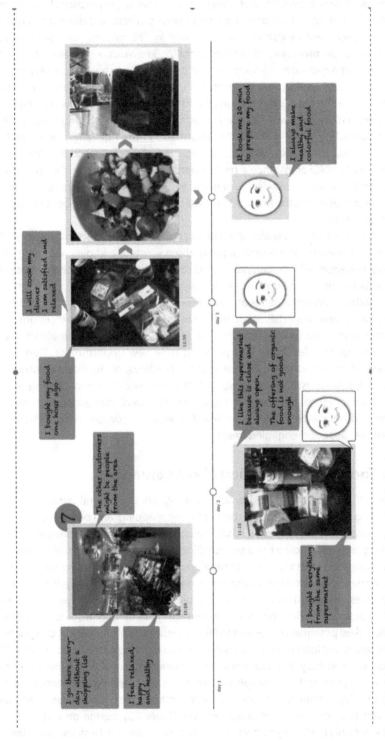

Figure 8.3: An example of the outcome of a generative tool.

Second, designers' *knowledge brokering* practice—that is, the practice of transferring market and technology knowledge acquired in prior projects and different industries to current projects and industries—helps greatly in FFE information management. Through knowledge brokering, designers mobilize knowledge domains apparently unrelated and not regarded as relevant. This not only increases the chances of detecting untapped opportunities, but also reduces FFE uncertainty since designers' (positive) experience in other industries is regarded as valuable grounding for new directions.

Once information has been gathered, designers support its sharing through their practice of *translating*, namely, converting information from one language to another (e.g., verbal to visual and vice versa; tacit to explicit and vice versa) so that it is usable by a broader audience. For instance, a mood board exemplifies the translation of verbal language—that is, the text description of brand values and market information—into a visual language that makes the information easier to use for generating better and more innovative concepts.

Designers' practice of *condensing* information is also important for the FFE. FFE-related information can be unstructured, disconnected, and overwhelming, thus challenging managers' information processing capabilities and generating uncertainty. Given their familiarity with complexity, designers can help companies sort insights, highlight key data, and combine them into relevant knowledge. Additionally, designers use visualization and materialization to communicate the condensed information in an engaging manner (*animating*), thus facilitating knowledge internalization and subsequent usage for identifying and pursuing truly new opportunities. Due to the effectiveness of the condensing and animating practices, an increasing number of design agencies, such as XPLANE, JAM, and INK, are specializing in offering visuals (e.g., infographics, animations, posters, digital visualizations) for condensing complex information (e.g., market intelligence, technical knowledge, company information) into comprehensive, clear, and engaging images.

Design Tools for Information Management

For enacting the above-mentioned practices, designers use several human-centered tools, including context mapping, customer journey mapping, and personas.

Context mapping is a qualitative design research method to uncover deep insights on how users experience a product or a service (Sleeswijk Visser, van der Lugt, & Stappers, 2007). Such insights deepen the FFE information base and help in finding more innovative solution spaces. In context mapping, participants are provided with generative tools (e.g., prototypes, photo cameras or recorders, diaries) to map their experience with a problem or a product/service category in an engaging manner. For instance, an organic food retailer asked potential consumers to fill in a verbal and photographic diary of their grocery shopping and food preparation habits to discover innovation opportunities. The simple and engaging diary tasks made respondents more aware of their experiences and enabled them to provide deeper insights on their grocery shopping behavior during the subsequent in-depth interview. Since the context mapping exercise usually generates lively answers and compelling visualizations (see Figure 8.3), design professionals use the exercise not only to identify opportunities, but also to engage business practitioners with their potential customers' life.

Customer journey mapping is a tool for mapping all the stages a customer goes through when using a product or a service (Stickdorn & Schneider, 2012). By covering customers' emotions, goals, interactions, and frustrations, the journey map provides a thorough view of the customer experience, highlights untapped opportunities, and stimulates idea generation. Figure 8.4 shows the customer journey of a train traveler.

Developing informative customer journeys requires time, effort, and the involvement of cross-functional teams with complementary competences (including design professionals for their intrinsic human centeredness). The following steps should be undertaken:

1. Define the subject of the journey (i.e., the type of customer) specifically.
2. Use a horizontal timeline to chronologically map all the activities a customer goes through when using a certain product or service, or when completing a certain task, including the before and after. Taking the customer's point of view is essential to keep the focus on activities rather than physical touch points.
3. Characterize the identified activities by describing the customer's aims, emotions, frustrations, challenges, and satisfactions. Both this characterization and the previous mapping could be supported by qualitative research such as in-depth interviews with customers.
4. Discuss the customer journey with different stakeholders (including customers) to identify opportunities related to the mapped activities.

Personas (see also Chapter 3) are fictional representations of current or potential customers describing and visualizing their behaviors, values, and needs (Pruitt & Adlin, 2010). In the FFE, design professionals use personas for summarizing and communicating the findings of market research in an engaging manner, for developing a shared user focus across different stakeholders, and for generating new ideas and concepts. The main strength of personas is their cognitively compelling nature, due to the fact that they give a human face to otherwise abstract customer information. Figure 8.5 offers an example of a persona ("Anna") for public transportation travelers.

The design professionals developed Anna for making the innovation team aware of authentic user needs and for embedding the user perspective in key decision moments (e.g., for discarding innovation directions that were not user centered).

Design professionals usually follow the steps below for creating personas:

1. Broadly define the type of customer for which personas should be developed (e.g., public transportation travelers).
2. Collect customer information from different sources (e.g., market research, expert interviews, desk research).
3. Based on the data, identify key characteristics that create differentiation within the selected type of customer (e.g., likes/dislikes, needs, values, interests). Normally, demographic characteristics are not considered at this stage.
4. Identify, name, and characterize three to five different personas within the selected type of customer.
5. Visualize each persona through pictures (e.g., their face, their activities, visual elements from their context), demographics (e.g., age, education, job, family status), representative quotes, and affective text.

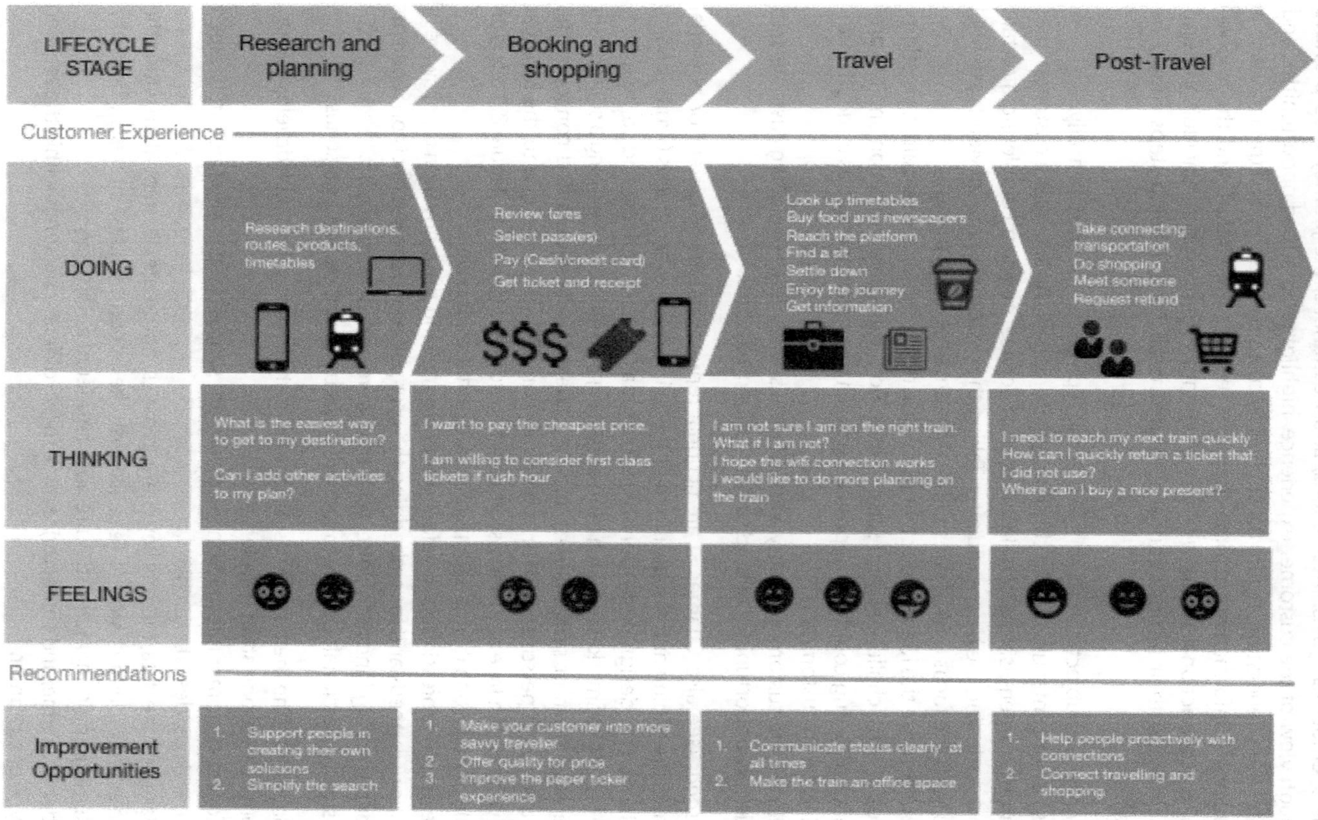

Customer Experience

LIFECYCLE STAGE	Research and planning	Booking and shopping	Travel	Post-Travel
DOING	Research destinations, routes, products, timetables	Review fares Select passenger Pay (Cash/credit card) Get ticket and receipt	Look up timetables Buy food and newspapers Reach the platform Find a sit Settle down Enjoy the journey Get information	Take connecting transportation Do shopping Meet someone Request refund
THINKING	What is the easiest way to get to my destination? Can I add other activities to my plan?	I want to pay the cheapest price. I am willing to consider first class tickets if rush hour	I am not sure I am on the right train. What if I am not? I hope the wifi connection works. I would like to do more placing on the train	I need to reach my next train quickly. How can I quickly return a ticket that I did not use? Where can I buy a nice present?
FEELINGS				

Recommendations

| Improvement Opportunities | 1. Support people in creating their own solutions
2. Simplify the search | 1. Make your customer into more savvy traveler
2. Offer quality for price
3. Improve the paper ticket experience | 1. Communicate status clearly at all times
2. Make the train an office space | 1. Help people proactively with connections
2. Connect traveling and shopping |

Figure 8.4: An example of a customer journey.

"I like working on the train, so I go for those I think are less crowded."
"I don't use public transport during the weekends."
"I need to be on time!!!"
"I wish it would be easier for my parents to use public transport to visit me."

Figure 8.5: An example of a persona.

Name	Anna	Interests	Likes sport (morning runs, tennis, yoga). Loves dinners with friends in restaurants. Can't live without her mobile.
Age	31	Home life	Lives with her boyfriend and a dog, commutes four days a week, does not want a car, has a salary allowing for two long exotic trips a year.
Working	Brand manager in a multinational		
Background	Got her master's degree at 25, still thinking about doing a PhD.		

8.4 Design Practices and Tools for Assisting in Stakeholder Management

Design Practices for Stakeholder Management

As shown in the bottom row of Figure 8.1, design professionals engage stakeholders in the FFE by continuously *inspiring* them through new perspectives, insights, and approaches to FFE challenges. Thanks to their future orientation, their openness to exploring new ideas, and their use of compelling visualizations for working and communicating, design professionals help business practitioners suspend risk-averse judgment and embrace new innovation directions.

Design professionals also use *co-creating* as a practice for generating and maintaining stakeholders' commitment to the FFE over time. Specifically, in all the FFE activities in which stakeholders are involved, design professionals stimulate their active participation and frequent interaction. Through co-creation, design professionals encourage stakeholders to consciously devote cognitive effort to FFE activities, ensuring that they develop ownership of the project and of its innovative outcome.

Personal interests and hidden agendas might inhibit an effective FFE, particularly when many different stakeholders are involved (e.g., in network innovation projects or in innovation projects for the public sector). Design professionals can help clients align different perspectives (*integrating*), by leveraging their outsider and expert status and by pushing the user perspective as a decision-making criterion for achieving agreement during the FFE. By immersing themselves in user experiences, business practitioners are less likely to base their decision making exclusively on their own perspectives and interests, thus becoming more open to alternative solutions with more market potential.

Design Tools for Stakeholder Management

Tools supporting designers in stakeholder management include storytelling, early prototyping, generative sessions, and stakeholder mapping.

Storytelling (see also Chapter 7) refers to the use of visual and verbal narratives for conveying information (Beckman & Barry, 2009). Communicating through storytelling offers a more compelling and effective way of delivering information than using "dry facts," thus helping designers develop trust and commitment across stakeholders. Storytelling can focus on explaining use and usability challenges (*informative story*) or on simply creating emotional connections between customers and stakeholders (*inspiring story*). Both types of stories may help stakeholders generate ideas and agree on interesting innovation directions. Figure 8.6 shows an example of a storyboard used by design professionals to convince a provider of public transport services that a more user-centered travel information website could substantially improve travel experience.

Early prototyping (see also Chapter 7) permits the testing of different ideas and concepts in a rapid and iterative fashion. Through early prototyping, design professionals provide tangible artifacts that allow stakeholders to experience more vivid manifestations of the future and eventually develop commitment to new directions. Designers for a consumer electronics manufacturer used early prototypes to develop digital services for their high-end coffee machines, and to persuade business stakeholders to transform their revenue model from selling high-end products to selling product-service systems. Due to the "hands-on," iterative working style, the use of early prototypes increased business stakeholders' sense of ownership and commitment, which were essential to take the digital service innovations to the market.

Generative sessions are used in connection with context mapping and usually entail inviting users to share their experiences and engage in activities in which they express their views on new product ideas and concepts. Design professionals use generative sessions also with FFE stakeholders in order to stimulate them to share their experiences, views, and opinions and break the silos that might prevent a fruitful collaboration. To prepare for these sessions, participants are given a specific task and generative tools (such as cameras or diaries) to allow them to record specific events, feelings, or interactions. These tasks and creative facilitation techniques during the sessions help stakeholders reflect on their ideas and motivations, and to open up to the discussion. The results of a generative session are never definitive, due to the high amount and the raw form of the generated insights. Thus, while the objective of creating stakeholders' agreement and commitment should be achieved during the session, its outcomes are normally further fine-tuned ex-post by the design professionals.

Stakeholder mapping is a tool for visualizing the stakeholders involved in a project and their interests, relationships, and interdependencies. Design professionals use stakeholder maps in the FFE as the cornerstone for building a common agenda. Although interests, relationships, and interdependencies may be highly dynamic and nontransparent, a stakeholder map provides an initial overview for the early detection of stakeholder-related opportunities or obstacles to effective FFE. Additionally, many design professionals use stakeholder maps dynamically, by introducing game elements to monitor the evolution of stakeholders' interests, relationships, and

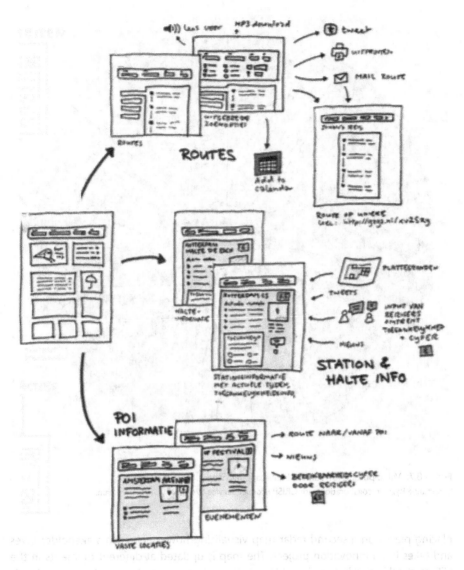

Figure 8.6: An example of a storyboard.

interdependencies over time. Value pursuit is an example of a dynamic stakeholder map using two radar maps to respectively identify the stakeholders and monitor them throughout a project. The first step in building this map is identifying the most relevant stakeholders. Each stakeholder is then placed in one of the sections of the radar map. The identified stakeholders are subsequently described in terms of their expectations, contributions, and struggles they might experience in the innovation project. This first step is visualized in Figure 8.7, where the numbers listed at the outer rim of the map (1 to 7) each represents a stakeholder. Stakeholder positions are represented by

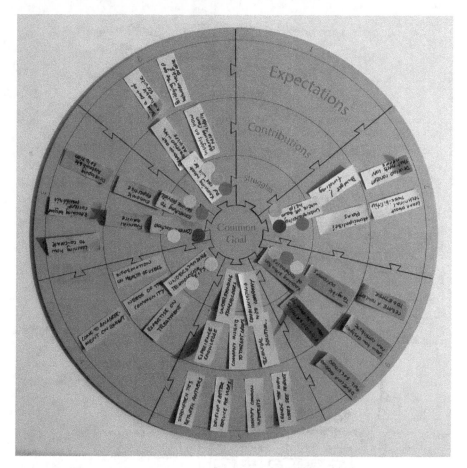

Figure 8.7: Value pursuit for stakeholder mapping: Step 1.
© Karianne Rygh, in collaboration with CRISP Product Service Systems 101 research team.

placing pieces on a second radar map visualizing how much each stakeholder gives and takes in an innovation project. The map is updated at different moments in the FFE, to check if and how stakeholder positions change over time. Design professionals usually co-create this map with the stakeholders, thus leveraging the mapping process itself for facilitating stakeholders' mutual understanding.

8.5 How to Integrate Design Professionals in FFE

In the prior sections, we showed how integrating design professionals and their practices and tools in the FFE helps to address some key FFE challenges. However, design professionals' strategic integration in the FFE is still not the norm but occurs only in companies with a design-oriented corporate culture. In this section, we describe some tactics developed by business practitioners and design professionals to achieve this strategic integration. The principle underlying these tactics is that design practices should complement rather than substitute for business practices in the FFE, and that design tools should be applied together with business practitioners to co-create the key outcomes of the FFE (i.e., new ideas and new concepts). Thus, in the FFE, design professionals should work with (rather than for) business practitioners. The identified tactics include:

Building a long-term, trustful relationship between business practitioners and design professionals. The chances of successfully integrating designers' practices and tools in the FFE increase if there is a long-term, trusting relationship between business practitioners and design professionals. After repeated, satisfactory transactions, business practitioners can assess the quality of the design professionals' practices and tools and progressively involve them in strategic innovation activities such as helping to identify *and* select opportunities. Under conditions of uncertainty (like in the FFE), team composition is often driven by personal trust based on prior experience. Once established, experience-based trust enables reciprocal and enduring relationships.

Developing mutual understanding. Both business practitioners and design professionals should invest in empathizing with each other's way of thinking and acting. To build the trust needed for playing a central role in the FFE, design professionals need to quickly develop a deep, authentic understanding of the innovation project and of the needs, objectives, and challenges of the business practitioners with whom they are working. Designers' empathizing efforts (e.g., generative sessions, asking the right questions, adjusting to the specific work environment in terms of, for example, language or clothing used) help to develop this mutual understanding and trust. However, business practitioners should be open to the intuition-driven practices and tools of the design professionals. As business practitioners are more familiar with analytical, linear, and quantitative tools, they might be skeptical about the appropriateness and effectiveness of designers' tools and practices in the FFE.

Preparing the ground. This refers to practices that explicitly or implicitly prepare business practitioners for undertaking FFE with an integrated intuition–rational approach. As noted earlier, this is often needed, as business practices and tools tend to be based on rationality rather than intuition. Practices to prepare the ground for an integrated intuition–rational approach should be planned and implemented at the beginning of or even before the FFE, and include activities like conversations and workshops aimed at activating business practitioners' creative and intuitive side. Another way to prepare the ground is to participate in promotional workshops where design practices and tools are showcased and experienced firsthand by the business practitioners. These workshops (commonly termed *jams*) are normally offered by design consultancy firms, but more and more often also by internal design departments attempting to promote their innovation approach to other departments. During these workshops, participants engage in the solution of a hypothetical case through design tools and methods. These cases could focus on problems of common interests (e.g., sustainability, community problems, personal health) or on company-specific problems if organized by internal design departments. Jams last three to four hours, take place at inspiring facilities, and involve 20 to 30 participants, including company owners and senior and middle managers. On the basis of these jams, business practitioners get a taste of what strategically collaborating with designers implies.

8.6 Conclusion

Design professionals are progressively establishing themselves as multifaceted and strategic sources of expertise for the FFE. Despite the growing number of companies integrating design professionals into the FFE, there is still limited knowledge on why and how to effectively implement such integration. In the previous sections, we elaborated on the "why" by showing how design professionals can use their practices and tools for addressing three key management challenges in the FFE. We also elaborated on the "how" by describing some key tactics, developed by both business practitioners and design professionals, for effectively integrating design practices and tools in the FFE. Our insights are summarized below:

1. Design professionals can help solve key challenges in the FFE by means of specific design practices and tools listed in Figure 8.1.
2. A critical step in the FFE is the appropriate formulation of the innovation challenge (*problem definition*). Designers can help to effectively take this step by reframing initial problem definitions and thinking holistically, using design tools such as mind mapping and metaphors.
3. Management of information for reducing FFE uncertainty can also be addressed by involving design professionals. Relevant design practices include sensing future trends, brokering knowledge from different fields, and making information more graspable by translating, condensing, and animating it. Useful design tools to address the information management challenge are context mapping, customer journey mapping, and personas.
4. The third key challenge in the FFE is getting and maintaining stakeholder support. Design professionals address this challenge by inspiring key stakeholders, by co-creating for maintaining stakeholders' commitment, and by aligning and integrating different stakeholder perspectives. Common design tools to help designers to get and maintain stakeholder support are storytelling, early prototyping, generative sessions, and stakeholder mapping.
5. Routes by means of which business practitioners and design professionals can work together on a more strategic level in the FFE include, among other things, establishing long-term, trusting relationships; empathizing with each other's way of thinking and ways of working; and "preparing the ground" to undertake the FFE with an integrated intuition–rational approach.

It is important to emphasize that the full potential of integrating design practices and tools in the FFE is achieved only when design professionals and business practitioners see each other as innovation partners, recognizing and building on each other's strengths. Design tools and practices complement rather than replace business practitioners' tools and practices for addressing FFE challenges. Thus, reciprocal recognition of each other's contribution in the FFE of innovation is essential.

References

Beckman, S. L., & Barry, M. (2009). Design and innovation through storytelling. *International Journal of Innovation Science*, 1(4), 151–160.

Buzan, T. (1996). *The mind map book: How to use radiant thinking to maximize your brain's untapped potential*. New York, NY: Plume.

Hey, J., Linsey, J., Agogino, A. M., & Wood, K. L. (2008). Analogies and metaphors in creative design. *International Journal of Engineering Education*, 24(2), 283.

Koen, P. A., Bertels, H. M., & Kleinschmidt, E. J. (2014). Research-on-research: Managing the front end of innovation—Part II: Results from a three-year study. *Research-Technology Management*, 57(3), 25–35.

Paton, B., & Dorst, K. (2011). Briefing and reframing: A situated practice. *Design Studies*, 32(6), 573–587.

Pruitt, J., & Adlin, T. (2010). *The persona lifecycle: Keeping people in mind throughout product design*. Burlington, MA: Morgan Kaufmann.

Reid, S. E., & De Brentani, U. (2004). The fuzzy front end of new product development for discontinuous innovations: A theoretical model. *Journal of Product Innovation Management*, 21(3), 170–184.

Sleeswijk Visser, F., van der Lugt, R., & Stappers, P. J. (2007). Sharing user experiences in the product innovation process: Participatory design needs participatory communication. *Creativity and Innovation Management*, 16(1), 35–45.

Stickdorn, M., & Schneider, J. (2012). *This is service design thinking*. Amsterdam, Netherlands: BIS Publishing.

About the Authors

Giulia Calabretta is Assistant Professor in Strategic Value of Design at Faculty of Industrial Design Engineering, Delft University of Technology (Delft Netherlands). Giulia earned her PhD from ESADE Business School (Barcelona, Spain). Her research interests are in the area of innovation and design management. Currently, her research is focused on understanding how design skills and methods can be effectively integrated in the strategy and processes of companies, with a particular interest on the role of designers in innovation strategy and early development. Her research has been published in such journals as the *Journal of Product Innovation Management, Journal of Business Ethics*, and *Journal of Service Management*. Correspondence regarding this chapter can be directed to Delft University of Technology, Landbergstraat 15, Delft, Netherlands or g.calabretta@tudelft.nl.

Gerda Gemser is Full Professor of Business and Design at RMIT University of Technology and Design (Melbourne, Australia). Gerda earned her PhD degree at the Rotterdam School of Management (Netherlands). She has conducted different studies on the effects of design on company performance (in cooperation with the European governments and design associations). She has held positions at different universities in the Netherlands, including Delft University of Technology and Erasmus University (Rotterdam School of Management). She has been a visiting scholar at the Wharton School, University of Pennsylvania (United States), and Sauder School of Business, University of British Columbia (Canada). Her research is focused on management of innovation and design in particular. She has published in journals such as *Organization Science, Organization Studies, Journal of Management, Journal of Product Innovation Management, Long Range Planning,* and *Design Studies.* Correspondence regarding this chapter can be directed to College of Business, 445 Swanston Street, Melbourne VIC, Australia, or Gerda.gemser@rmit.edu.au.

9

THE ROLE OF DESIGN IN EARLY-STAGE VENTURES: HOW TO HELP START-UPS UNDERSTAND AND APPLY DESIGN PROCESSES TO NEW PRODUCT DEVELOPMENT

J. D. Albert

Bresslergroup

Introduction: An Emerging Start-up Culture

The huge boom in hardware development is largely tied to an emerging entrepreneurial culture—a growing trend of inventors and professionals dedicated to making their own functional products. Just as, years ago, anyone with a website could launch a new business, today anyone with a 3D printer or a few electronic development boards can create a new product. The range and affordability of new production technology and access to funding has made creating new products cheaper than ever, and greater cultural acceptance of entrepreneurs and innovation has built public support for new devices.

Major corporate tech and innovation leaders have caught on to the culture of innovation and the appeal of new, physical products, too. They are setting up semisecret research and development (R&D) labs, where "intrapreneurs" are tasked with developing new innovations. Google, for instance, launched its GoogleX lab in 2010 to develop the self-driving car, and Amazon founded Lab126 in 2004. Lab126 has since produced hardware and software for devices including the Kindle Fire HDX, Kindle Paperwhite, Amazon Fire TV, and the Amazon Fire Phone. Even Nike has its own R&D product incubator—the Nike "Innovation Kitchen."

With more products flooding the market, entrepreneurs and intrapreneurs need good counsel. Though the settings are different, they face many of the same challenges in taking a product from concept to successful market launch. The old saying, "If you build it, they will come," is not a guarantee. Research by Booz and Co. shows that of the 50,000 new consumer packaged goods (CPG) products introduced every year, an estimated 66 percent fail within two years of introduction. According to Product Development Institute Inc., an estimated 46 percent of the resources that companies devote to the conception, development, and launch of new products go to projects that do not succeed. They either fail in the marketplace or never get there in the first place.

This chapter will outline the best practices for optimizing new products using design thinking to help early-stage ventures understand product design and development processes. Two real-life case studies will demonstrate how these processes can vary when applied to very different products.

9.1 The Basics

Research: An Overview of Different Types

If there is one thing entrepreneurs have plenty of, it is novel ideas. In this sense, entrepreneurs often feel like they're ahead of the game, but it would be a mistake to bypass the beginning of the typical design phase, which involves user research to identify product opportunities and other types of research to hone in on your product concept.

User research (Figure 9.1) entails talking with users and customers to understand what they are looking for and determine whether a product suits their needs. Less formal user research can take a few days. Other times, companies spend two years trying to develop an understanding of what users need from their products (e.g., Gillette spent over 3,000 hours studying consumers to create its latest razor for India).

Competitive research might identify and assess business rivalry (e.g., Porter's Five Forces, SWOT [strengths, weaknesses, opportunities, and threats], or 5Cs Analysis), pull from their best practices, learn from the competition's mistakes, and determine how to position and brand the new product in the target market. *Background research* should study the problem or need being met by the new product as well as the history of products in the target market.

Market research helps entrepreneurs understand the markets their products are entering. Entrepreneurs should do their homework and understand the market, its history, competitors, and business models that have and have not worked. If all the similar products on the market are priced at $20, why will a new product sell at $40? If two companies manufacture all of the products in a given market, entrepreneurs trying to break into those markets should know the companies and spend time gaining an understanding about them.

Figure 9.1: A designer at Bresslergroup digesting the results of some persona-based user research.

None of this entails big expenses—it's more about doing your homework by using what's out there to understand the landscape of the marketplace. Talk to customers, probe pricing dynamics, and get a good sense of whether what you're working on is unique and what's unique about it. The Internet is one source, but interviewing and talking to customers is critical. Conjoint analysis is a great tool to help understand the features and cost trade-offs from the customer perspective.

Defining and Refining the Product

The more a start-up can hone in on exactly what they are looking for, the more time and money they will save. Invariably, there will be open questions that require user research and testing, but there will be logical times to answer these questions during development.

Once the product concept is selected and some boundaries to work within are set, the first task is to expand the product vision and to consider the ideal user experience and product form, envision how the product will function, and consider how it will be produced. This vision, along with physical prototypes, should allow investors to "see" what they are investing in.

Intellectual Property to Protect and Drive Innovation

Start-ups cannot afford to cut corners in background research on intellectual property (IP). Generally there are existing IP patents that need to be maneuvered around, and trying to patent new ideas can be tricky. It is extremely beneficial for entrepreneurs—if they are able—to work with patent attorneys. If hiring an attorney is not an option, online research may suffice. Google Patents and the U.S. Patent and Trademark Office (PTO) are both good search tools. (See Chapter 21 in this book.)

Getting a trademark and the rights to name a new product are another piece of the puzzle. It is best to secure trademarks and naming rights as soon as possible. What an entrepreneur does not want is to have to change the name of a product after it has already hit the market, built brand identity, attracted customers, and potentially even cut the name into tools and hardware. Entrepreneurs should view product development holistically and think of designing the branding (both physical logo and communication approach) and packaging at the same time they consider designing the product itself. Done correctly, the brand, packaging, and product design language all reinforce one another.

9.2 The Process

Loop de Loop: The Winding Path from Idea to Product

One of the most common questions entrepreneurs ask is, how long it will take to develop an idea into a real product? Of course, the answer varies greatly depending on the complexity of each product, but development typically flows through a three-part design cycle that includes definition and design, engineering, and production "loops." The goal is to move gracefully from one loop to the next, but more often than not, a product has to run through a loop more than once.

In the *definition and design loop,* the product is refined through an iterative process of creative thinking, sketching, building 3D models on a computer, and making prototypes. Entrepreneurs who have little experience with the definition phase might want to check out the iDea Fan Deck, a tool meant to direct brainstorming and concept generation. It's well worth spending time on this formulation of the raw idea, whose quality is a reliable predictor of the product's ultimate success.[1]

The *engineering loop* is about taking the product designs and turning them into parts that can be manufactured. Engineers try to make the product light, sustainable, or cost-effective, depending on the clients' priorities. At the end of this loop, engineers build prototypes that look and work like the product. These are tested and improved. Finally, engineers create refined preproduction prototypes that offer a last gate for them and the product's design team to assess design, function, and user acceptance before production.

The third and final loop, *production,* often takes longer than entrepreneurs expect. Generally speaking, it is hard to develop a new product in less than 6 months. Nine to 15 months is more common, and many products take longer—medical devices in particular. Production separates a good idea from a great product. There is a huge amount of work that goes into getting the right product back from a manufacturer, and entrepreneurs need to diligently monitor quality.

[1] http://journals.ama.org/doi/pdf/10.1509/jmr.12.0401

Case Study: KidSmart Smoke Detector

An Alarming Success from a Couple of Biz School Grads

Major challenges: Designing around patents; production scale-up issues
Development Process: 20 months

In October 2003, a couple of entrepreneurs fresh out of business school wanted to develop a smoke detector that would allow parents to record their own voice message to wake sleeping children in the event of a fire. Research at Victoria University had proven that children wake up to smoke alarms with familiar voices much more easily than they do to smoke alarms with a generic beep.

Definition and Design

The entrepreneurs behind KidSmart (Figure 9.2) patented the idea of a device that allows parents to record their voice and then started on the product development course. The research from Victoria University and the work securing a

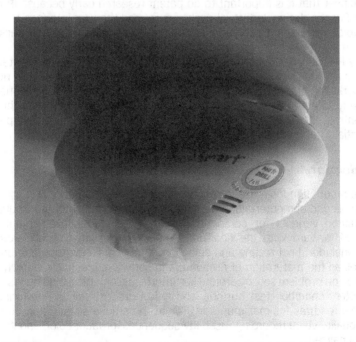

Figure 9.2: Kidsmart allowed parents to record their own voice message to wake sleeping children in the event of a fire.

patent gave these entrepreneurs a jump start on the design cycle because they had already done some concept generation. They continued into the

(continued)

design cycle by mapping out performance requirements, usability strategies, and product cost targets. It was critical early on to identify component suppliers and production strategies for all major components, including the actual smoke sensor.

Unlike traditional smoke alarms, this one needed high-quality recording and playback features. Without these features, the smoke alarm would not qualify for necessary safety certifications, and retailers would not carry it. To make sure the standards were met, a highly regarded acoustics expert was brought in to help specify a speaker and internal speaker chamber. The team came up with a few concept choices before settling on one that allowed the speaker to point toward the sleeping child's pillow to maximize the alarm volume.

Engineering

During the engineering phase, the electronics team modeled the main electronic and mechanical components and casework parts in SolidWorks. Electronics team members focused on nailing down the component specification and developing top-level firmware and software. The mechanical design team members worked out details for the major mechanical features and were forced to design around several patents. To do this they detailed new battery loading and changing functions and invented a new ceiling mount feature— a reminder that it is important to do patent research early because IP issues can influence design. Industrial designers refined the user interface to enable the voice record and playback features, and the team designed packaging with an eye toward differentiation on the shelf.

By February 2004, the first functional models were ready for limited preview at a trade show. Those early prototypes gave KidSmart what they needed for marketing purposes, and they were sent to Underwriters Laboratories (UL) for early safety and compliance testing in the UL's "smoke box." In this way, the entrepreneurs were able to get multiple uses out of their prototypes and essentially get more bang for their prototype bucks.

Production

In the production loop, the third loop of the development process, the bill of materials and computer-aided design (CAD) files were distributed to qualified production vendors for competitive quoting. Still, the development team pressed forward with the finalizing production design. Critical electronics tasks included final testing and selection of key smoke-sensing components as well as the maturation of firmware and software. Mechanical engineering focused on problem solving around key interface elements including resolution of battery chamber, test buttons, and indicators, as well as fine-tuning the internal features for maximum air flow to the smoke sensor, and output of high-quality voice recording. Usability sequences, product color, and graphics were finalized.

The entrepreneurs decided to have the smoke detectors manufactured abroad, so the team toured several facilities in China. They selected a manufacturer and started to work through the transition plan toward final production.

Prototyping and testing resulted in a list of changes to interface, electronics, mechanics, and packaging design, and the team worked with the Chinese production company to apply those changes.

By early May 2005, the production partner had created the tools and first preproduction boards. The parts were inspected and tested for functionality to iron out the remaining design and software bugs. As so often happens, there were some surprises, including a design flaw in the battery holder area and a struggle to optimize the code to work within the constraints of the microprocessor capability and battery life requirements.

From concept to production, the development process took 20 months. While all of this was going on, the KidSmart entrepreneurs were working in parallel on a public relations campaign. The campaign resulted in endorsements from firefighters, prominent media coverage across the country, and demand for the product that created a modest but manageable back order.

Prototyping: Increasing Fidelities for Different Benefits

Smart start-ups invest in prototypes. Different levels offer different benefits along the process journey. Their level of fidelity, or quality, increases as the product evolves. The prototyping process is a physical manifestation of the growth mind-set that informs design thinking: the information gleaned from testing each prototype—and each prototype's particular weaknesses—informs the next evolution. What follows is an accounting of the process journey via the prototypes an entrepreneur would do well to develop along the way.

Prototype for Initial Exploration

It can be very effective to catalyze the product definition phase by crafting exploratory prototypes. For example, a start-up led by serious racquet sport enthusiasts wanted to develop a new training aid. Once they settled on a concept, they built a crude "proof of principle" model, or "low-resolution" model, to test out the idea. After a few weeks of work, they had a first prototype that gave them the look and feel of the physical product they had in mind. It was hacked together using an Arduino (modular, adaptable electro\mechanical kit), quick and dirty CAD, additional development or "dev" boards, and some crude 3D-printed cases.

The initial configuration didn't work at first, so they switched things around and did some more testing. The next prototype was a huge leap forward in terms of understanding what the real product would need to do. This first-round prototype was instrumental in increasing the team's confidence that the idea was sound, and it laid a foundation for writing an initial specification document to guide the continued design and engineering of the product.

This kind of first-cut prototype is now easier to make than ever before and its importance should not be underestimated. With it, entrepreneurs can attempt to learn if a product works well, if it is easy to use, if it looks well designed, if it's easy to assemble, if it is affordable to manufacture, if it's robust, if the product is easy to change and customize over time. Early prototypes should be tested, documented, and iterated upon. Small loops should expand to include higher-level fidelity prototypes with more features. Prototypes that fail are ripe learning opportunities.

Prototype to Determine the Appropriate Direction

There is often a point in product development where there are multiple potential directions and not enough information to definitively choose one over another. It can help to develop a prototype for each leading option and compare them to determine which is the better product or which is more cost-effective, sustainable, or realistic—depending on an entrepreneur's priorities and constraints.

Sometimes the variable is the type of technology. For instance, a medical device start-up was developing a product and knew that there were two competing pumping technologies that would work. One was aimed at the higher end of the market and the other at the lower end, so that start-up created two prototypes to learn about the different solutions and to gain a better understanding of what would be required to produce one over the other. These prototypes were not fully designed products, but they were learning vehicles for performance and testing that allowed the start-up to choose one technology over the other.

Entrepreneurs should not to be afraid to break their prototypes. There is a time to develop a polished, presentable prototype to show investors or share at a trade show, but it is just as important to learn from rapid prototype iterations leading up to design lock.

Prototype to Attract Funding

As the design process advances, the prototypes become more refined and changes become subtler. This is the point at which most start-ups take their prototypes on the road to share them with investors. Investors are more swayed by preparedness than passion.[2] Part of being prepared is having the wherewithal and foresight to have developed compelling prototypes. People's attitudes about a product change when they engage with the product concept. (Have you ever seen an entrepreneur brave the "Shark Tank" without a prototype?)

If you don't have a prototype, make sure you have a great story, excellent team, and a proof-of-concept model to show. While having the best possible prototype available during a pitch session is best practice, even a less refined prototype will work toward helping investors understand where you are in the process. For example, when E Ink—the company that makes the electronic paper displays for Kindles and Nooks—sought funding, the "ink" barely worked. It took years for the technology to reach that point. But E Ink had a compelling story about electronic books and newspapers replacing paper. This coupled with a proof-of-principle prototype was enough for investors to connect the dots and see what was possible.

---

[2] http://amj.aom.org/content/52/1/199.short

Prototype to Garner Feedback

Entrepreneurs often cycle through the product-development process as they are raising money. Typically, they will pause at different phases to complete fundraising rounds or to use the prototype to get customers and companies interested in a more refined version.

This can lead to essential feedback that informs the next phase of design. It is much easier to respond and provide input when you have a physical prototype. Handling a product raises questions about usability, materials, form, and interactivity that would never come to someone's mind from looking at a rendering, no matter how well it's done. Either through user feedback or testing, prototypes often expose weak points. From a design standpoint, this helps you perfect your product, and from a funding standpoint, revealing weaknesses can help investors understand what further resources are needed.

Prototype to Define Patents

To file a patent, entrepreneurs need a solid understanding of what they are trying to do. Going through the process of building and understanding a prototype can help them write stronger patents because they will be able to state exactly how and why their product works. The product specification and definition phases set the stage for formal patent applications. Prototype development can also define new opportunities to broaden the patent suite. With prototypes, entrepreneurs may be able to envision future changes or additions and get claims on those as well.

Prototype to Facilitate the Manufacturing Process

Quality prototypes can help pave the way for a smooth manufacturing process. Ultimately, how good a prototype is and how deeply it has been debugged will determine the end result. When a start-up moves into the manufacturing phase, it is critical to be able to show a manufacturer what qualifies as an acceptable finished product. This is especially true when start-ups work with overseas manufacturers, but no matter where the product is made, eventually it is important that the prototype be as close to the final product as possible.

Testing, Testing: Assess Your Product Early and Often

Too many start-ups decide they can't afford research, which adds a tremendous amount of risk and reduces variable inputs for making improvements. Initial testing should help answer questions such as: Is it durable enough? Is it easy to use? Is it well designed? Do people like how it looks? How much are people willing to pay?

Some testing truths:

1. **Don't wait until the product is "perfect."**

    Early testing helps determine if the product is ready and leads to the most promising sales channels, but many times entrepreneurs don't want to hear criticism. This is a fear start-ups need to face head-on. It is both easier and cheaper to correct problems early in development rather than later on during production, so it is important to be open to early criticism.

Initial feedback can come from existing networks of people such as friends, family, and even online social networks. Entrepreneurs and start-ups should consider offering these first guinea pigs a range of possibilities, such as different styles and colors, and asking their opinions about how the product could work better and what features could be added. Most importantly, entrepreneurs need to remain open-minded about the feedback they receive and careful not to get locked into a product concept too early.

2. **Practice excessive abuse.**

Entrepreneurs need to realize their product will be used in ways and contexts that might terrify them. If it is possible to insert something upside down, people will insert it upside down. Laptops tested in the tight confines of a commuter flight, freezing temperatures, and abusive conditions imposed by kid use can reveal weaknesses as well as opportunities. At a certain point, entrepreneurs must determine if the product is robust enough for the real world.

For entrepreneurs testing the limits of their products, the question becomes, what abuse is common, standard, and excessive? Does the device need to be waterproof, or would dunking it in water be considered excessive abuse? Ultimately, entrepreneurs need to decide how much risk to assume versus how much risk is the users' responsibility. (And no matter what, accept that there will always be a couple of Amazon reviewers stating, "I ran this over, and it broke. This thing sucks.")

3. **Find the testing sweet spot.**

While learning during product development is safer, it is common for companies to not really "get it" until they do performance testing in the actual marketplace. Entrepreneurs need to work toward that sweet spot between development learning and product launch learning. Ultimately, this is a balance of time and money and how much a start-up can afford to figure things out postlaunch versus prelaunch.

For better or worse, public support and feedback can help determine when a product is ready to launch. Kickstarter, other crowdfunding platforms, and social networks are making it easier for entrepreneurs to stay in touch with customers, who can help demonstrate that a market is ready for a product to be introduced. If the product is to be tested by a very specific market or industry, start establishing connections early. It will take a lot of time to build this kind of network.

Production and Supply Chain: Solving for Complex Equations

Once a product is defined, iteratively prototyped, and tested, it is time to move on to production. Each product requires its own unique manufacturing strategy and entrepreneurs must consider how something will be made. Will it be a do-it-yourself (DIY) process, or will a contract manufacturer lead production? Will production be local, domestic, or offshore? Deciding which method to use is a complicated equation influenced by the product complexity, the production volume, how refined the design is, how hard it is to assemble, whether the product is made using established processes or new processes, and the materials used. There is no cookie-cutter approach.

DIY production means the entrepreneur or start-up will be the last point in the production chain. Often in DIY production, entrepreneurs will source parts and assemble the product themselves. In some cases, they might outsource assembly as well. This works best for products that will be made in low volumes, and it gives the entrepreneur or start-up a chance to look over each product before it ships. Sometimes DIY production is used as a starting point that gives way to another strategy.

With *contract manufacturing,* an independent manufacturer oversees sourcing and assembly. Using contract manufacturers does not mean entrepreneurs can kick back and relax, though. It just means the entrepreneurs have a production partner. Contract manufacturers can be especially efficient when a product is being produced with a tried-and-true manufacturing method, when the production volume is large, or when the design is refined and clear.

Regardless of the production method, costs are going to add up. To generalize, entrepreneurs and start-ups generally spend more on tooling, production, and building sales momentum than they do on defining, prototyping, and testing a product. Entrepreneurs should be optimistic yet conservative about production volume numbers and tooling costs. Tooling can add up really quickly, so it is important to be smart about budgeting and strategizing for tooling.

It is often difficult for entrepreneurs to be patient, but this is exactly what they need to do during the production phase. The time required for production typically surprises entrepreneurs. Finding vendors, documenting the design, waiting for tools, debugging the parts, and any required regulatory compliance always takes longer than they would like. There is a huge amount of work that goes into getting the right result back from a manufacturer, and entrepreneurs have to diligently monitor quality no matter where or how something is made.

Product Launch and Everything Else: Branding, Packaging, Certifications

Sometimes entrepreneurs get so caught up in product design that they don't think about all of the other essential business and design elements, like branding, packaging, certifications, and so on. If product branding and naming happen too late in the process, entrepreneurs risk missing out on opportunities to generate support and potential customers. Entrepreneurs must be able to articulate what their brand is about and lock down logos and product names before they think about releasing the final product. Unique packaging solidifies the brand identity—but for that to happen, the brand must have an identity.

Certifications are another key component. There are a lot of different agencies that need to stamp new products and give their approval. Final product (or in some cases prototypes) may need to be tested to any one of a number of UL, CE (European conformity), and Federal Communications Commission (FCC) standards. Those tests can be expensive and time consuming, but often the challenge is figuring out which standards are applicable.

It is best to identify which certifications will be necessary early in the process because the specifications of the various certifications do sometimes shape design, and sometimes adding a feature will add a new certification requirement. Ideally, a project manager will be available to oversee all of these "extra" but essential considerations. More often than not, that project manager is the entrepreneur/CEO/product manager who is running the show.

Case Study: Clean Cut Towel Dispenser

Redesigning an Existing Product at a Lower Price Point

Major Challenges: Miscommunication with overseas manufacturers, high cost to make
Development Process: 12 to 15 months

An entrepreneur in Michigan had set up an operation producing touchless paper towel dispensers out of his garage and selling them for $300 a piece. SMART Venture Concepts, a company with a history of creating new products and setting up companies behind them, saw an opportunity to improve the paper towel dispenser's design, lower the price point, and mass produce at a greater volume (Figure 9.3). To do this, SMART Venture entered the dispenser into the product design cycle.

Figure 9.3: Smart saw an opportunity to improve upon an inventor's touchless paper towel dispenser.

The touchless paper towel dispenser was unique in that it could dispense and cut any commercially available paper towel without human contact. The core technology enabled it to cut paper towels to any length, all through a gestural interface. This made it an ideal product for messy kitchens and workrooms.

Definition and Design

The initial device was mostly made out of sheet metal and machine components, and it required a lot of manual assembly. SMART Venture saw the potential for the product to be commercialized. If they could make a single device for around $25, they might be able to mass produce it, and they might have a winner.

At first, SMART Venture took their idea overseas to a manufacturer in China to both design and manufacture the device, but the manufacturer did not maintain the core technology and the final product was disappointing. They decided to essentially start from scratch and to partner with a U.S. design and engineering consultant. This is a trade-off that many entrepreneurs face. There are manufacturers overseas who will sometimes design and produce a product at an affordable rate, but this does not work for all products. If entrepreneurs choose this route, they must have an especially strong working relationship and excellent communication with a production partner who understands their mission.

Since the endeavor here was to redesign an existing device, the concept generation did not take long. The design phase started with the product definition and feasibility phase. Here, SMART Venture did an engineering analysis to understand the existing product and identify the core technology and functionality they wanted to maintain. They also did a cost analysis in which they took the device apart, counted the parts and fasteners, and tried to understand where the costs were. They built a cost model to see which costs came from sheet metal, fasteners, labor, and electronics and then identified the opportunities to design out some of those costs.

In the preliminary design phase, they came up with a rough architecture for the product. They decided it would be best to design the bulk of the product out of plastic molded products that would allow them to get rid of the fasteners by combining a lot of separate sheet metal parts into a single molded piece that would be easier to mass produce. They also did some industrial design development to assess ergonomics, ease of use, work flow, and user-device interactions.

Through this design and research, they determined that it would indeed be possible to produce a refined device in the $20 to $30 range while maintaining core technology.

Engineering

Next, the engineering loop consisted of concept refinement, engineering development, and alpha prototyping. One hurdle they faced was that a lot of the automatic paper towel dispensers require that light reflect off of a user's clothing. If a user is wearing dark clothes, the dispenser won't work. To work around this, the SMART Venture team implemented a through-beam technology that was not dependent on light reflections.

When it came to initial prototypes, the team used foam mockups to get a sense of size and shape and to test things like the visibility of the paper towel roll. In the original device, the paper towel was hidden inside the device, and there was no way to tell how much paper towel was left. The reengineered device features strategically placed windows.

The design loops and first engineering loop produced the concept definition, refinement, and feasibility work which took about 30 to 45 days. Then, in another engineering loop, the team built and tested three alpha prototypes. Each of the prototypes essentially looked and worked like the device, but they were made with components that could be manufactured in low volumes. Building these three prototypes and getting the device ready to enter production took 12 to 15 months. At the end of the process, SMART Venture had a device that could be produced in the $25 to $30 range and sold for about $75, rather than $300 like the original device.

9.3 Troubleshooting Common Mistakes

Simplifying the Product and Vision

Wanting to build everything from scratch and innovate around all aspects of the product are common missteps. Entrepreneurs need to consider how to reduce risk and maximize their chances of success by zoning in on what is critical. Following are a few key takeaways that can help entrepreneurs do just that.

Repurpose

While making everything yourself can be very satisfying, it is often smarter to find other products or companies and build on their devices or components. Doing so can save both time and money—limited resources in the new product development world.

For instance, a start-up wanted to build a custom tablet for use in the hospitality industry, but the start-up didn't have a lot of product development experience. They walked through the pluses and minuses of customizing a tablet for their application, and after laying out the development costs and risks, they decided to make a custom case rather than a custom tablet (Figure 9.4). The custom case fit around commercially available tablets and required a much simpler development process that saved considerable resources.

Often, solutions for a product in one industry can be pulled from a completely different industry. For instance, there is a commercially available garden sprayer that uses the guts of a carburetor to optimize its performance. (One advantage of working with

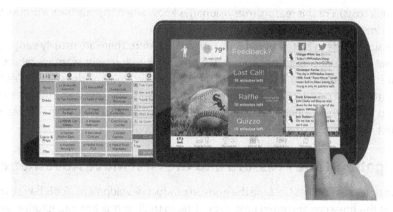

Figure 9.4: This start-up started out wanting to make a custom tablet but saved considerable resources when they decided to make a custom case instead.

product development consultants is their exposure to a wide range of products that gives them this broad base of knowledge.)

Entrepreneurs must be cautious when selecting off-the-shelf components to use in their new products. They risk putting themselves at the mercy of whoever makes the components, so it is critical to choose wisely and make sure the supplier is dependable. Think about whether that other company might view you as a threat and how you would recover if the product, component, or subsystem became unavailable. Have a production plan that doesn't depend on one supplier.

Keep It Clean

Entrepreneurs tend to struggle to limit their product's features, but added features often make products harder to use and may cloud what an entrepreneur is really trying to accomplish. Here, it is important to keep it clean, streamline, and pare down.

Prioritize the most important features and constantly question why something needs to be included. Does the device really need an app? Does it really need to function on different platforms? Is that feature actually necessary? It might help to strip functionality from the product and test whether it still works. If it does, it might be best to keep it simple. Test it with consumers—again, conjoint analysis can help companies make design trade-offs.

Each addition requires another layer of design, testing, documentation, and debugging. That includes factors and variants such as color and size. In a perfect world, an entrepreneur will launch with a minimum viable product and add more features and variants as the product attracts more customers.

Kill Your Darlings

At times, it is necessary to scrap a project and start over. When this does happen, it is often too late, and by then a great deal of money, time, and energy have already been lost (sunk costs). For this reason, true visionaries know when to scale back a product line. Don't throw good money after bad.

Passion needs to be tempered with rational thinking. There are usually early indicators that a product should be scrapped. If product development team members start to question the direction a product is headed or something just doesn't feel "right," it is wise to really consider whether to move forward. In some cases, an entrepreneur or start-up will learn so much from working through the production process—even if the product never goes to market—that it is worth it to continue.

Navigating Time Pressure and the First Move Advantage

Insane time pressure has become the norm in product development. Tools have evolved to meet this pressure, and rapid prototyping, rapid tooling, quick manufacturing, and air shipping allow unprecedented velocities during product launch. But there is still plenty of risk in going fast. Entrepreneurs are more likely to rush a decision, not test something fully, or miss an opportunity for a design breakthrough if they are too focused on delivery dates. Speed can be costly. Overnight shipping alone can add thousands of dollars to prototyping costs.

The flip side to that is that delivery dates can sometimes mean all the difference between launching first and getting the spotlight versus being a follower. Since it is easier than ever for entrepreneurs and start-ups to create new products, these new products are flooding the market. With so many people creating products along the same thought lines, being the first to launch a new product can provide a real advantage.

Design Thinking for Success

Launching a hardware start-up is not easy, but grounding a product development process in design thinking will help the rising number of relatively green entrepreneurs and intrapreneurs navigate the unfamiliar and avoid hiccups. As stated at the beginning of this chapter, there has recently been a huge boom in entrepreneurship stemming from access to prototype-making devices and other technologies, democratization of access to funding, rising awareness of design, and a cultural fascination with innovation.

All of this excitement and passion around new product development can obscure the fact that, in order to succeed, there needs to be a method to the madness. Knowing the tactics and specific skills—prototyping, testing, gaining feedback, learning from feedback and iteration—to follow through on a design thinking mind-set will set up entrepreneurs and intrapreneurs to succeed as they encounter investors and in the marketplace.

About the Author

J. D. Albert is Director of Engineering at Bresslergroup, a product development firm in Philadelphia. As a veteran of two start-ups and an experienced consultant, J. D. has fulfilled his own and his clients' visions many times over and across multiple industries. After receiving his bachelor of science in mechanical engineering from MIT, he co-founded E Ink, whose technology is used in e-reader devices including the Kindle and Nook. Later, he became CEO of the solar roofing start-up, SRS Energy. He has been granted more than 60 patents and is frequently called on to speak and write about product development in entrepreneurial settings, including at SEGD, Intersolar, Wharton, and Harvard Business School.

10

DESIGN THINKING FOR NON-DESIGNERS: A GUIDE FOR TEAM TRAINING AND IMPLEMENTATION

Victor P. Seidel
Babson College

Sebastian K. Fixson
Babson College

Introduction

Design thinking provides a tremendously powerful set of tools for designers and non-designers alike. However, non-designers face the difficulty in learning the tools and mind-set of design thinking while lacking the long training period that experienced designers undertake as part of their education. For example, when General Motors began working with a design consultancy to improve their own vehicle innovation processes, they faced a challenge that some of their engineering and marketing staff were to start using design thinking methods, but these staff had little formal training in how to use design thinking techniques. The increasingly widespread training of design thinking for teams of *non-designers* raises questions of how relative novices can learn effective methods given realistic time constraints for training such teams.

Indeed, ambitious efforts are under way across firms to get more of their staff involved in design thinking approaches. For example, IBM has opened a 50,000-square-foot "Home of IBM Design Thinking" in Austin, Texas, that they described as part of a "new approach to reimagining how we design our products and solutions," and Infosys has plans to train 30,000 of its employees in design thinking. While non-designers can relatively quickly learn the basic concepts behind design thinking and a user-centered approach to innovation, experience shows that not all product development teams trained in design thinking are successful. While the theory can be learned easily, the actual practice of design thinking comes with many practical challenges for implementation. To counter this problem, there are three important strategies for training teams of non-designers:

1. Encourage "dual-mode debate" of not only ideas but also processes.
2. Manage design thinking transitions of key mind-sets.
3. Adapt tools under changing team membership.

For each of these strategies we provide two points of specific implementation guidance. Our guidance is applicable to those implementing design thinking training programs as well as members of design teams hoping to improve their effectiveness. Before embarking on the strategies and implementation guidance, it is important to consider what non-designers will need to learn as part of adopting a design thinking approach.

10.1 What Do Non-Designers Need to Learn?

Training non-designers in the design thinking process can take many forms, from relatively brief lecture-based overviews of the underlying concepts to more hands-on experiential training sessions built around either simulated or actual projects. In general, teams need to learn design thinking tools as well as key mind-sets to improve their performance (Seidel & Fixson, 2013). Figure 10.1 presents a broad, three-part categorization of design thinking tools that teams typically are exposed to as part of their training: need-finding tools, brainstorming tools, and prototyping tools.

Considering the specific tools in each category helps to understand the range of tools that those new to design thinking will be exposed to—one guide lists 101 tools

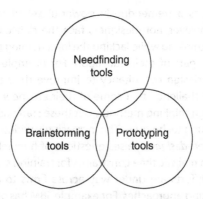

Figure 10.1: Three main categories of design thinking tools.

that aspiring users of design thinking might consider (Kumar, 2012). For example, teams being trained in need-finding tools may learn how to better interview for empathy, following a step-by-step guide to developing an interview protocol to promote wide-ranging responses. A second tool may be in how to build a "journey map" of a user experience, helping to chart emotional highs and lows across a user's experience with current products that may trigger insight into a new opportunity for innovation. In training on brainstorming tools, teams may learn how to apply "How might we" questions as input to an ideation session that might start with a few minutes of individual ideation before proceeding to group brainstorming. Another brainstorming tool is to adhere to specific rules that promote variance in ideas. In applying prototyping tools, teams may learn how to develop "low-fidelity" prototypes using cardboard or foam core to quickly test out ideas, or they may learn how to incorporate potential users into a prototyping process. Some tools apply to multiple categories, as represented by the Venn diagram in Figure 10.1, such as a brainstorming approach that involves the use of physical prototype materials.

In addition to specific tools, non-designers need to learn key mind-sets that are typically used by designers, such as encouraging a climate of debate, developing a sense of empathy, and promoting respect of different viewpoints. Learning the formal tools of design thinking along with relevant mind-sets gives design teams a set of capabilities for innovation, but the design thinking approach can also lead to challenges for non-designers, which we discuss next.

10.2 Challenges Teams Face with Design Thinking

Design thinking provides a tremendously powerful set of tools, but teams with non-designers face three main challenges in making use of design thinking. A first challenge they face is in understanding that the use of design thinking tools is dynamic and requires adaptation over time. Design thinking tools often have to be introduced to novice teams in a linear fashion, or illustrated as a journey through well-defined phases of analysis and synthesis (Beckman & Barry, 2007), but there is no "one-size-fits-all" way for the tools to be applied in any design scenario. While this point is made in almost all publications on design thinking—including the overview of design thinking found in Chapter 1 of this book (Luchs, "A Brief Introduction to Design Thinking")—this dynamic aspect of design thinking is particularly difficult for novice teams to adopt. Many teams may be expecting to learn a well-defined linear process, and so it is difficult for them to learn that as teams they need to decide which tools to use in an emergent, nonlinear, and iterative fashion.

The tools of design thinking are applied within phases of a product design process that can take many labels. For example, IDEO product development has described their range of tools as occupying three "spaces" of inspiration, ideation, and implementation (Brown, 2008). The d.school at Stanford University (2013) groups tools within five "modes"; Chapter 1 of this book includes four modes across two phases; and a high-level view of product development separates out two primary phases of concept generation and concept selection (Ulrich & Eppinger, 2012). A comparison of

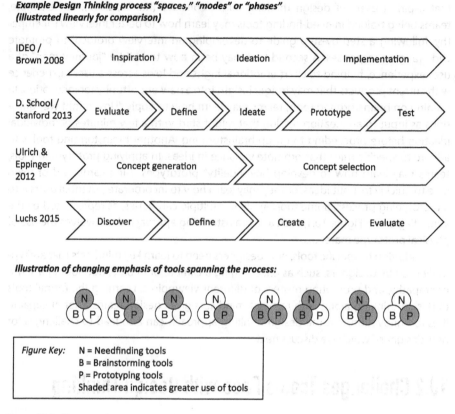

Figure 10.2: Example design thinking process phases and illustration of changing emphasis of tools over time.

these approaches is given in Figure 10.2, in which the nonlinear process of design is by necessity represented linearly for comparison purposes. This figure also includes a representation of the use of design thinking tools over time. Certain categories of tools—need-finding, for example—may have more emphasis at different points in the process (represented by shading in the figure), depending on how the innovation process has unfolded. Consider, for example, a team that was initially developing a range of concepts around new luggage. After the team presented three concepts at the end of their concept generation phase to a review panel, they realized one concept they had concerning a means to provide cell phone charging within the luggage was going to require additional need-finding activities, rather than further prototyping as they had initially planned. New data in the form of feedback reviews, technical milestones, or competitive offerings can change the decision rationale of which design tools are relevant as a next step. It can be a challenge for teams to know how to proceed in a process that is not overly prescriptive about which steps to take in every situation.

A second area of challenge for design teams is in knowing how and when to encourage different design thinking mind-sets. Debate over ideas is encouraged in a design thinking context, and once mastered this can be alluring to team members who relish the opportunity to engage in spirited discussion over options. However, as the well-known advice to avoid "feature creep" in engineering projects illustrates, too much debate over new options can slow the progress of projects without an increase in innovation. Jeff Hawkins was famous at Palm Computer in insisting that the main challenge for devices was to know when to limit new features. Knowing when to draw on a divergent mind-set seeking out new ideas versus when to focus on executing a course of action is critical for teams.

A third area of challenge is that team membership may change over the course of a project. This is the reality of innovation in modern corporate contexts, where individuals are often moved from one project to another. This contrasts to the more stable environment in which design is taught in academic environments or in corporate team training sessions, where stable membership during the process is usually assured. Taken together, these three main challenges provide a context where teams may easily become frustrated with their progress or are prevented from realizing their full potential. How can teams of non-designers adopt the most efficient strategies for success?

10.3 Three Team Strategies for Success

While teams with non-designers face challenges in adopting design thinking, there are concrete strategies they can take and specific guidance for implementation that facilitate success. Table 10.1 presents three areas of challenges for design teams discussed above, along with team strategies and guidance for implementation, which will be addressed in turn.

Encourage "Dual-Mode Debate"

Encouraging design teams to vigorously debate ideas is well established as a means to facilitate innovation; past studies have shown that active debate of different design concepts can lead to a wider variety of ideas for teams to consider, and that more questioning and debating ideas within a team is related to more product development

Table 10.1: Design Thinking for Non-Designers: Challenges, Strategies, and Implementation		
Challenges	**Team Strategies**	**Implementation Guidance**
Non-designers are often unaware that use of tools is dynamic and requires adaptation.	Encourage "dual-mode debate" of both ideas *and* design process.	Add metrics for idea and process debate. Conduct scenario exercises to analyze process options.
Non-designers learn benefits of key mind-sets but not phase dependency.	Manage design thinking transitions of key mind-sets.	Enforce discussion of process phase check-ins. Specify a "contract" of ideas no longer open to debate.
Team membership changes can cause lack of common understanding of design journey.	Adapt tools under changing membership.	Proactively document and share past elements of design journey. Purposely use brainstorming for onboarding of new members.

success (Hoegl & Parboteeah, 2006). While encouraging active debate of new design *ideas* is common, learning to actively debate the next step of the design *process* is relatively rare. This process debate can focus on whether the next step is to the next phase, to iterate back to a different phase, or which specific tool to apply.

Teams that engage in "dual-mode debate," in which they actively debate both ideas and the next step of the design process, apply a critical skill. As an example of encouraging "dual-mode debate," one high-performing team developing a device to help facilitate spine surgery was very open not only to new concepts early in the process but also to the tools that they would use. One member commented afterward that the team was "open-minded," stating, "We kind of thought that everything had a reasonable chance." This team was noted for debating both ideas and which design tools to use early in the process. In contrast, a low-performing team at the same time reported they were more focused on being "efficient" than in taking the time to engage in debate over their process.

Experienced designers know that the next step of a design process is always subject to debate, but non-designers may need help learning this. Teams of non-designers may have debated ideas during brainstorming sessions, but they may not know to spend time debating whether the results of a brainstorming session should be followed by additional need-finding or with rough prototyping, for example. Learning how to debate the next step of a process can be very uncomfortable for some team members, and so executing this strategy can be helped by two points of implementation guidance, as outlined in the final column of Table 10.1: (1) add metrics for both idea and process debate and (2) conduct scenario exercises to analyze process options.

A first point of guidance is for teams to add metrics for both idea and process debate. It is important to note that what is labeled idea or process "debate" here falls within the larger category of what social sciences sometimes call conflict. While debate over ideas or processes can be beneficial, personal conflict almost always leads to poor team performance. To help teams assess their ability to foster debate while avoiding personal conflict, teams can be asked to fill out surveys at various points in the process to evaluate their team dynamics. On five-point scales, team members can be asked to what degree are they debating ideas and the design processes they are using, and team facilitators can use this information to assess whether interventions are required, as we outline in the accompanying text box ("Add Metrics for Idea and Process Debate").

Add Metrics for Idea and Process Debate

Teams need to know how well they are engaging in active debate. One simple but effective survey tool that has been used with teams in the past has been to ask them the following two questions on a scale of 1 to 5, with 1 representing "to a little extent" and 5 representing "to a great extent": (1) I feel that we debate ideas within our team, and (2) I feel that we debate the process to follow within our team. These metrics should be collected in at least two points. First, the survey should be run in the heart of concept generation, such as after a team has completed at least one brainstorming session and is developing a range of potential product concepts. Second, the same survey should be conducted after concept selection, when a team has narrowed down to one main concept and is further refining the concept. How much debate is ideal? The circumstances of the industry and teams may change how individuals rate a level of debate, and so the ideal level of idea and process debate can vary by context. The scores of new teams can be compared with successful teams in the past, allowing managers to determine if teams need to be reminded of how to encourage debate in the early stages of the process. Also, if scores among individuals within a team vary greatly, that can be a signal that intervention is needed to see why team members have such disparate perspectives on the level of debate—such as debates being dominated by one or just a few members of the assigned team, for example.

A further point of implementation can be to embed exercises in debating a design process as part of team training, as we outline in the accompanying text box ("Conduct Scenario Exercises to Analyze Process Options"). For example, a case-study scenario can be given on a design process that is under way, and team members can engage in debate as to what their next step would be: Do they stay within the same phase, do they proceed, or should they iterate back to an earlier phase? Do they need to consider a different tool? Reference texts can be provided to demonstrate that many alternative tools might be appropriate (e.g., Kumar, 2012), and team members can assess and debate which might be most beneficial as well as if iterating back to a prior phase is beneficial. Rather than blindly following a process of how design was done in a textbook case, non-designers can learn how to debate and evaluate the process and the range of tools available to them.

Conduct Scenario Exercises to Analyze Process Options

When adopting design thinking, it can be difficult for teams to understand how the tools of design thinking can be applied in many different phases and that the specific order of tools can change with each project. Conducting a case-study scenario exercise where team members can learn how to debate different process options can help team members to be open to debating a range of options. Sources of case-study scenarios can include articles from *PDMA Visions* magazine, cases and articles from the Design Management Institute, and cases purposely written for training teams within a certain industry. For example, a team that is working on new service design in the food and beverage industry might be given a case write-up describing the development of a new bever-

(continued)

age bottle. The case may describe, for instance, the need-finding activities that identified the potential for a product that added recommended doses of vitamins to a water bottle of the user's own water and the initial ideation sessions that led to different means to inject and mix vitamins. However, the case may note how user reactions to initial prototypes identified a concern with cleanliness. How should the team proceed? In this scenario setting, teams can debate the merits of further brainstorming about methods, of prototyping the delivery, and of further involving users in understanding their requirements for "cleanliness," describing what further information they would need to gather to make their decision. Encouraging debate in the safety of a scenario can help teams better explore options in their own projects and help them learn the benefits of spirited debate about next process steps.

Manage Design Thinking Transitions

Even design teams that become good at debating both ideas and processes are not immune from difficulties along their design journey. Some teams are markedly better than others in managing design thinking transitions in key mind-sets, as outlined in the second row of Table 10.1. The specific "design thinking transitions" are those concerned in making the shift from a high-debate divergent mind-set to an emphasis on execution within a focus mind-set.

One high-performing team in the medical device field was very good at making the transition from debate to execution and focus. One member reflected on their decision to use a special glue instead of another attachment as part of their design, stating: "Once we knew that we were going to go ahead with glue, we did not revisit. It was not like, okay, at every meeting you are going to think about the need criteria [again]." Medtronic has been lauded as one company that has been good at controlling too much ideation in the latter stages of development. In contrast, Sony's Playstation products have at times been cited as experiencing delays due to the inability to focus on a few innovations at a time rather than trying to incorporate many last-minute additions.

Skilled designers have years of experience in how the practice of debate and a divergent mind-set needs to be moderated over time, an insight that novice teams may lack. As design teams move from concept generation through to concept selection and final refinement, it is critical for design teams to transition from active debate to a focus on design execution. This can be a great challenge for teams, and we offer two points of implementation guidance, as outlined in the accompanying text boxes.

First, team leaders or facilitators should enforce a discussion of where the team is in their process at regular points. For a project of six months' duration from initial ideation through to final prototype, there could be biweekly check-ins where members can discuss the degree to which they need to start to finalize key features of their concept, so to be able to implement a final prototype on schedule. An example "process phase check-in" guide is given in the accompanying text box ("Enforce Discussion of Process Phase Check-Ins").

Enforce Discussion of Process Phase Check-Ins

Whether called "phases," "stages," "milestones," or otherwise, the relative position of a team in the development process has implications for what behaviors will be more successful. Should a team be in a divergent mind-set or a more focused convergent mind-set? Either mind-set may have a place at different process phases, depending on whether the team needs to explore new areas or needs to focus their efforts, though as deadlines approach a focusing effort will be necessary. Many teams struggle when some members are in a divergent mind-set while others are focusing. By engaging in a process phase check-in, the team can consider whether they should be encouraging debate—and if so, what type—or if they should be focusing their effort on execution of the design. The following questions can help to inform a process phase check-in, though the specific form can be tailored to the context:

1. Where in the development process does the team view itself, and do all members agree?
2. Is the team currently needing to seek out diverse ideas and information, or does the team need to select among ideas currently at hand?
3. Would an iteration back to another process phase be helpful at this point?
4. Would an iteration of a specific design thinking tool be helpful at this point?

Second, teams should consider developing a "contract" of what ideas are decided on and no longer open to debate, as we outline in the accompanying text box ("Specify a 'Contract' of Ideas No Longer Open to Debate"). Too many teams get caught in the trap of revisiting decisions over and over again. Teams need a procedure to help them resist the urge to revisit decisions already made, unless there is compelling new information available to them. As mentioned, Medtronic has for years been very careful in documenting aspects of their designs as they proceed, working to make sure that the evolving design is well communicated among all team members. This type of commitment and discipline is especially relevant for the product development velocity of complex products, ranging from consumer electronics to automobiles.

Specify a "Contract" of Ideas No Longer Open to Debate

During an innovation project, it can be difficult to know when to narrow focus to a few ideas central to the product, as the iterative process requires holding many options open while needs are being continually reassessed. However, at a certain point, teams need to decide which new ideas that can be incorporated as product features are well enough settled to be treated as fixed. At one major

(continued)

electronics company, they referred to a set of fixed design criteria and related product ideas as being "in the box." Once they had enough data to support a certain feature of the final product, they put this criteria metaphorically "in the box," so all engineers and designers on the program knew these did not merit revisiting unless some fundamentally new information came up. The "in the box" listing served as their contract of which ideas were no longer open to debate. To successfully specify a team contract of ideas no longer open to debate, teams can consider the following steps:

1. Name the contract, such as "The Box," "The Vault," or "The Design Contract," and make sure all design team members are aware of this contract at the start.
2. At each major milestone, be sure relevant team members can nominate ideas or features to be included in the contract.
3. Specify the voting process by which ideas or features can be added to the list.
4. Make the contract readily accessible and visible to team members.

Adapt Tools under Changing Membership

While student design teams and design consultancies may have stable team membership during a project, real-world design teams in corporate settings often face a challenge of changing membership. Sometimes people are moved midway through the project to other projects that appear to have more pressing needs; at other times, new people may join the team, coming from other projects that have ended. These shifting team compositions are challenging for two reasons. The first reason is practical, resting on a question of information transmission. Whenever a new member joins a team, this new member needs to quickly learn what the existing team already knows. The second reason is emotional, resting on a matter of trust. In other words, a relationship needs to form between existing and new members of the team that enables productive collaboration.

There are two specific points of guidance for implementing this strategy, as outlined in the third row of Table 10.1. Both points of guidance for improved onboarding of new team members make use of design thinking tools and their outcomes. First, one tenet that underlies much of design thinking is to work visually (Liedtka & Ogilvie, 2011). For example, during need-finding, using tools such as interviews and observations, design teams are expected to display their data on available work surfaces to allow for producing new associations and insights. Vertical surfaces in front of which teams can gather are particularly well suited for these activities (Doorley & Witthoft, 2013). Subsequently, these data are then transformed in visual tools such as journey maps. Since the goal of these visualizations of the user and his pain points is to create alignment within the team and between the team and other stakeholders (such as clients, executive management, etc.), these tools capture the essence of many hours of the team's work. For that reason, these visuals are also an effective tool to help new team members quickly develop an understanding of key aspects of the project, and the resulting guidance is that it is important for teams to proactively document and share past elements of the design journey. Design team leaders must ensure that their teams have access to appropriate workspace and materials, as outlined in the accompanying text box ("Proactively Document and Share Past Elements of Design Journey").

Proactively Document and Share Past Elements of Design Journey

Complex situations composed of many different kinds of information are best understood and interpreted when displayed in their complexity on work surfaces sufficient in size so that they allow making connections and associations. Most industry "war rooms" have this aspect as their underlying idea, whether set up in a major design consultancy like Continuum or a major automotive manufacturer such as Toyota's "Obeya" rooms. In ideal situations, design teams occupy project rooms whose walls they can use to lay out their data from interviews, ethnographic observations, drawings, storyboards, and findings. In situations where design teams do not have access to permanent project space, it helps to make "movable walls" available to which complex sets of information can be attached but are themselves movable. Such movable walls can range from 24" × 36" posterboard to 4' × 8' foamboard. The task for team leaders and managers is to make available the materials as well as the storage space for the boards (which must be very quickly accessible). For 3D materials and prototypes, project boxes (or "cubbies") have proven very helpful, keeping items at hand that reflect the design process already undertaken.

Second, in addition to using the outcome of design tools as an onboarding device for new team members, some of the tools themselves can also take on this purpose. For example, being part of a structured collaborative brainstorming session is an intense and highly social experience, and this process can help team members learn what knowledge is available within the team and what lines of thinking have been explored. As one designer with the firm IDEO put it, "Brainstorms teach us what designers and clients know, and how to fit it together" (Sutton & Hargadon, 1996, p. 696). As a consequence, managers and design team leaders can use the tool of brainstorming not only to generate ideas but also as a way to integrate new team members faster, as we outline in the accompanying text box ("Purposely Use Brainstorming for Onboarding New Members").

Purposely Use Brainstorming for Onboarding of New Members

Brainstorming is a particularly well-suited tool for integrating new team members. The underlying reason is that well-facilitated brainstorming sessions create a well-defined and protected space for idea exchange and generation.

(continued)

The typical rules for brainstorming such as "one conversation at a time," "generate many ideas," "crazy ideas are welcome," and "suspend judgment" create such a space, and in this environment newcomers face lower entry barriers for participating. In other words, a second purpose of a brainstorming session can be to serve simultaneously as an icebreaker. One team working on the development of a medical device purposefully used several brainstorming sessions to help new team members get "ramped up to speed" with a complicated project in surgical products.

10.4 Conclusion

The three key team strategies and associated implementation guidance can help teams best adopt tools to the real-world experience of design thinking as non-designers. Members of design teams will increasingly come from ranks of professionals who do not have the luxury of academic design preparation, and it can be possible that the challenges of design could put them off a sustained use of design thinking. Design thinking encourages us to be flexible in how we view a given situation, and to that end these team strategies help teams to engage with design thinking in an adaptive way that gets to the heart of viewing design thinking as a valuable but flexible toolkit for innovation.

References

Beckman, S. L., & Barry, M. (2007). Innovation as a learning process: Embedding design thinking. *California Management Review*, 50(1), 25–56.

Brown, T. (2008). Design thinking. *Harvard Business Review*, 89(6), 84–92.

Doorley, S., & Witthoft, S. (2013). *Make space: How to set the stage for creative collaboration*. Hoboken, NJ: Wiley.

d.school. (2013). Stanford d.school "bootcamp bootleg." Retrieved November 15, 2014, from http://dschool.stanford.edu/wp-content/uploads/2013/10/METHODCARDS-v3 -slim.pdf

Hoegl, M., & Parboteeah, K. P. (2006). Team reflexivity in innovative projects. *R&D Management*, 36(2), 113–125.

Kumar, V. (2012). *101 design methods: A structured approach for driving innovation in your organization*. Hoboken, NJ: Wiley.

Liedtka, J., & Ogilvie, T. (2011). *Designing for growth: A design thinking tool kit for managers*. New York, NY: Columbia Business School.

Seidel, V. P., & Fixson, S. K. (2013). Adopting design thinking in novice multidisciplinary teams: The application and limits of design methods and reflexive practices. *Journal of Product Innovation Management*, 30(S1), 19–33.

Sutton, R. I., & Hargadon, A. (1996). Brainstorming groups in context. *Administrative Science Quarterly*, 41(4), 685–718.

Ulrich, K. T., & Eppinger, S. D. (2012). *Product design and development* (5th ed.). New York, NY: McGraw-Hill.

About the Authors

VICTOR P. SEIDEL is on the faculty of the F. W. Olin Graduate School of Business at Babson College, where he teaches product design and development. He is also an associate of the Harvard School of Engineering and Applied Sciences, where he teaches the design and innovation process at the undergraduate level, and he is a visiting scholar at Said Business School, University of Oxford. He received his PhD from Stanford University. He has worked with corporate clients including Dell, L'Oreal, Raytheon, and General Motors. His research interests include organizational practices supporting innovation, the role of online communities in innovation, and the use of design methods.

SEBASTIAN K. FIXSON is Associate Professor of Technology and Operations Management at Babson College, where he teaches innovation, design, and operations related courses on various levels (undergraduate, MBA, and executive education programs). His research is concerned with how organizations build innovation capabilities through design and innovation practices, tools, and incentives. He has worked on these issues with both entrepreneurial start-ups and established large organizations such as Alcoa, BBVA, Boeing, Delphi, Ford, GM, Harley-Davidson, Raytheon, and Spirit Aerosystems. He holds a Dipl-Ing (MSc) in mechanical engineering from Technical University in Karlsruhe, Germany, and a PhD in technology, management, and policy from MIT.

11

DEVELOPING DESIGN THINKING: GE HEALTHCARE'S MENLO INNOVATION MODEL

Sarah J. S. Wilner

Wilfrid Laurier University

Introduction[1]

Designers are trained in their field's logics, but many organizational members with a stake in new product development have had little, if any, exposure to design thinking's precepts. Yet if design is important to firms' value creation, we must consider ways in which its practices might be instilled beyond the design department. To provide guidance and inspiration, this chapter looks inside one maverick internal design studio's efforts to embed design thinking within one of America's oldest (and among the world's largest) companies: General Electric (GE). Tracing the development and implementation of the studio's Menlo Innovation Ecosystem, I examine the process, challenges, and outcomes of creating an internal design thinking innovation program.

Thomas Edison's Menlo Park research laboratory is often credited as being the first industrial research and development lab. Several important innovations emerged from within it, including incandescent bulbs and phonographs. More than a century later, the legacy endures: invention and innovation continue to be central to GE's culture and operations, a strategic perspective captured in the organization's slogan: "Imagination at Work."

[1]With thanks to Bob Schwartz, Mark Ciesko, and Rebecca Bhanpuri; sincere gratitude and admiration to Menlo's leaders: Doug Dietz, Emil Georgiev, and Megan Wimmer; and appreciation to all of the participants of the May 2014 discussion group.

11.1 GE Healthcare's Design Organization

The design studio at the heart of this story is part of GE Healthcare, a strategic business unit headquartered in the United Kingdom and employing more than 46,000 people worldwide. GE Healthcare's early innovations included X-ray tube technology, and today the organization continues its groundbreaking work, specializing in medical imaging and information technologies, diagnostics, patient monitoring systems, drug discovery, biopharmaceutical manufacturing technologies, and performance solutions services.[2]

GE Healthcare's Global Design group is a cross-functional organization with offices in the United States, France, China, India, and Japan. The group's practice includes more than 60 professionals across multiple disciplines: industrial design, interaction design, design research, innovation, human factors, ergonomics, cognitive psychology, visual design, surface design, and technology architecture. The unit whose practices are described in this chapter is global design's largest group, and its work is distributed across all product groups to enable product branding alignment, and innovation.

Evidence of the design studio's success includes 10 International Design Excellence Awards (IDEA) acquired in just the past three years, placing GE Healthcare in league with Apple and Samsung in corporate design recognition. The team's design philosophy is summarized by the phrase, "The Magic of Science and Empathy."

> Whether a product, user interface, or environment, our philosophy is to enrich that experience with technology, delight, hope, and understanding of human needs.... Our design values [include] authenticity; empathetic design; shared intelligence and trusted relationships; imagination at work; essential expression; and the science and mathematics of beauty.
>
> *GE Healthcare document, "Global Design/User Experience" (2012) for internal distribution only.*

Such values have improved both business performance and patient outcomes, and one of the key means of transmission has been an initiative called "the Menlo Innovation Ecosystem" (hereafter, "Menlo"). Menlo was established to disseminate design thinking in team-based environments, and has triggered a number of strategic and cultural conversions—the latter, arguably, the biggest challenge of adopting design thinking. While Menlo's scope continues to evolve, its central activity is a series of multiday workshops that help internal teams solve their business challenges by encouraging both individuals and teams to shift how they think and act.

11.2 The Menlo Innovation Ecosystem

Established just over five years ago, Menlo is evolving, and workshops are dynamic works in progress. However, several components stay constant, having emerged from a unique blend of organizational needs, history, and skills. Menlo's leaders have tested a range of workshop structures and time frames, settling on a five-phase approach delivered over the course of approximately 10 workdays (the workshop is suspended midway for field research; see Table 11.1). However, not every team engagement includes all phases.

---

[2]www3.gehealthcare.com/en/about_us

Table 11.1: Phases of the Menlo Innovation Workshop

	Stage	Purpose	Roles of People Involved	Key Activities	Duration
1	Exploratory	Get brief from business leader on goal Dig into team challenges, identifying struggling teams	**Non-Menlo:** "Sponsors"; ~3 key managers **Menlo:** One or two of the program leaders	Briefing by sponsors Determination of key problem to address in Menlo Determination of the duration and framework of the workshop Definition of participant team's makeup	1 day
2	Boot camp	Trust building; learning to divert from less helpful corporate culture norms Team building Build empathy via immersion Ideation and rough prototyping First pass at a research question Ends with team returning to their areas with a new mind-set and frame for future	**Non-Menlo:** Whole team **Menlo:** Facilitator and coaches. Menlo advises on composition of team of people attending this phase (like "ensemble casting").	**Initially:** Trust and interpersonal learning **Next:** Adjacent or similar (non–health care) design challenge **Then:** Introduction to research and plan development	4 days: −3 days boot camp −1 day of "What we wish we knew …" research planning
3	Implementing the Research Plan	Take learning/new mind-set back to daily task Work on research for actual challenge	**Non-Menlo:** Whole team may not be involved but is strongly encouraged to participate **Research Specialists:** Serve as guides to core team during research activities and are responsible for overall research strategy and data capture/consolidation **Menlo:** Coaches serve as integrator and facilitator back at home area	Data collection, etc. May involve outside suppliers	1–3 months

(continued)

159

Table 11.1: (Continued)

4	Innovation Camp: Ideation + report out	Move from research to opportunity: key innovation stage A deep dive into the problem and potential solutions Report out: Present concept/solutions to top management	**Non-Menlo:** Whole team + may involve customers (internal or external); + top management for report out **Menlo:** All	Research summary Active listening to customers, if present Writing opportunity statements (individually, then grouped and voted on) Ideation based on findings Prototyping Solution iteration High-level solutions (3–5 concepts) Identification of intellectual property filing opportunities Report/presentation	5 days
5	Follow Up *(new phase: currently under development)*	Ongoing support to teams that have participated in Menlo workshops to ensure that new mind-sets and behaviors are put into action after the workshops are over.	**Non-Menlo:** As needed **Menlo:** Core coach	Under development, but may include: Open houses with multiple teams/workshop alumni Scheduled check-ins to monitor progress Embedding a coach in the project team	TBD

Depending on the challenge, some teams might, for example, complete their Menlo engagement within three to five days.

Phase 1: Exploratory

Among the most critical phases is the first meeting with a team's sponsors—who initiate the workshops—known as "Exploratory." Exploratory is analogous to writing a project brief; initiating a dialogue between team leaders and Menlo staff, this initial phase provides a forum for determining key issues and helps ensure that the resulting workshop process will allow participants to succeed in addressing them. It also reflects the consultative and empathetic processes of design thinking.

Menlo coaches listen carefully to sponsors in order to diagnose the cause of their challenge. Often, however, they must also dig deeper, seeking clues to complicating factors that might inhibit or undermine possible solutions. This process reflects a critical assumption in the Menlo model: the problem a team is experiencing is likely not a function of its raw ability, but rather a symptom of more subtle issues. Accordingly, Menlo leaders inquire about business problems, but also probe further to uncover possible impediments. The team may have an inability to communicate effectively, or may not have found an effective solution because they haven't really understood the customer's point of view.

During Exploratory, Menlo staff also preview their approach to the sponsors, including working through sample activities, to expose the sponsors to Menlo's principles and environment as well as manage expectations and achieve consensus on the workshop's objectives and basic structure. At the end of Exploratory, staff will have gathered enough information to begin formalizing a delivery plan for the full-team workshop, including customizing its content or duration for a team's specific needs. Decisions made at this stage include determining:

- The specific program framework that can best address the business challenges.
- The team and individual skills that warrant development.
- Key activities to foster and build those skills.
- Duration and timing of the workshop(s).
- Plans for offsite trips or special guests.
- Possible additional support from team sponsors going forward.
- Participants, in terms of both functional area/expertise and leadership level.

At Menlo, the last item is known as "ensemble casting" and is approached as a director might assemble a dramatic group, by considering how each actor's unique skills, personality, and point of view can contribute to the group's perspective and performance as a whole. For example, if a group is having difficulty developing a new product, the team's manager might assume that engineers alone should attend. However, Menlo staff might also suggest that representatives from marketing, human resources, or even an individual from another business area entirely who is not working on the project be present to add new sources of information and bring fresh insight.

Phase 2: Boot Camp at the Menlo Innovation Lab

The next phase is known as "Boot Camp" because it is an intensive preparatory experience intended to build trust, develop basic skills, and facilitate the team's cohesion for the innovation "battle" to come. Like military boot camp, participants are sent to a new, unfamiliar site for training: the Menlo Innovation Lab. The open-space lab is designed to foster communication and collaboration. Furniture is mobile, adaptable, and comfortable. Bright orange, part of the Menlo graphic identity, accents the lab area, where it signifies energy and stimulates creativity. JWD-Creative, the agency that worked on the Menlo lab identity, notes that they were inspired by Edison's own words: "Hell, there are no rules here; we're trying to accomplish something."[3] In this sense, the Menlo boot camp is the antithesis of military training because the objective is not for innovation team members to surrender their individuality, but rather to function productively as a cohesive unit.

The next clue that the design thinking workshops are not "business as usual" is the request that participants surrender their mobile devices. As one design coach said, "This is what we used to call a wet lab: [rather than a device, we want you to] use your head, use your anatomy—use your brain" (Lawrence "Murph" Murphy, chief designer and facilitator, May 2014). He recounted an occasion when participants were distracted by a significant reorganization happening while the team was away in the workshop training. Sensing the stress, the Menlo staff showed a video by Honda automotive articulating its "Kick Out the Ladder" philosophy. The phrase describes a situation in which an individual is striving and climbing and, just as he gets to a high rung, the ladder is kicked out from under him. The metaphor conveys that valuable change cannot be achieved by moving safely and incrementally (rung by rung); instead, innovation happens when *not* innovating is not an option. After showing the participating team the video during a lunch break, the coaches erected a five-foot ladder in the lab and directed participants to leave their devices—and the tickertape of unsettling news being broadcast there—on its steps.

During boot camp, a team is guided through exercises that foster trust, lateral thinking (De Bono, 1992), empathy, creativity, improvisation, and collaboration. It is an experiential onslaught intended to break conventional modes of interaction, remove masks of professional personas, and create a team that not only functions together, but functions better as a unit.

For example, if Exploratory found that the team's interpersonal communication skills were causing an innovation bottleneck, the facilitator might lead the team in an exercise called "Pile of Rocks." Everyone sits on the floor around a large pile of rocks; each participant is told to select a rock, look at it, and then place it back on the pile. They are then told to retrieve their rock, and the difficulty of doing so quickly conveys the importance of focus and attention to seemingly trivial details. Next, each is paired with a partner, to whom they must describe "their" rock so that their partner can retrieve it. Having to characterize an object that could easily be dismissed as indistinguishable requires precision of specification: "gray and round" is insufficient, and success comes only to pairs who can

[3]www.jwd-creative.com/work.php#/branding-campaign/1/menlo_lab/1

effectively communicate. An important part of the exercise is participants' discovery that not everyone shares identical concepts of vague descriptors such as "large" or "smooth."

In this activity, as in many conducted by the Menlo team, the first impression can be one of absurdity—why pull advanced engineers away from projects to pick rocks out of a pile? But dismissing these as "silly games" would be a mistake. Instead, one lesson to draw from this is the importance of disruption. The Menlo team speaks of the importance of "scraping the GE off" participants. This phrase is not meant as a slight to GE, but rather is a way of expressing that innovation requires departing from the norm. GE is populated by very bright people accustomed to achievement who want to be shown a system so that they can master it. Yet so-called "wicked problems"—the very kind that can catalyze radical innovation—defy quick mastery. Introducing an entirely new form of playing field disrupts existing operating modes and forces participants to experiment with new problem-solving approaches.

> [For Pile of Rocks], everybody's sitting really close to each other. Their eyes are locked together, because they've got to listen to every word that this person is saying to go be able to find that rock. And they learn that we're not always great at communicating to one another; we usually don't listen very well at all. We have tons of other crazy stuff that we do, and it's all the product of thinking about immersive activities that we can do to get a participant in the spirit of noticing where these problems are probably going to show up. After they've experienced an issue with team members in an activity, and they still have that feeling, we ask them, "How does same issue this show up at work?" It's really easy to make that quick connection when it's still raw.

> **Doug Dietz, Innovation Architect, member of core Menlo team, and Menlo Facilitator, personal interview, May 15, 2014.**

A second premise and lesson to be drawn from the Menlo model: experiential learning is exponentially more powerful than passive information transfer. The eccentric activities are not intended to be literal skill training, but rather the means to foster memorable experiences, which in turn precipitate emotion that can be connected to a larger concept. Being told to "pay attention" and "listen carefully" is far less effective than having a visceral memory of finding an effective strategy to help a colleague find one specific rock among dozens.

Reflection is a critical step in transforming experience into meaningful learning, and at Menlo, time is always reserved to debrief at an exercise's end. Before moving to a new activity or phase, the subteams return to a single group to consider and answer a single, critically important question: "What just happened?" This part of the process can sometimes take as much time as the activity itself. Reflecting on when a similar interpersonal dynamic occurred during a typical project back at the office, noticing how and which emotions were provoked during an exercise, or explicitly articulating the specific connections participants make between their expectations and their experiences—all serve to cultivate "group genius," the synergistic benefits that high-functioning teams develop. Moreover, although reflection is encouraged directly after activity completion, facilitators also structure the workshop's arc so that the "payoff" of a given exercise doesn't appear until later in the process. The skills that accrue in the course of multiple active listening, collaboration, and creative problem-solving exercises are intricately woven into a team's newly forming ability to move as a unit toward an objective. It is often not until a later stage that individuals can be made aware of how differently their team has begun to function, compared to when it began the workshop process.

This conversion is particularly transformational for employees of large companies assigned to cross-functional, sometimes geographically dispersed, work teams, for each exercise provides a highly personal, meaningful introduction to colleagues who until now may only have been a name at the bottom of an e-mail. Whether competing against other small groups to erect a camping tent in silence, describing yourself as an app, or introducing oneself through a collage that answers the question, "Where do you feel the most and least creative?" Boot Camp activities not only ignore traditional functional skills, but are designed to be challenging for *all* participants, an equalizing process that focuses talent at the team, rather than individual, level.

A third lesson of the Boot Camp process—and perhaps a critical explanation for the failure of some companies who have tried to adopt a design thinking orientation—is that enabling or activating a design *mind-set* is critical before any other modes of practice can be introduced. The Menlo facilitators emphasize that readiness is key to a team's ability to successfully learn, practice, and implement design thinking principles.

> You might not get to a traditional design-thinking exercise for days, but that's a version of getting them to "fail early." The traditional design thinking exercise might not happen until the middle of Day Two. You need to spend the entire previous day more on team activities or higher-level principle work to get the team working well.
>
> **Mark Ciesko, Manager of GE Healthcare Americas Studio and Menlo Coach, personal interview, May 15, 2014.**

Indeed, readiness can be understood as providing the *experience* of design thinking principles within the product development team—prototyping, failing early and often, customer-centric empathy—before labeling them as such.

> It is a very important factor, internal empathy for the team. For every organization that is team-based, most of the development and innovation happens within teams. Without this empathy and without setting the team up for success, everything else that you do will either be subpar or will not work at all.
>
> **Emil Georgiv, Senior Menlo Innovation Strategist and Menlo Facilitator, personal interview, May 16, 2014.**

Fostering group genius—teams that work in concert and synergistically to allow relevant and powerful solutions to emerge—is a core value at Menlo. Indeed, although sponsors may have articulated a project-based objective during Exploratory, that project is not directly addressed during Boot Camp. According to Menlo philosophy, attacking the problem with existing perspectives, group dynamics, project history, and personal "stakes in the ground" can result in only incremental results. Boot Camp is an opportunity to immerse the team in a non–health care setting to help them learn how to see and solve problems with fresh perspective.

For example, a team that engaged Menlo for a problem with poor workflow assumed their problem was software related. The team was given a (seemingly) unrelated assignment: to redesign the fast food restaurant drive-through experience.

As one facilitator explained, "We needed to help them see that workflow wasn't about a specific tool—software—but about the critical relationships among information, people, and needs." Subgroups ideated concepts, which were then presented in a series of skits (itself an introduction to the concept of consumer journey prototyping). One team's solution called for the menu to be projected onto the car dashboard so it would be easier to read; the employee taking the order became a "health concierge," and the customer's car was washed while waiting for the order to be filled.

One Menlo facilitator noted that making early design thinking trials removed from the actual business problem provides freedom and space for creativity: "It's not health care related, so they can be free to see how [re-imagining work flow] *feels*." The resulting solutions can even surface issues that are relevant to the specific healthcare issue facing the team—how can value be added during experiences like waiting that would otherwise be experienced as pain points? What happens when employees start with the consumer's perspective?

The "design a better drive-through" example highlights other important Boot Camp components. These include opportunities for problem immersion, ideation, prototyping, and iterative refinement. Contextual immersion fosters empathy, both for teammates and for customers. As one facilitator asserted, "Empathy isn't transferable," meaning that if a team doesn't share common experiences, it is unlikely to engage in meaningful interpretation and action. The workshops' inherent "learn-do-reflect" philosophy provide opportunities for participants to collectively *experience* design principles like "fail early, fail often" as a team.

It is hard to overstate the significance of conveying these principles through practice. In an engineering-led culture such as GE's, the reflexive response to problems is more likely to focus on system development and control than on generating potentially messy, irrational, or emotional ideas. "[Managers] often want to feel control," noted a coach. "It's hard for them to let the team find its own way to address the challenge. But not letting go strangles creativity; it usually just doesn't work well."

Menlo leaders have created a variety of worksheets to allow participants to document their learning in ways that they can refer back to and be inspired by when daily responsibilities insinuate themselves again. For example, a worksheet entitled "Design Thinking: Take It Home," features three columns. Colored hexagons, each containing one of the fundamentals of design thinking practice (e.g., "Empathy," "Define," "Ideate," etc.), are literally at the center of everything, dotting the length of the page like stepping stones across a river. Each is captioned with useful prompts (e.g., the words *listen* and *inquire* underlie "Empathy"). The top of the otherwise unmarked left side of the page is labeled "Personal Behaviors"; the column running along the right side is labeled "Business Challenges." The message is clear: individuals cannot reap the rewards of design thinking unless their own behaviors are synchronized with the objectives they are working on for their business.

On the fourth day of Boot Camp, teams focus their efforts on "What We Wish We Knew," an activity that asks participants to consider the kind of information they need to move their project forward. The list that results forms the basis of a research plan to gather the information that can make proposed solutions more likely to succeed.

Phase 3: The Research Plan

The phrase "design research" describes "any number of investigative techniques used to add context and insight to the design process."[4] Once a team has determined what they need to know to advance their desired innovation, Menlo's leaders help them identify the research methods that can best address their questions. These might include observation, contextual inquiry and cultural probes, interviews and focus groups, or techniques to reveal user emotions. No two projects are identical, so the number and combination of methods employed vary by project.

The teams are matched with design researchers within Global Healthcare Design, who guide them through data acquisition and analysis in preparation for the next workshop phase. While the design researchers are specialists and external suppliers are sometimes brought in to assist, the Menlo philosophy is grounded in experience, so facilitators strive to maximize team participation during the research phase, which is the longest portion of the Innovation Workshop model and can last up to three months. Whenever possible, the plan includes methods requiring engagement, a critical step toward fostering empathy. Facilitators have found that when participants observe or interview customers firsthand, they develop new appreciation for the perspective of those for whom they are developing a solution. Just as exhorting workshop participants to pay attention is less effective than placing them in a setting in which they will fail if they don't, telling those involved in developing a product to be mindful of user experience is not nearly as effective as engaging with the technicians or patients for whom a device or process is being developed. Health care is, after all, a context where design can have life or death consequences. Menlo facilitators report witnessing deep transformation in perceptions and behavioral patterns of workshop participants who are afforded the opportunity to not only learn about, but also develop empathy for, customers by personally conducting design research.

The research findings provide the foundation on which solutions are evaluated.

Phase 4: Innovation Camp

The fourth phase of the Menlo program is Innovation Camp. There, the research results are presented, and work begins toward solving the initially identified business issue. Co-creation is important for success, so customers are frequently invited to this stage of the workshop to talk about their experiences and challenges, while participants practice active listening. A new experience for many managers and their customers, the results deepen relationships within the team as well as with external stakeholders. And, now that the team has considered the research results in light of their customers' input and feedback, they are ready and able to (re)define the original business objective.

Participants are taught to write "opportunity statements," brief summaries of the problem to be solved. Such statements, prefaced by the phrase, "wouldn't it be nice if ... " prepare the ground for the ideation that follows, so participants must identify the need to be addressed, rather than determine a specific type of product that would fill a gap. Coaches work to ensure that statements are neither so broad that they are too

[4]www.uxbooth.com/articles/complete-beginners-guide-to-design-research/

general to work from (e.g., "Wouldn't it be nice if we made a better health care experience for patients?"), nor too narrow, which would presuppose or only provide a very limited solution ("Wouldn't it be nice to give every patient an iPad to customize their room lighting?"). A better opportunity statement might read something like, "Wouldn't it be nice if we created a more comforting patient environment?" because it identifies the goal but allows for multiple possible solutions.

Individuals are encouraged to develop six to eight opportunity statements. The statements, each written on a sticky note, are then aggregated on large whiteboards and clustered into common themes. Next, participants vote for their preferred opportunities by placing sticker dots on those they believe will be most important to develop further.

With a new level of focus, the group begins ideation exercises, inventing possible solutions for the opportunity statements while learning how to accept rather than dismiss others' concepts and build on them. Rejection emerges organically, the by-product of a "build to learn" philosophy in which simple, rough prototypes are constructed and tested. Those that fail are improved and the revised prototype is, in turn, retested. Such iterative improvement helps focus the group on meaningful, empirically evaluated solutions and minimizes failure once resources have been invested. Importantly, customers are frequently included in the prototyping stage of the project, where they are invited to critique and co-create emerging concepts. According to senior Menlo innovation strategist Emil Georgiev, multiple patent filings have emerged from Menlo innovation workshops that included customers as co-inventors.

Of the many solutions developed and prototypes tested, the team selects three to five concepts, presenting them—with collective pride and conviction—to senior leaders on the final day of the workshop.

Phase 5: Follow-Up

From its inception, Menlo has emphasized its own continuous improvement. So while early workshops were successful at shifting mind-sets and driving new behaviors, over time the Menlo group realized that the transformation individuals and teams experienced during the workshops was sometimes not sustained once they returned to the old habits of their normal work environment. Early attempts to stave off relapse focused on having participants complete exercises like the "take it home" worksheet described above to remind them of important lessons and principles. While still important, these efforts are now supplemented with more formal processes to transition teams back to their units as well as provide support when they tackle their business challenge with full resources and accountabilities.

The structure for this newest phase is still in development, and like the entire Menlo program, its implementation is customized for specific teams. Nevertheless, Menlo leaders have been experimenting with a range of formats to address the needs of current teams as well as the growing group of workshop alumni, who form an important organizational learning and support network. Current plans include hosting open houses for Menlo participant-alumni to share new design techniques, challenges, and ideas; scheduling a series of checkpoints so that the team can touch base with its facilitator; and having a coach embedded within the team during its further project development.

11.3 The Significance of Design Thinking at GE Healthcare

The technology required for magnetic resonance imaging and other similarly sophisticated machines traditionally drove product development at GE Healthcare. Prior to establishing Menlo, design's role had been inscribed within primarily ergonomic and styling concerns, a task sometimes derisively characterized as "colors and covers" (Dietz, personal interview, September 8, 2009). Indeed, a culture that privileged incremental engineering innovation over user experience dominated at GE Healthcare less than a decade ago.

In an organization as old and as large as GE, cultural change is unlikely to result from a single memorandum. One early catalyst to focus on the "human side of the equation" through design came in the form of a project initiated by Doug Dietz, a 25-year veteran of GE Healthcare's design team. In 2008, Dietz was visiting a children's hospital to check on a magnetic resonance machine he had worked on. As he spoke to a technologist about the machine, he was satisfied and proud: the device was functioning and serving the radiology department well. But his visit was interrupted by the need to leave the room because a patient was being brought in for scanning. The little girl was crying, terrified of the massive machine and the unknown procedure to come.

Encountering the child was a turning point for Dietz, who began a campaign to change the health care experience for some of GE Healthcare's youngest and sickest consumers. Dietz and a radiology team took the issue of customer experience as the problem focus for an early design thinking workshop, and the result was "Adventure Series," a dramatic redesign of the radiology imaging experience for pediatric clients and their families. An immersive experience in which storytelling and imagination transform pediatric radiology from a frightening, anxiety-laden experience into a Disney-like themed adventure, Adventure Series[5] illustrates the power of design to positively influence and accomplish multiple objectives for a range of stakeholders.

The series has been profitable, in part because the redesign helped key customers for the machines—pediatric hospitals—to differentiate their institution in a meaningful way to *their* stakeholders: parents. Where once the imaging equipment had been purchased by hospital procurement as needed, mixed and matched across brands, the series has strengthened loyalty for GE's suite of imaging devices, which are fully integrated within the service delivery concept. Perhaps most importantly, Adventure Series has increased comfort and compliance for patients, which has resulted in a cascade of positive outcomes including lower sedation rates, lower treatment costs, fewer complications, and increased satisfaction for patients' families, who are an integral part of the treatment process when the patient is a child.

[5] www.gehealthcare.com/promo/advseries/index.html

The impact of the Adventure Series's development went beyond market and medical success. It was also an early example of what allowing designers to do more than style machines might mean for the GE Healthcare organization. And while the pediatric context was an accident of Dietz's initial site visit, it became an important advantage to fostering support within GE. For example, the project was initiated as a pilot with the Children's Hospital of Pittsburgh, and the hospital's enthusiastic participation highlighted the benefits of co-creating with key customers. Senior designers Dietz and Murphy modeled best practices by conducting extensive observational research and working closely with stakeholders including radiologists, technologists, child life specialists, nurses, patients, and families to better understand the experiences of users at every touch point. The Adventure Series project also dovetailed with the arrival of a new general manager for the Global Healthcare design practice, Bob Schwartz, from Procter & Gamble in 2007. Having come from a consumer products firm, Schwartz was passionate about the importance and influence of experience in each point of the sales channel, from retailers to buyers to end consumer. Not long after Schwartz's arrival and the initial development of the Adventure Series, a "product experience" group was formally established in the design practice.

It is difficult to underestimate the importance of "buy-in" when proposing a design-based initiative, and to that end the Adventure Series became an important public relations tool across the organization. When GE launched *Healthymagination* as a sibling to its original *Ecomagination* campaign, leaders throughout the organization looked for illustrative examples to help them adopt the initiative. Dietz proudly recalls the excitement and engagement Schwartz heard from leaders from another organization, then a prominent GE subsidiary: NBC Universal, where television executives, engaged by the colorful images and stories of kids' transformational experiences, were suddenly interested in learning more about radiology.

Success Factors

Menlo's developers are quick to credit those from which they have drawn guidance and inspiration in developing their innovation lab model. Among these are Stanford's d.School; P&G's Clay Street initiative; The Creative Problem Solving Group, and Matrixworks' Mukara Meredith and Sean Sauber. The following factors are common to Stanford, P&G, the Creative Problem Solving Group,[6] Menlo, and other successful innovation labs:

Physical factors: Having a separate physical space that can both signal the end of "work as usual" and provide a safe environment in which to be vulnerable while learning and experimenting is vital. The space should be conducive to creativity, with no corporate boardroom or classroom-style meeting rooms, comfortable and adaptive furniture, and ample materials for expression.

Cultural conflict: Deeply entrenched organizational norms; function-based thought worlds and national culture can each create barriers to understanding, stifle creativity, and contribute to discord. Menlo's leaders have found that recognizing

[6]www.cpsb.com

the sources and symptoms of stress and dysfunction is imperative to ameliorating it. However, these underlying tensions are rarely addressed directly, but instead are coaxed out in the course of activities designed to surface—and eventually resolve—them organically.

Autonomy: The importance of a "skunkworks" level of independence cannot be over-stated in creating an effective design thinking program: bureaucracy is innovation's kryptonite. At Menlo, the leadership is largely supportive of the training program (some top managers, like Mark Ciesko, Manager of the Americas Design Studio, are also facilitators), but Menlo programs are conducted separately from the design studio's usual projects.

Challenges to Overcome

Menlo has not been without its challenges, and new hurdles regularly appear. These include:

Resource allocation: It can be difficult for programs without known outcomes—like design thinking—to secure sustainable funds. Not only is leadership support imper-ative, but it is equally valuable to develop new business models. Menlo, for example, generates some of its funding by operating as an internal consultancy.

Growth: Success breeds opportunities, but growth can strain resources. At Menlo, there are two areas currently exerting (welcome) pressure on the program: the first is hav-ing adequate staff to conduct workshops while maintaining quality and consistency for the program's core ideas, skills, and values, and the second is GE's global presence. Menlo has addressed the former with a "train the trainers" model in which partici-pants who demonstrate enthusiasm and acumen for the curriculum are encouraged to apprentice as coaches. Meanwhile, Menlo has just begun adapting its successful curriculum to a range of different cultural contexts, including the design studios in Europe and Asia.

Resistance: New work modes can be threatening to those whose comfort and success is derived from the status quo. Because the Menlo program is so heavily team based, resistors can limit the momentum available to a group working on difficult or com-plex problems. Dietz (personal interview, December 1, 2010) explains:

> When you bring people in, does everybody get it 100%? No, you're still going to have cynics. I love them; they've just been at GE so long that they've got this crust over them. We can get through that crust, but it's going to take some time. Usually by about three-quarters through the session, you'll see them start to take a few more risks. If you can build them up, you'll see them do something that's really unexpected. Before you know it, they've changed.

Menlo leaders have consciously embedded periods during workshops in which cus-tomers and GE executives participate in and validate the work the team is doing in order to illustrate and reinforce the benefits of the process. As a result, only two groups have chosen not to complete the workshop, and both were instances of departmental change that had to take priority.

Lessons Learned

While every organization has a unique culture and strategic objectives that would influence the development of an internal design thinking program, it is worth reiterating the hard-won lessons of GE Healthcare's Menlo Innovation program leaders, including:

1. **Seek information and inspiration.** Be alert to ideas from other innovation groups; investigate training techniques and new research from an array of sources. Dietz reads widely and also often credits his work with teens in the community for helping him develop new workshop activities. He reasons that if he can get reluctant adolescents to become vulnerable to new ideas, managers can't be much more difficult to engage.

2. **Buy-in takes time.** The process of developing the Menlo program did not happen overnight. Large organizations have short-term goals, nested commitments, and turbulent markets competing for their attention. Menlo's leaders persevered in their quest to teach their colleagues design thinking because they believed in its benefits. Just as designers are trained to prototype and iterate, the Menlo workshops have benefited from small wins and a commitment to continuous improvement.

   Similarly, incubation is vital. Dietz uses the phrase "going vertical" to refer to cases where a project stuck in neutral suddenly jumps forward as ideas click, risks pay off, and solutions emerge. Managers expecting a steady stream of incremental gains as the measure of success must be taught to be patient for a payoff.

3. **Bigger isn't always better**. Menlo's leaders might have been tempted to create franchises throughout the company to diffuse its curriculum, but they have focused instead on building a firm foundation for the program before scaling up. Moreover, the "train the trainer" model means that Menlo's reputation is carried by word-of-mouth rather than a managerial mandate to participate. Sponsors who request workshops have heard from trusted sources that the resource investment is worthwhile; they are therefore more likely to be committed to the process.

11.4 Conclusion

Given the profusion of magazines and books touting the benefits of "design thinking," one could be excused for believing that implementing a program within a large organization is easy. GE Healthcare's Menlo Innovation model illustrates the advantages of such a program, but also the challenges. Merely getting managers to put down their cell phones, laptops, and project schedules is ambitious, let alone asking them to jettison professional comfort zones in the name of as-yet-unknown team achievement. Many articles that promote design thinking focus on the promise of breakthrough innovation without acknowledging how difficult the process is. For the teams that prevail, however, the experience can produce profound transformation. Menlo Innovation Ecosystem workshops build caring and productive relationships among participants and forge meaningful understanding between teams and their customers, levering each group's shared scientific and emotional intelligence to imagine better, valuable solutions to wicked problems. In the final analysis, Menlo's achievement has been to make the alchemy of design—its unique combination of science and empathy—accessible to all. Magic, indeed.

References

De Bono, Edward (1992). Serious creativity: using the power of lateral thinking to create new ideas. New York: HarperBusiness.

Dietz, Douglas (2014). Personal interview, September 8, 2009.

Dietz, Douglas (2014). Personal interview, May 15, 2014.

Georgiev, Emil (May 16, 2014). Personal interview.

About the Author

DR. SARAH J. S. WILNER is a Professor of Marketing at Wilfrid Laurier University (Canada). Her research interests include product design, development, and innovation; managers' interpretations of their consumers' needs, desires, and behaviors; and the intersection of design (including industrial, service, and graphic) with consumer culture. Her scholarship has garnered multiple awards and has been published in top-tier publications. Her papers have been presented at the Product Development Management Association's Research Forum, the International Society for Product Innovation Management, the American Marketing Association, and the Academy of Marketing Science, among other conferences. Dr. Wilner dedicates this chapter to Peter Lawrence, founder of the Corporate Design Foundation, whose groundbreaking work and mentorship ignited her passion for design.

12

LEADING FOR A CORPORATE CULTURE OF DESIGN THINKING

Nathan Owen Rosenberg Sr.
Insigniam

Marie-Caroline Chauvet
Insigniam

Jon S. Kleinman
Insigniam

Introduction

This chapter explores the critical impact of corporate culture on design thinking and is organized in five sections. In the first three sections, we highlight the critical impact of corporate culture on design thinking, our perspective on culture, and the forces in any large corporation that work to undermine the principles and practice of design thinking. In the last two sections, we provide insights and practical applications to evaluate, design, and shift or transform the corporate culture through the four pillars of innovation and the four stages of implementing a culture of design thinking.

12.1 The Critical Impact of Corporate Culture on Design Thinking

Empathy. Ideation. Collaboration. Iteration. These are not the typical terms in the everyday conversations of executives at a large corporation. Instead, top executives are likely to be focused on the top and bottom lines, market share, return on investment, share price, and employee retention. But *empathy?* Not in most big companies.

Yet empathy, ideation, collaboration, and iteration are critical aspects of design thinking. For executives who want to install design thinking as a source for their companies' successes, knowing and understanding terms like this, and the practices and processes behind them, are also important to achieving those other key measures of success.

Roger Martin, the former dean of the Rotman School of Management at the University of Toronto and one of the founding fathers of design thinking (along with the Institute of Design's Patrick Whitney and Stanford d.School's David Kelley), once wrote that incorporating design thinking into a large company

> … is not as simple as hiring a chief design officer and declaring that design is your top corporate priority. To generate meaningful benefits from design, firms will have to change in fundamental ways to operate more like the design shops whose creative output they covet. To get the full benefits of design, firms must embed design *into*—not append it *onto*—their business (Martin, 2005, page 5).

Embedding design thinking into a business means embedding it into the company's strategy, corporate culture, processes and practices, systems, and structures. For too many large companies, their corporate cultures are obstructive to design thinking at best, and at worst are destructive of this important new business and management method. However, we believe that an enterprise that embeds design thinking in its corporate culture—in the enterprise's everyday ways of working, its shared practices, beliefs, and values—can gain a competitive advantage over those that do not adopt design thinking.

To gain this edge, organizations will need to reevaluate the organizational context in which they currently operate. And here is the Gordian knot: a company's context is transparent to the people who work in the company.

Culture as Context

The context in which people are working in an organization is primarily the corporate culture. The organizational context influences, shapes, emphasizes, diminishes, or distorts everything that happens in an organization. It reinforces the choices executives make to pursue some strategies and discard others. It virtually chooses the tactics by which managers execute. It encourages employees to behave in specific ways and rewards them for that particular behavior and often can discourage them from, or even punish them for, acting in different or new ways. Context can be a potent force for change—pushing organizations to continually look for new opportunities—or for stagnation—encouraging organizations to stick to the *status quo*, failing to recognize changes in the market. Corporate culture can drive innovation to significant value in the market or kill a great idea. Indeed, corporate culture can have one company see an opportunity and another miss or dismiss the same one. In this way, corporate culture is a singular determinant of organizational effectiveness.

When a company chooses to implement a radical or fundamentally new initiative, like embedding design thinking, the success of that initiative is not simply going to be a product of training and education, nor of management telling people what to do and following up, nor even a product of some new compensation or reward system. Implementing a discipline like design thinking will be successful only if it fits in with the corporate culture, even when that means supplanting some elements of the current culture with elements drawn from design thinking.

But transforming a corporate culture is complex, difficult, and fraught with risk of failure. Arguably, organizational transformation is one of the most difficult initiatives that a company can undertake, the equivalent of an experienced mountain climber scaling Mt. Everest: a long, complicated, challenging journey that is not to be undertaken lightly.

Default Culture

In most companies, culture is not intentional or purposeful. Typically, culture has evolved organically from the company's founding days, from the personality and likes and dislikes of the founder(s). Like Topsy, it " … just growed."

Corporate culture is likely to *default* to reinforcing what has worked in the past and avoiding what has not worked, especially avoiding significant failures. It is a relic from the past that powerfully shapes both perceptions and actions and limits possibilities. It often occurs as a given; corporate culture is *the way it is around here.* That can be true even when culture had been an intentional creation.

At Ford Motor Company, Henry Ford shaped the culture from the firm's earliest days to avoid a previous traumatic failure and to cause the company's success. So important was his influence on the firm's culture that in the early 1980s—more than three decades after the founder's death—Ford's ghost was said to be walking the halls of the company. Although the auto industry and the methods of manufacturing had changed drastically, Ford's culture had not.

By way of example, in 1985, a colleague of ours was conducting a training session with both older and younger employees of the Body and Assembly Division of Ford. Each participant was asked to write down something very significant that had happened in their tenure at Ford, something that they had put away in their "silver box of memories."

An older participant shared that in his first year at Ford, he was in the lunch room eating the ham sandwich that his wife had made, when Henry Ford sat beside him. The employee said, "Mr. Ford was somewhat of a health nut. He had special bread baked every day, and when he traveled, he had his bread flown to his location. In this instance, Mr. Ford said to me, 'John, you know that stuff you're eating is bad for your health. You shouldn't be eating it,' and then he rose and walked away."

Our colleague asked, "What happened next?"

Looking a bit surprised at the question, the gentleman said, "I've never eaten ham since that day."

A culture that made Henry Ford a quotable and inspirational industrialist and gave his company a huge competitive advantage had become a barrier to Ford Motor Company's success in a much changed marketplace. Fortunately, his successors, Donald Peterson and Harold "Red" Poling, led the cultural transformation to recover Ford's competitiveness—the first-known intentional transformation of the culture of a large corporation.

In too many large organizations, corporate culture is a barrier to design thinking and potent innovation. And transforming corporate culture is a particular challenge for executives hoping to move their organizations toward design thinking.

Why do we say this? Design thinking is human-centric—focused on the customer, the consumer, or whoever the end users may be. Design thinking requires a high degree of empathy for the end user, as well as big doses of risk taking, prototyping, and failing. Therefore, the practice of design thinking is likely to be antithetical to the corporate culture of most large companies, where data-driven decisions, rigid organizational hierarchies, well-established rates of return on investment, and a high cost of failure are often the preferred business-as-usual ways of operating.

Impact of Corporate Culture on an Organization's Ability to Innovate through Design Thinking

Companies that want to embed design thinking would do well to evaluate their current culture first. They should identify which of the company's shared patterns of perception, thinking, and acting may be at odds with design thinking and which ones are in harmony with design thinking's key principles of empathizing, defining, ideating, prototyping, and refining. The "Distinctive Elements of Corporate Culture" (Figure 12.1) can be used as structure for making this critical assessment.

Said another way, design thinking cannot simply be wedged into an organization whose values are at odds with design thinking's principles and practices. As Jeremy Utley, the director of executive education at Stanford's d.school, a leader in design thinking, has said, "It's a fool's errand to try and go against the culture. You have to find the elements of your business culture that support [design thinking's] kind of working and thinking mind-set."

If an organization is to embed design thinking, it must reveal, confront, and take responsibility for all aspects of its current culture. Then it must design a culture that leverages design thinking for success in the marketplace of the future. (Not coincidentally, the application of design thinking can enable this step.) Finally, the firm must rapidly make the needed changes. Any other process risks simply dressing up the old culture without changing it. And that can result in the *new* culture unwittingly inheriting aspects of the old culture—aspects that undermine the advantages of design thinking.

12.2 What Is Corporate Culture?

Every organization of any significant size—whether a commercial enterprise, a non-profit charity, or a governmental agency—operates within its own distinctive culture. Because it influences, shapes, and distorts the actions, perceptions, and thoughts of the people within the company, corporate culture is a singular determinant of an organization's effectiveness and can be an arbiter, or at least a critical factor, in long-term success or failure.

Distinguishing Corporate Culture

Corporate culture is the particular condition in which people perceive, think, act, interact, and work in a particular organization; it acts like a force field or an invisible hand. It shapes, distorts, and reinforces the perceptions, thinking, and actions of the people

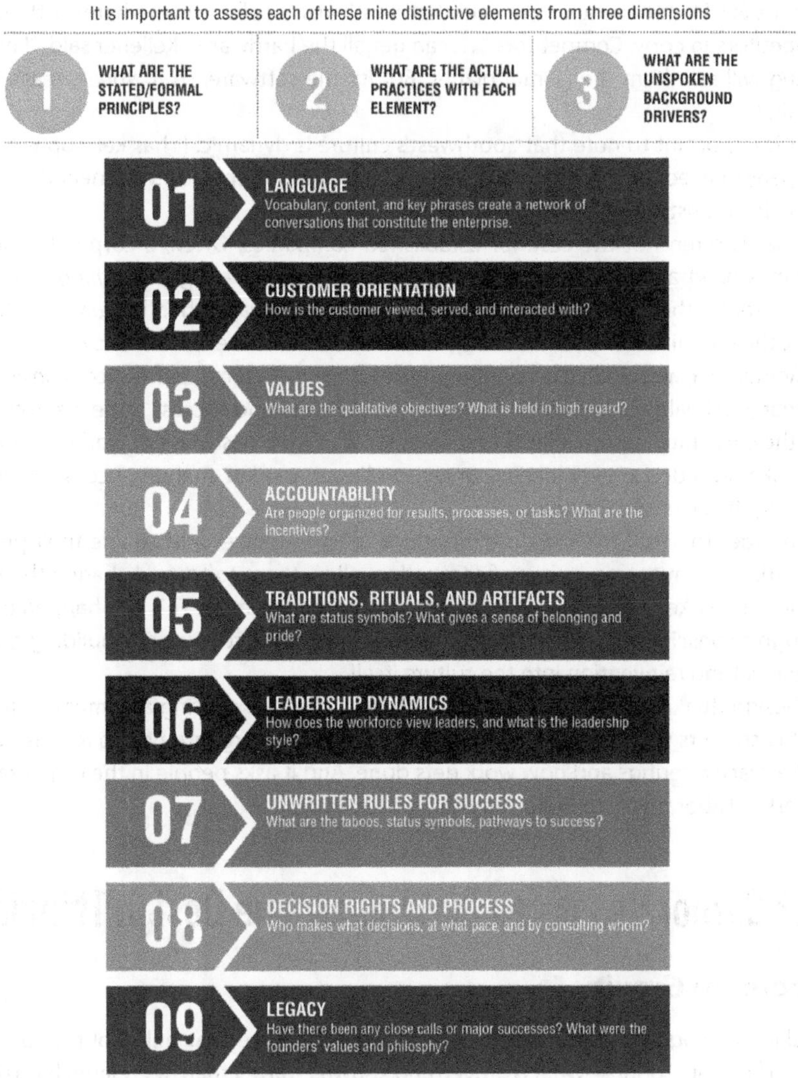

It is important to assess each of these nine distinctive elements from three dimensions

| **1** WHAT ARE THE STATED/FORMAL PRINCIPLES? | **2** WHAT ARE THE ACTUAL PRACTICES WITH EACH ELEMENT? | **3** WHAT ARE THE UNSPOKEN BACKGROUND DRIVERS? |

01 LANGUAGE
Vocabulary, content, and key phrases create a network of conversations that constitute the enterprise.

02 CUSTOMER ORIENTATION
How is the customer viewed, served, and interacted with?

03 VALUES
What are the qualitative objectives? What is held in high regard?

04 ACCOUNTABILITY
Are people organized for results, processes, or tasks? What are the incentives?

05 TRADITIONS, RITUALS, AND ARTIFACTS
What are status symbols? What gives a sense of belonging and pride?

06 LEADERSHIP DYNAMICS
How does the workforce view leaders, and what is the leadership style?

07 UNWRITTEN RULES FOR SUCCESS
What are the taboos, status symbols, pathways to success?

08 DECISION RIGHTS AND PROCESS
Who makes what decisions, at what pace, and by consulting whom?

09 LEGACY
Have there been any close calls or major successes? What were the founders' values and philosophy?

Figure 12.1: Distinctive elements of corporate culture.
Used with permission from Insigniam.

within the company, whether realized or not. Corporate culture is the unwritten rules for success inside the corporation, creating unseen walls and boundaries. It is the corporate paradigm. In short, it is whatever is reinforced within the organization. Analogously, it is like a company's personality.

In many cases, that invisible hand offers a company a huge competitive advantage in markets, where the differences between competitors are limited. Southwest Airlines, long lauded for its unique culture and for being the most consistently profitable U.S. airline over the past three decades, is a good example. The company's co-founder and former CEO, Herb Kelleher, once said that Southwest's culture is the hardest thing for competitors to copy. Competitors " ... can get all the hardware," Kelleher said. "I mean, Boeing will sell them the [same] planes. But it's the software, so to speak, that's hard to imitate."

It is important to note that Southwest's culture is dynamic. It has kept up with the company's incredible growth over the decades, helping keep Southwest near the top of the airline industry.

That is often not the case for established companies. Often, a corporate culture becomes fixed and unquestioned, the absolute view of reality, *how things are (and ought to be)*, rather than simply a way to work—*the right* way rather than *a* way. In those cases, the organization loses flexibility, waste increases, and execution slows.

Remember, we said that corporate culture is the unwritten rules for success *inside* the company. In healthy companies, the arbiter of behavior and success is the marketplace, and the corporate culture adapts to market forces. When past ways of working and culture take precedence over leading or responding to market change, success becomes pleasing the boss and fitting in.

In order to avoid this trap, organizations must empower and enable their people to continually invent new ways of competing, allow them to try and change the rules within the marketplace, as well as inside the corporation itself. This can happen either through a conscious and methodical cultural reinvention process or by building a spirit of renewal and reinvention into the culture itself.

Design thinking promises to do just this. It challenges existing assumptions about what customers want and need. It constantly pushes the organization to reconsider its marketplace offerings and how work gets done. And it asks people in the organization to work collaboratively to build something new.

12.3 Corporate Forces that Undermine Design Thinking

Corporate Gravity

Speaking analogously, corporate gravity is a hidden force that pulls your employees back to familiar ground—what is proven and known—rather than freely launching them toward innovation. The pull of the corporation is greater than the pull of the consumer and the marketplace. Corporate gravity is a product of the world view and concomitant processes, systems, and structures that protect the legacy business model and core products or services. Corporate gravity pulls resources to maintaining and improving what-is-perceived-as the source of corporate success.

To achieve success, ultimately, any organizational transformation must be led by the chief executive. Having said that, an antidote for corporate gravity is to appoint a chief change officer or a chief innovation officer (CIO). This executive is seen as the hand and brain of the CEO, has a budget and a department under her or him, and has the accountability and commitment to embed design thinking in the corporate culture. This CIO has the power and authority to initiate, lead, and manage change. For companies moving toward a design thinking model, the change can be facilitated by an executive who can bridge the gap between top management and the design and innovation teams—someone who can bring their two worlds closer together.

Corporate Immune System

Your body's immune system rejects and fights foreign substances. It is an involuntary response. In an analogous way, organizations can seemingly reject and fight changes to their culture, even when the leaders are actively leading change.

There is no such thing as inherent resistance to change. People will not make changes if they are threatened by those changes. People will rapidly adapt to and/or cause change when they can see an opportunity for themselves.

All successful employees possess a key item of knowledge: how to make their managers happy. Do bosses demand that the current product development process be executed exactly as laid out in the corporate product development manual? Well then, how likely are employees to apply design thinking to that process, potentially disrupting it or completely reinventing it? How will they know how to make the bosses happy then?

This is one of leadership's toughest challenges. Effective corporate change demands conversations—lots of them. Managers must tailor their message to individuals or, in a large-scale change process, tailor it division by division. What will inspire the scientists and engineers? What will move the marketers? Managers and executives need to interview and observe people in those divisions. They must then design a conversation that will open up opportunities for their people. In other words, executives and managers need to bring the designer's tools and methods to their own work. That way, not only are employees engaged in the change, but they also see design thinking in action.

Corporate Myopia

There is a joke:

Question: How many designers does it take to change a lightbulb?
The Designer's Answer: Does it have to be a lightbulb?

The power of design thinking is that it asks those engaged to think in different ways—about the product itself, the way the consumer will use it, the way it will be made. Everything is up for contemplation and a shift in perspective. That can be a problem for companies that are undergoing a transformation to design thinking while also operating an ongoing business.

Corporate myopia keeps executives from seeing value in innovations, including new methods like design thinking. Successful executives think that they know what the consumer wants and what will succeed in the marketplace. A breakthrough innovation in a product or a process may threaten an executive's sense of his worth or not fit her understanding of what is valuable or not conform to the corporate strategy. In some companies, anything that does not hit financial hurdle rates or show well on forecast sales volume evaluations never makes it to market.

On several occasions, Nestlé executives tried to kill the now-successful Nespresso coffee system. Nestlé was in the food business, not the kitchen gadget business. Executives were skeptical of a technology developed by research and development (R&D) that did not fit the mass-market business model of the time and was a major departure from most of Nestlé's lines of business. The Nespresso System survived and thrived when it was established as a separate company, in a different building, and an *outsider* was brought in for new perspectives and ideas.

The antidote to corporate myopia is design thinking itself. Rather than executives determining the value of an innovation, design thinking prototypes are held and actually used by consumers or users. Consumers and users determine value and how to improve or add value.

12.4 Four Pillars of Innovation for Enabling Design Thinking

No one, not even expert mountain climbers or the Sherpas who live in the Himalayas, just shows up one afternoon and starts climbing Mt. Everest. The effort takes years of experience and months of preparation and requires that many things go according to plan.

In the same way, a corporate culture cannot be formed around design thinking without taking the time to build a stable base on which that change will rest. There are four critical pillars on which a shift or transformation to design thinking must be built, as illustrated by figure 12.2.

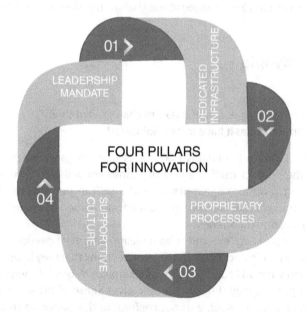

Figure 12.2: Four pillars for effective creativity and innovation.

Pillar 1: Leadership Mandate

The top executives have to commit to innovation using design thinking as a corporate priority; this is the corporate equivalent of Agamemnon hauling his ships onto the beach and burning them. The requirement for design thinking has to be baked into the corporate strategy. The executives have to learn and practice design thinking. They have to be committed to leading its adoption throughout the organization.

They have to design and communicate the case that innovation and design thinking are critical to the future of the organization. That mandate needs to be loud and clear and relevant to employees across the enterprise. They also must give clear permission to do fresh thinking and back this up with funding, people, time, and space.

To illustrate: When A. G. Lafley, Procter & Gamble's CEO, set out to remake that company around design thinking in 2001, he said, "We will not win on technology alone. Therefore, we need to build design thinking into the DNA of P&G." And he backed those words up with his own actions.

Mr. Lafley was a regular attendee at workshops focused on design thinking, where designers were paired with senior managers so both could share the principles of design thinking. He also routinely received input from an external design board (those fathers of design thinking, among others), set up to critique P&G's design decisions. He met regularly with the design executive whom he had tasked with overseeing the transformation, Claudia Kotchka, Vice President for Design, Innovation, and Strategy, so that he could stay informed about the latest steps in the process. On virtually every business trip, he took time to go into consumers' homes and observe how they lived. In those ways, Mr. Lafley communicated and led, in both words and actions, the vision and commitment to make P&G a company built around design thinking.

Pillar 2: Dedicated Infrastructure

A dedicated infrastructure organizes people, resources, budget, timelines, space, and metrics. The dedicated infrastructure always mirrors the seriousness of the mandate. If there are visible resources invested in supporting the mandate, it is taken seriously. If not, it can be seen as lip service. It can include specific organizational roles, such as the office of innovation, a function that goes beyond new products or setting up self-managing teams.

To illustrate: A few years ago a large, successful health care enterprise held a typical executive offsite. The top executives discussed good news. Revenues were steadily increasing. So, too, were margins. They discussed a strategic plan that had been developed to ensure continued growth. But then came a surprise.

"We realized," says one top executive, "that what had gotten us this far wasn't going to get us where we wanted to be in the future. We took a real gut check and asked, 'What do we need to change to achieve the strategic goals we had set?' and 'Are we going to make the investment in those changes?'"

The answers were eye opening. It was decided that the corporate culture needed to be redesigned with a focus on patients. The corporate culture had been centered on fiscal discipline, a heritage that was valued and honored. Centering on patients would require a transformation. It was also decided that this transformation was worth the investment of money, time, and the risk involved in changing the operational values of a financially stable enterprise because the executives were smart enough to realize that the risk of staying the same was even higher.

When an organization takes that kind of risk, though, it does not take it lightly. In this case, they executed systematically, step by step.

Step 1: Conduct a cultural assessment.

A cultural assessment of custom-designed questions was used for interviews and surveys of employees from every level, function, and geography in the company to reveal the current culture. This process is not unlike the ethnographic tools that design thinking relies on to reveal the needs of consumers who designers hope to help. The assessment was conducted against the "Distinctive Elements of Corporate Culture" (Figure 12.1). The analysis and report from the survey highlighted key aspects of the culture, the aspects that would support the strategy and those that would inhibit it, and recommendations to initiate the transformation.

Step 2: Set up an office of transformation and a transformation leadership team.

They called this team "the leadership coalition." It was composed of 40 people from across all divisions of the organization. While not every team member was dedicated full time, they allocated a set amount of time for their new roles, as if part-time jobs. A full-time transformation executive was appointed and acted as team leader. The CEO's opening statement on the coalition's first day was that the group was to leave their titles at the door as they drafted a new vision, a new mission statement, new corporate values, and new operating practices for the company.

Step 3: Set a budget and some deadlines.

Cultural transformation is neither quick nor cheap. It has hard costs and the need for sustained effort—a test of commitment, all of which should be considered carefully in advance. The company had an 18-month window for its first, most important phase of cultural transformation and allocated a specific dollar amount to make that change happen. The leadership coalition set a timeline working back from the close of the window.

Step 4: Create a team to win over the team.

An enrollment team that drew employees from all levels and functions of the organization was asked to inspire and engage the workforce to adopt the company's new direction, even as it was being developed. These team members were trained in design thinking, effective communication, and enrollment. They designed both the message and delivery methods to fit the company and its people.

Step 5: Create a big commitment, new capabilities and increased capacities, and lots of project teams.

This company also created a 30-person Keystone Project Team that, as its name suggests, was responsible for developing and executing a key project that would move the company to its new customer-focused goals. A keystone project is a multi-year commitment to producing critical results that can only be accomplished in the new culture. The Keystone Project Team commissioned project teams to move the keystone project forward and to deliver the intended results.

At the same time, 18 of the company's top executives were engaged in a year-long leadership development initiative that would give them the skills to work in an environment where collaboration was emphasized far more than it had been. As part of the leadership program, each executive designed and led a leadership project.

As one executive said, "We developed a real focus on teams. There was much less interest in executives saying, 'How can I get this done?' and more on, 'How can we get this done?'"

Within the 18-month window, the initiative delivered notable change and produced remarkable results, moving the company from the middle of the pack to near the top of its industry in the three key metrics used to measure success.

Note: The first rule of management is that you tend to get what you reward; obviously, everyone knows that a reward structure is needed to support the transformation. However, the second rule of management is that you tend to get what you measure; not so obviously, as part of the infrastructure, you have to put in place a scoreboard to measure the value generated by design thinking. Otherwise, the value of design thinking is lost in the mix of overall business results.

From the time he joined Clorox in 2009, Wayne Delker successfully drove new product development as head of research and development and then as chief innovation officer for the entire company. Dr. Delker invented metrics to measure the value derived from innovation that helped sustain executive management's investments in innovation, creating a virtuous cycle.

Pillar 3: Proprietary Creative Process

For an enterprise's innovation and design thinking process to complement its culture, infrastructure, and mandate, it must be *their* process, meaning it must be proprietary. The process needs to reflect the unique business and assets of the company, as well as its corporate culture; ideally, the process should evolve over time.

Learning from leading-edge businesses and educational institutions can and will provide value, so why not just cut and paste their process into your organization? When an organization tries to wedge another entity's process into its own business, the background and organizational context that allowed the process to be successful in the originating company is lost. The implementing company often finds that the *off-the-rack* solution does not integrate with the other elements of their enterprise. Simply put, company X's process will not fit company Y's infrastructure or culture—organizational context—because it was not designed to fit. Context trumps content; context is decisive.

Consider this statement once made by Norio Ohga, the former chairman and CEO of Sony, a company that has effectively utilized design thinking. "At Sony," Ohga said, "we assume that all products of our competitors have basically the same technology, price, performance, and features. Design is the only thing that differentiates one product from another in the marketplace." Design, implement, and utilize a proprietary innovation and design thinking process for your company.

Pillar 4: Supportive Corporate Culture

A supportive corporate culture is friendly to new ideas, ranging from incremental to transformational, and not just those from the top down. The culture has to avoid breeding a fear of risk and failure. Risk management is healthy; risk avoidance is deadly. A supportive culture limits corporate gravity, inoculates against the enterprise immune system, and fights corporate myopia, the three forces we discussed earlier in this chapter.

Compare the corporate cultures of Boeing and Airbus, a duopoly of commercial airplane builders, and you will find that the cultures are not even remotely similar, while their businesses are essentially the same. Think of the difference in corporate cultures at General Motors and Toyota.

This is why culture can be a huge advantage to some companies and a huge disadvantage to others. Remember the quote from Herb Kelleher at Southwest Airlines from earlier in this chapter about culture being a differentiator.

Cultures are not one-size-fits-all, and neither cultural transformation nor any serious design thinking endeavor can be a one-size-fits-all solution. These must be specifically designed with the existing culture and the corporation's purposes and strategic intentions in mind. It must take into account the organization's history, leadership, and the mandate for change. Any plan put in motion to move an enterprise on an innovative path toward the future must first begin by recognizing and revealing where and what the organization is today—for better or worse.

12.5 Four Stages of Transforming to a Culture of Design Thinking

Okay, what if we have done a good job and convinced you that you need to transform your corporate culture with design thinking embedded in it? What if you have realized that a designed culture with design thinking embedded in it would give your enterprise a competitive advantage? Beyond the five steps and the four pillars outlined above, there are four stages to move your organization through for a successful cultural transformation (or any organizational transformation, for that matter).

Stage 1: Reveal

- What are the current aspects of your strategy, culture, processes and practices, systems, and structures that enhance or inhibit design thinking?
- What are the hidden assumptions and deeply held beliefs that operate as an invisible force in the organization, telling people what is possible and not possible?
- What are the unwritten rules for success?
- How are new products and services brought to market? Is the company driven by internal decisions or customer insights?
- How do past failures, as well as successes, determine people's thinking about the business, market dynamics, the competition, and the customer?

- Are you innovating or just keeping up with the competition?
- Is your company's relationship with the marketplace generative or reactive?
- Assess the current culture against the Nine Elements of Corporate Culture.

Stage 2: Unhook

- What interpretations and beliefs cloud your view of the facts?
- To what degree do you blame forces outside of your control for your results, for example, "It's the economy," "Marketing's data is flawed," "R&D can't deliver on our customer needs"?
- Are you listening to what your customers are *actually* telling you, or do you already know what they are going to say?
- To what degree have your ways of doing things become the only way of doing things?
- What was said in the past and has now become *the way it is*?
- What are the sacred cows that need slaughtering?
- What assessments and judgments were made and are now related to as facts?
- Take responsibility for all of those conversations, stop relating to them as reality, and put them aside.

Stage 3: Invent

- What will be the marketplace in the future?
- What kind of company would thrive and be wildly successful in that marketplace?
- What will be the purposes and ambitions of your organization that will inspire, challenge, and excite the people who are your organization?
- What values will support your commitments? What are the fundamental principles that will inform people's thinking and working?
- What is your leadership mandate for design thinking?
- What will be your proprietary innovation and design thinking process?
- How do you need to design your strategies, processes and practices, systems, structures, and teams to leverage design thinking?
- What rewards and recognitions will reinforce and support design thinking?
- How will you measure the value generated by design thinking?

Stage 4: Implement

- Is leadership aligned with the future that is being created?
- In what new conversations will you engage the people of your enterprise?
- How are you going to get people to work across functional lines?
- How are you going to get the customer present in virtually every conversation and in every day of work?
- What projects and initiatives will utilize design thinking?
- Do you have a communication strategy that is sufficient to support the culture change (communication increased by a factor of 10)?
- What education and training will make a difference in empowering and enabling what people within the company?
- Are people held accountable for behaving and acting consistent with the new culture?

12.6 Conclusion

The lesson of this chapter? Design thinking is a powerful new approach to business. Design thinking would likely be a source of competitive advantage, if it were embedded in the corporate culture, as well as the company's strategy, processes and practices, systems, and structures. Those companies that have embedded design thinking have found that it produces great results for customers, employees, and their organizations as a whole. To embed design thinking means both a strategic and cultural transformation for most large corporations. But cultural transformation is complex, difficult, and fraught with risk. To use design thinking to achieve competitive advantage, the corporate culture must at least align with and, ultimately, pull for design thinking. By installing certain structures and by working on specific elements of the culture, executives can achieve this level of performance for their businesses.

References

Martin, R. Incorporating Design Thinking into Firms. Rotman: The Magazine of the Rotman School of Management, Fall 2005, Toronto, Canada.

About the Authors

NATHAN OWEN ROSENBERG SR. co-founded Insigniam. His 29 years in consulting have catalyzed measurable breakthrough results and valuable innovations and transformations for 17 percent of the world's 1,000 largest companies and the more than 89,000 people with whom he has worked directly. Nathan is a board member of Boy Scouts of America, Educate Girls Globally, and the United States Air Force Academy Rugby Excellence Fund, and is a trustee of the Committee for Economic Development. This chapter is dedicated to his brother, Werner Erhard, who is the inspiration for much of what has been written.

MARIE-CAROLINE CHAUVET is a partner in Insigniam and a member of the firm's design and innovation team. She has two decades of corporate experience in financial management and corporate strategy. She received her MBA at the Université Paris-Dauphine. Marie serves on PDMA's Board (2014–2017). She resides in Toulouse, France.

JON KLEINMAN brings more than 15 years of experience in consulting with Fortune 500 companies. As the partner leading Insigniam's design and innovation team, Jon has researched, designed, and executed new interventions for Insigniam that provide extraordinary value for clients. Jon is on the board of the Jewish Community Center of Southern New Jersey.

13

KNOWLEDGE MANAGEMENT AS INTELLIGENCE AMPLIFICATION FOR BREAKTHROUGH INNOVATIONS

Vadake K. Narayanan
Drexel University

Gina Colarelli O'Connor
Rensselaer Polytechnic Institute

Introduction

Design thinking has emerged as the next frontier in the competitive landscape of many industries and firms. Design thinking has been defined as combining *empathy* for the context of a problem, *creativity* in the generation of insights and solutions, and *rationality* in analyzing and fitting various solutions to the problem context (Kelley & Kelley, 2013). Its principles and practices are directed toward intractable human issues, so-called *wicked problems* for which an optimal solution or even a knowable solution may not exist (Buchanan, 1992). It is a method of creative action and experimentation, focused on solving complex problems.

How to best handle knowledge for large intractable problems for which optimal solutions are not knowable is a tantalizing arena for purveyors of knowledge management experts to explore. Typically, organizations structure their knowledge management systems to cumulate knowledge, experience, and expertise in certain market and technology domains, and leverage that knowledge repeatedly to enhance their efficiency and effectiveness, thereby outpacing competitors and maximizing profitability for their shareholders. Partly as a consequence, knowledge management (KM) approaches have mostly been applied to these routine facets of an organization's operations, including product development, market orientation, customer relationship management, and others, where known markets and known technologies are leveraged for success, and the challenge is being more efficient and effective than competitors. But the application of KM principles to design thinking is also a fruitful arena for consideration.

KM had its origins in information technology (IT), but recognizing that decisions involve information and knowledge, not merely data, the KM that many organizations institutionalized came to be viewed as a function dealing with *acquisition, utilization*, and *dissemination of knowledge*. Indeed, in some facets of product or business development, KM tools have been valuable. However, the "fuzzy front end of innovation" did not and indeed could not benefit from traditional KM tools. Some suggest that this situation is beginning to change.

This chapter will focus on the perspectives, principles, and practices required in KM to address the world of *breakthrough innovations*—innovations that demand design thinking because they are characterized by high levels of uncertainty and address big, complex, and sometimes intractable problems. We will discuss the shift in perspective from *intelligence leveraging* to *intelligence amplification* as a key characteristic of KM to address the arena of breakthrough innovations. This shift emphasizes the role of insight development over leveraging available knowledge.

The chapter will deal with the tools for embedding KM in design, with examples from several large but leading-edge companies and other organizations such as incubators, idea labs, and consulting organizations. The tools will be described within the framework of discovery, incubation, and acceleration, three capabilities necessary for breakthrough innovations (O'Connor & DeMartino, 2006; O'Connor, Leifer, Paulson, & Peters, 2008). The material in this chapter is targeted at people in corporations and other organizations (such as incubators) who engage in innovation of a nonincremental nature, and who find that the more they leverage what is known, the further removed they are from the possibilities that their opportunities enable. We hope this chapter provides a frame of reference to help readers recognize that they are engaged in a different sort of innovation altogether, and that the use of knowledge must occur in a different way than typifies most organizational practices today.

13.1 Designing Amidst Uncertainty

Much of the world of design for new product development (NPD) has been fitted within the traditional Stage-Gate process (Cooper, 2001). Ideas are generated, screened, and approved by a gate review board, and the project follows a sequence of steps designed to ensure cross-functional involvement through project scoping, building the business case, detailed design and development, testing and validation, and launch. While the team may not have access to the necessary information, it is easily accessible using traditional tools. This approach works well for incrementally new products that leverage past designs, technologies, and customer loyalties.

But firms also introduce breakthrough innovations. These opportunities arise in a couple of ways. One way is to engage customers directly to understand a deep-seated problem, find solutions, develop really new products and services, and get to market first. A second approach is based on identifying applications for advanced, emerging technologies that enable new solutions to interesting problems. Companies that invest heavily in research and development (R&D) develop deep technical expertise and can gauge shifts in technology to meet known and unknown needs in the marketplace. Firms that adopt the first way are sometimes labeled Need Seekers, and those following the second approach Technology Drivers (Jaruzelski, Loehr & Holman, 2013). Both of these approaches lack complete technical or market expertise, respectively, in the domain in which they are innovating, and so they operate in domains of uncertainty. But they do produce *breakthroughs*. Regardless of which approach, both can benefit from a KM framework and tool set that helps managers expand beyond current knowledge base. Technology Drivers do not use their current customer base as a referent for future innovation. And those engaging deeply in the market to identify unarticulated needs do not necessarily draw on known solutions as they forge a new product for a deep seated need they've uncovered. In fact, they may work with customer partners to co-develop solutions through experimentation.

These groups' innovation experiences compare with those using the Stage-Gate® process for incremental innovation as more ambiguous; forecasts, business cases, operating models, market reactions, and production systems are all unknown. Following O'Connor and DeMartino (2006), we define three basic stages required for breakthrough innovations (BIs): *discovery, incubation,* and *acceleration*:

Discovery: Involves creating, recognizing, elaborating, and articulating potentially breakthrough opportunities. Discovery activities can include invention and lab research, hunting inside and outside the company for ideas and opportunities, partnering with universities and licensing technologies or placing equity investments in small firms that hold promise.

Incubation: Matures breakthrough opportunities into business proposals. A business proposal is a working hypothesis about a technology platform, potential market space, and a business model. Incubation is not complete until that proposal—or, more likely, a number of proposals, based on the initial discovery—has been tested in the market, with a working prototype. The skills needed for incubation are *experimentation* skills. Experiments are conducted not only on the technical front but also for market learning, market creation, and testing the business proposal's match with the company's strategic intent.

Acceleration: Activities ramp up the fledgling business to a point where it can stand on its own relative to other business platforms in the ultimate organizational unit (SBU) in which it will reside. Whereas incubation reduces market and technical uncertainty through experimentation and learning, acceleration focuses on building a business to a level of some predictability in terms of sales and operations. Acceleration activities include investing to build the business's physical infrastructure, focusing and responding to market leads and opportunities, and developing repeatable processes for typical business functions such as manufacturing scheduling, order delivery, and customer relationship management. Scaling a business involves uncertainty on a variety of dimensions as the opportunity is faced with many degrees of freedom and the company's and market's reactions to choices made are extremely malleable.

Discovery, incubation, and acceleration differ markedly from the conventional Stage-Gate process for NPD, where markets and solutions are drawn from the existing stock of knowledge and expertise held within the company. For the breakthrough innovation stages, the firm cannot rely on its past storehouse of knowledge. It must instead rely on its ability to *amplify* what it knows. The company will be engaged in creating new knowledge together with market agents as the opportunity is enlarged in discovery, and then incubated and accelerated into a full-fledged new business.

Table 13.1 provides a summary comparison of the differences between incremental and breakthrough innovation and the challenges of KM for each. We take this theme up next.

13.2 Knowledge Management Tasks for Breakthrough Innovation: From Intelligence Leveraging to Intelligence Amplification

KM as practiced in organizations has relied on a number of tools that enabled *intelligence leveraging*: garnering existing data and knowledge to create efficiencies in organizational processes. Broadly, these tools enabled codification and dissemination of information, and identification or establishment of social networks. These tools (e.g., FAQs,

Table 13.1: Differences between Incremental and Breakthrough Innovations		
	Incremental	**Breakthrough**
Characteristics of the process		
Knowledge of customers	High	Low
Knowledge of technologies	High	Low
Characteristics of the process	Sequential/Stage-Gate	Iterative
Level of ambiguity	Low	High
Utility of design thinking	Moderate	Critical
KM characteristics		
Perspective	Intelligence, leveraging	Intelligence, amplification
Objective	Embed KM tools to increase the efficiency of the process	KM tools for (1) insights, (2) inventions, and (3) experiments

data mining), organizational mechanisms (e.g., communities of practice), and analytic approaches (e.g., social network analysis) have come to represent the technical core of KM; they enable embedding best practices in many organizational processes. During the past two decades, spurred by the IT revolution, many corporations have institutionalized a KM function. Recognizing that decisions involve information and knowledge, not merely data, KM came to be viewed as a function dealing with *acquisition, utilization, and dissemination of knowledge*. Over time, KM scholars and managers have developed a number of tools currently in use in organizations. These tools enabled *intelligence leveraging*: garnering existing data and knowledge to create efficiencies in organizational processes. Table 13.2 provides a select set of KM terms and tools in use today.

As noted in the previous section, in the case of incremental innovation, relatively low levels of ambiguity characterize the discovery, incubation, and acceleration stages. Thus, the individuals or teams that are involved in the discovery are familiar with the markets, customers, and the dominant designs (and the technologies undergirding them). Their tacit knowledge and attendant interpretive frames are robust enough to wade through the databases and knowledge depositories. Similarly, incubation can be modeled as a structured process (e.g., Stage-Gate process) using well-understood tools to gather information about customer preferences and business models, and acceleration has some chance of success because reasonable predictions can be made about cash flows and returns on investment. In incremental innovations, KM, as intelligence leveraging, can enhance the efficiency of the process by enabling the search of market and customer data through knowledge depositories, deriving best practices though the creation of communities of practice, whereby product development teams are brought together for sharing their experiences, and building social networks for the transfer of personal knowledge.

Breakthrough innovation, however, requires that KM functions—acquiring, disseminating, and utilizing information and knowledge—take on a hue that is different from

Table 13.2: Knowledge Management Tools for Intelligence Leveraging	
1. Codification and dissemination of explicit data or data that can be digitized	These tools include data warehousing, knowledge engineering, and FAQs. Corporations like Siemens employed KM to enhance the effectiveness of the bidding process in the market, transferring knowledge gained from developing economies to emerging economies.
2. Transfer of tacit data	This kind of data (especially best practices) cannot be digitized, but has to be transferred from individual to individual. The tools include mentoring or communities of practice where individuals involved in the specific set of practices (e.g., new product development) form a learning community to share practices and learn from each other. Some organizations like NASA addressed the scarcity of talent (e.g., program managers) due to retirement by investing in training and mentoring a new generation of project managers. Professional organizations such as, for example, the Project Management Institute, have instituted communities of practice for project managers.
3. Tools to identify individuals with knowledge	These tools enable accessing individuals within a firm or outside who have knowledge or expertise that may enable individuals or teams to perform their tasks. Social network analysis tools are an example. The consulting firm McKinsey and Company is reputed to have a system in place to access individuals with expertise for specific domain areas.

the case of incremental innovation because BI stages, especially discovery and incuba-tion, are characterized by high levels of ambiguity. Very often, the engineers involved in the discovery stage may be familiar with the technical details of the proposed solution, but neither they nor their marketing colleagues understand the markets and customers (who may not yet exist). Indeed, in the case of Technology Drivers, companies typically misjudge how the application markets unfold for the opportunity, where the products are likely to earn the highest returns, and how to meet the cost of capital objectives. In the case of Need Seekers, the individuals who are close to the markets and customers often are blind to the more effective solutions to the customer needs (see Christensen, 2000). Incubation requires market engagement more intensely than can be obtained through traditional market research techniques (Leonard-Barton, 1995), and both pro-totype and business model development involve "out of the box" thinking increasingly discussed in the literature. Even acceleration, with its focus on scaling the business, is fraught with unknowns regarding new processes, scaling issues, and market inquiries about new applications that require different tools than those that presume a grounding in what is known to be an adequate basis for judgment and decision making.

Breakthrough innovations can be facilitated by KM in the sense of intelligence amplification. By *intelligence amplification* we refer to the processes of discovery, imag-ination, and experimentation by which individuals, teams, and organizations expand their base knowledge, perspectives, and practices beyond what is currently available to them. Here, the focus is not on the efficiency of the process but on developing insights, unleashing imagination, and enabling experimentation—activities that are the key to success during discovery, incubation, and acceleration. This in turn requires a focus on interdisciplinary, peripheral, and sometimes speculative information and exceptions (Ruggles, 1997). For example, to develop insights, the individuals (or teams) involved in BI must gain exposure to information about potential markets or applications not in their areas of expertise; they may need to engage the customer more intensely, may need tools to unleash their imagination, and may have to be coaxed to experiment more frequently than in the case of incremental innovation. To unpack the enabling activities in insight development, imagination, and experimentation, we highlight five key functions that are part of knowledge amplification.

1. Information arbitrage

An information arbitrage function involves the deliberate movement of data and knowledge from one location to another to create value. In the context of breakthrough innovation, where problem–solution links are not obvious, this function allows for creative links of disparate elements of information. Information arbitrage addresses questions such as: How might we couple "problems" and "solutions" in a manner that yields highest returns? In other words, how might we move technologies to their most productive uses and/or identify the most effec-tive solutions to customer problems? Examples of information arbitrage at work in companies include technology fairs, for example, which 3M and others hold for internal personnel. Exposing members of business units or other global regions to the technological capabilities and discoveries that the company has allows 3M to increase the likelihood that someone in the company will articulate a new business opportunity because of their recognition of what is possible in the technology solution space. Another interesting version of information arbitrage stems from an HR policy. At Air Products, the director of New Business Development encouraged his staff members to attend multiple conferences each year to expose themselves to new markets, with the caveat that they could not attend the same conference two years in a row. They were required to learn about new markets each time. Each of these practices of information arbitrage unleashes imagination and enables insight development through specific managerial actions.

2. Customer engagement

This function involves deep engagement of the customers in the design and use of product. How might we engage the customer in the design process in a meaningful and productive fashion to conceptualize and prototype products? IBM's earliest attempts at developing an electronic book in the mid-1990s took place in partnership with Boeing. The first perceived application was a way to simplify the cumbersome manuals used by aircraft maintenance crews. So the IBM team loaded those manuals into an e-format and asked the maintenance crews to go about their work with the new devices. Three months later the IBM team engaged them to understand their reactions to this novel technology, and a number of iterations regarding the display, battery life, and storage capacity of the first e-book resulted from that experiment. This managerial practice of engaging an innovation team along with a customer partner in experimental learning enabled insight far beyond what any traditional KM technique might have surfaced.

3. Visualization

This function involves creating graphical (two- or three-dimensional) representations of products and business models that enable individuals (designers and customers) to get a nuanced understanding of the aesthetics and functions of products or the coherence, completeness, and value potential of business models. How might we enable individuals and teams to visualize their products and business models, both of which may be hazy in the beginning in the case of breakthrough products? The first presentation of a concept within Kodak for satellite-based photographs of earth to aid farming were simply dummied photographs that showed possible ways to analyze crop and field size and shape to maximize crop yield. The dummy photos sold Kodak's Venture Board on the idea by helping them imagine the possible.

4. Fail fast

This principal underscores the need for fast experimentation whereby an idea can be tested without much investment, and the unsuccessful ideas can be weeded out quickly. How might we devise a process to (a) allow individuals and teams to fail fast, (b) learn from the process so that early-stage investment of resources is low, the process is kept iterative and recursive, and thereby (c) enhance the probability of success?

5. Application migration

This is the phenomenon observed in technology push and breakthrough innovation environments in which a product is launched for one purpose but ends up being used for many others. How might we leverage breakthrough technologies into markets that were never imagined or planned? What are the best ways to speed the diffusion of a breakthrough innovation into niche applications? Analog Devices' commercialization of its accelerometer-enabling computer chip started with a killer app dream of replacing the detonation system for airbags, but in reality started with satellites, gyroscopes, niche video games, and many other smaller market opportunities before the killer app was realized. None of those first markets were predicted or planned, yet market experiments led to new insights.

Implicit in this set of five activities is the recognition that in order to be useful for BI, KM has to assist individuals and teams to "invent" rather than "embed best practices" identified. Each of these KM practices is a managerially controllable action that can be taken to enhance insight development, enable imagination, and encourage experimentation, all of which lead to the outcome of knowledge amplification.

KM can accomplish these through a variety of organizational mechanisms and technology enablers. Organizational mechanisms include individual roles such as knowledge brokers, facilitators, and transition managers and institutional mechanisms such as communities of practice (Wenger, 1988). Knowledge brokers serve as bridges of information by connecting people or teams who have different pockets of information and knowledge. Facilitators are individuals who are trained to guide a discussion group with specific goals. Transition specialists understand the challenges of transitioning between stages and help make the group or organization movement smooth. Technology enablers include physical resources, IT augmented tools, and knowledge depositories.

In what follows, we will illustrate how several knowledge amplification tools facilitate the discovery, incubation, and acceleration stages of BI. Table 13.3 provides a summary for reference.

13.3 KM and Selected Tools for Breakthrough Innovation

Discovery

As mentioned previously, the objective of discovery is to generate ideas or novel technologies and develop and elaborate them into robust potential business opportunities, which are then tested out in Incubation. We summarize three tools that amplify what is known: idea jams, technology translation tables, and technology market mind maps.

Idea Jams

IBM is known for hosting daylong meetings around the globe to which they invite experts across a wide variety of fields to participate in idea generation sessions. The objective is to consider big problems, technological progress, and social challenges and generate a plethora of ideas for new business platforms. IBM's first innovation jam took

Table 13.3: KM's Support for Breakthrough Innovation				
Stages of Breakthrough Innovation	Discovery	Incubation	Incubation	Incubation/Acceleration	Acceleration
KM Tasks	Information Arbitrage	Customer Engagement	Visualization	Fail Fast	Application Migration
KM Tools	■ Inventor experience ■ Technical conference presentations ■ Technology translation tables ■ T-A-P-M mind maps ■ Idea jams and idea capture tools ■ Idea combination tools ■ Ethnographic observation	■ Extended use trials ■ Customer immersion labs ■ Discussion groups/platforms ■ Co-design tools	■ Rapid prototyping ■ Visual simulation ■ Interactive simulation ■ Business model canvas	■ Learning plan and business experiments ■ Discovery-driven plan ■ Crowdsourcing ■ CRM and automated feedback systems	■ Ads in technical publications ■ PR ■ E-commerce tracking ■ Web analytics ■ Usage sensors ■ Social media dashboards

place in 2006, comprised of two 3-day sessions in which 150,000 IBM employees, family members, business partners, clients, and university researchers participated from 104 countries in online round-the-clock dialogue about potential growth opportunities (Bjelland & Wood, 2008). Every single post was captured and analyzed to discern the best new ideas for creating business opportunities from the technologies. IBM executives view the jam as valuable, and indeed *10* breakthrough projects were launched as a result. They continue to modify the process to make better use of the knowledge gained through subsequent jam sessions.

Technology Translation Tables

When inventors describe a new discovery, it is easy to get lost in the technical details that most excite them. A technology translation table is a useful aid when interviewing inventors and in subsequently researching and thinking about the possibilities that the invention may enable. The technology translation table is a three-part tool (see Appendix 1). The tool appears to be simple, though a thorough treatment of the translation requires deep thinking, conversation, and research to maximize the opportunity analyst's understanding of the scope of the opportunity's potential. Part I is a description of the business concept (Appendix 1a). It describes the technology, what it does, and why it is important, in just a few sentences. Second (Appendix 1b) is a table listing the key features of the technology, one by one, and how each differs from what is currently known. Whether value can be attributed to each of these differences is not important at this stage. We know that for breakthrough innovations, the feature that inventors and companies believe at the outset is the feature of value frequently turns out not to be the case; instead, another feature is considered valuable. Therefore, it is important to be as thorough as possible in describing each dimension of difference this technology offers over what is currently known or available. Finally is the technology translation table (Appendix 1c). Each key feature described in the previous table is broken down into the intellectual property claims that would be made in a patent application for the invention. Each claim is listed along with commentary about who might value that difference. The latter task requires creative thought and a broad scope of applications.

Technology Market Mind Maps

Another mechanism for helping the discovery analyst link technologies and markets to find breakthrough opportunities is a technology market mind map. This tool is a useful systematic process for either finding applications for technologies or finding technologies and product ideas for market needs. Several examples are described here, and the tool is arrayed in Appendix 2.

Each mind map is composed of four elements, arranged in different patterns for each case. The elements are M (market), A (application), P (product), and T (technology). The most conventional approach is to start with a market that we elect to serve, and ask, "What problems do you experience?" Those are the application areas. For each, we generate multiple products that may serve the problem in different ways. Each is executed via a different technology. In this pattern (shown in Appendix 2a), we search for a technology that solves a particular market need. For example, company X serves the energy market. Energy applications include in-home heat, mobile energy, and clean energy. For each we have multiple product/service offerings. For example, batteries and fuel cells provide mobile energy. Clean energy also incorporates fuel cells, so different applications can actually have a product in common. Each is fulfilled through a different technology.

A different approach, also common, is to start with a novel technology that we believe could generate a breakthrough innovation (shown in Appendix 2b). Just as the technology translation tool guides us, we use this mind map to ask ourselves about potential uses, or applications, for the technology. From there we ask who cares about those uses to uncover possible market segments. Each segment may need a different formulation of the offering, or product design. DuPont's biodegradeable Polyester branded as Biomax® was commercialized in this manner. The material could be used for many applications and could be designed to degrade in a prespecified period of time, depending on the user's need. Many potential applications were experimented with, including mulch, bags for harvesting bananas, diaper liners, and more. For each application there were multiple markets. Mulch, for example, could be targeted at home gardeners, farmers, or gardening communities. The product formulation differed for each one. It was made into sheets, scraps, pellets, or molded into plant containers. Similarly, diaper liners are marketed for babies, toddlers, and the elderly. Products vary by size and shape.

The critical KM amplification activities at this stage are information arbitrage and, to a lesser extent, visualization, as technology and markets are joined for an application. KM brokers who have access to a breadth of information can put individuals and teams in touch with needs and/or solutions associated with new markets and new solution domains. Knowledge brokers can induce the teams and individuals to look outside their habitual domains for needs/solutions by accessing knowledge outside the firm, including external consulting firms with relevant knowledge. A knowledge broker's breadth of information and willingness to reach out to new sources of information is a critical resource that can assist the teams and individuals in zeroing in on the more profitable venues for any specific BI opportunity.

Incubation

Traditionally, incubation required developing prototypes that may not have been ideally designed for maximizing customer interface, but whose purpose was to allow customers to get used to the breakthrough technology in an experiential manner. We know that customers will react negatively to most changes that they cannot compare with their current usage patterns. Computed tomography (CT) scanners failed focus group and concept tests because doctors believed that X-ray technology served the imaging purpose well. Households rejected microwave technology for cooking applications because the change in behavior upset too many norms regarding food preparation time and energy investment. Many other examples exist to reinforce the point that extended time with the new product is required before customers can provide valid feedback. So the methods that incubation managers have traditionally used required lengthy time periods before they were confident that the market reactions they were getting could be considered valid. These approaches are excellent for learning but can be costly in terms of time and money. Newer digital technologies help to alleviate some, though not all, of these challenges.

Customer Immersion Labs Including Visual and Interactive Simulation

Customer immersion labs are labs outfitted with equipment and digital technology that enables exposure of 3D, real-time depictions of product assembly, design, or servicing. They are used by incubation teams to gather reactions and data on new product designs directly from individuals as they are using them by providing a simulated experience. Such visual immersion can be useful for capturing reactions from multiple members of the value chain, including end users, maintenance and service personnel, assembly line crew, sales technicians, and account managers. In this manner, technology mimics reality as closely as possible. While extended use over time is not possible with immersion labs, immediate reactions regarding serviceability, usability, and ease of assembly can be easily captured.

Rapid Prototyping

Computer-aided design (CAD) software has become increasingly sophisticated and now can create 3D digital models to replace wooden or plastic mockups of products to ensure they can be produced. Craftsmen can work from these 3D models now rather than from 2D drawings. The models can be projected so that the designer and customer can examine a prototype together. Electric Boat, a defense contractor that supplies nuclear submarines to the U.S. Navy, now relies on 3D visualizations over its wooden mockups, thereby reducing time, expense, and cost of rework dramatically (Jaruzelski et al., 2013). Currently, the company is investigating the use of those models to generate holograms of the inside of the submarine to improve its design and layout.

Three-dimensional printing is rapidly gaining ground as the most effective tool for experiencing a product, both with internal collaborators and with customers. The key differentiating feature of 3D printing is the speed with which the mockup is made, which can be a matter of an hour. Changes can be made based on user response, and a modified mockup can be generated in a very short time to test the user's reactions and make necessary modifications.

Immersion-type labs, coupled with visualization tools, are also useful for building and evaluating business models for the product. Google, for example, runs boot camps where prospective venture teams are invited to work on business models for their start-ups, and they receive in-depth feedback from analysts.

The central KM activities for this incubation stage involve deep customer engagement, visualization, and fail fast. KM roles can facilitate deep but multidisciplinary sessions that often happen in customer immersion laboratories, and KM can either manage these internal labs or access external incubators. IT augmented tools are now available as technology enablers for visualization, whereas process tracking can enable fast failures by keeping an organizational memory of past trials, successes, and failures.

Acceleration

The acceleration stage requires developing and institutionalizing processes that will enable reliability, predictability, and scaling, all while the business is experiencing rapid growth. Conventional scaling tools include design for manufacturability, quality function deployment, and process development. All are important, and all take time and attention. While no substitute for some of these tried-and-true tools for managing growth and quality, new tools are emerging that can ease the financial and time pressures of an acceleration manager who is trying to grow a business and can help given the context of uncertainty of manufacturing process, organizational fit, deployment, and customer loyalty that he or she faces.

Usage Sensors

New, unintended uses may emerge as the market begins to incorporate the innovation into daily life. Baking soda is used to clean carpets and laundry and deodorize refrigerators. None of these uses were imagined by the original development team, who formulated it as a kitchen ingredient. Other similar examples exist that expand and elaborate the business opportunity beyond what the new business manager may have imagined. Recently, firms have adopted usage sensors to track the numerous and unintended applications of its new product that may turn it into a true business platform serving a broader array of needs than originally imagined. These tools enable companies to collect and analyze usage data directly from the user via automatic tracking technologies (Jaruzelski et. al., 2013). These are particularly relevant to web-based offerings, where click-throughs, time spent on certain pages, and prices paid are just a few of the elements of data that could be useful for expanding the business.

Although KM in the information-leveraging sense begins to be increasingly relevant for this stage, KM can underscore the need for product migration as the BI launch is completed. KM can provide the knowledge repositories from the previous stages and garner the tacit knowledge of the teams involved in the launch through knowledge engineering approaches to build other applications.

Design and development communities generate their own principles and approaches over time, partly as a result of learning from experience. Some of the deep expertise in this kind of multidisciplinary environment remains tacit and is not easily transmitted in organizations over time and locations. Here, the communities of practice and knowledge capture methods developed in traditional KM may be extremely useful. As in any innovation, this requires the support of senior management, who are ultimately the keepers of organizational culture.

13.4 Organizational Implications

In large firms, where new business growth is generated through breakthrough innovations, a KM amplification function is most useful. Especially in the early stages of a breakthrough innovation project, KM has the potential to influence the evolution of the project in value creating ways. To accomplish this, a firm has to develop an appropriate organizational architecture. As shown in Table 13.4, the organizational architecture may include setting up idea labs or knowledge broker roles. Since all these require commitment of resources, different firms may adopt different mechanisms.

A second requirement for implementation is the presence of a supporting culture. As also shown in Table 13.4, the organization and team leaders require a healthy attitude toward multidisciplinary thinking, ideas from outside, courage to undertake low-cost

Table 13.4: Implementation of KM as Intelligence Amplification

KM Function	Options for Organizational Architecture	Cultural Prerequisites
Information arbitrage	Data: Knowledge repositories; external sources Roles: Knowledge brokers for technology and applications	Receptivity to ideas from other perspectives
Customer engagement	Facilities: Idea labs Roles: Knowledge brokers for lead users	Tolerance for multidisciplinary inquiry
Visualization	Tools: Visualization tools Roles: Expertise in the use and interpretation of tools	Context of creativity or thinking out of the box. Ability to deal with complex problems.
Fail fast	Tools: Rapid prototyping Roles: Expertise in the use of rapid prototyping tools	Tolerance for failure Failure as an occasion for learning
Application migration	Facilities: Venture camps Roles: Facilitators, opportunity brokers	Nondefensive climate for authentic feedback; opportunism

experiments without fear of reprisals for failure, and, finally, the willingness to undertake disciplined inquiry beyond intellectual domains familiar to the firm. This means new business creation teams, their team leaders, and those to whom they report must be willing to admit they do not have the answer and that, in some cases, the answer is unknowable for now. To nurture this culture is the single most important leadership act in the implementation of intelligence amplification.

These firms typically encounter two pitfalls. The first is imposing intelligence leveraging as a KM approach for breakthrough innovations. Intelligence leverage relies on what is already known rather than on learning the new, and is an approach, as we have argued, that is appropriate for incremental innovations. Yet we see this occur all the time in companies, where admitting one does not know is viewed as weak and can result in a career misstep. As mentioned, this is where leaders must step up to enact the appropriate culture changes. A second pitfall is the use of intelligence amplification tools as silver bullets. These tools are to be viewed as facilitating a form of inquiry that is different from the ones typically conducted in organizations, and hence the shift in perspective, rather than the tools themselves, is the critical end point.

13.5 Appendices

Appendix 1: Technology Translation Tool

Appendix 1A: Introduction of the Technology, Its Points of Difference, and Problems It Addresses

The development of the liquid lens has been known for a while, but the fast focusing of water droplets by use of sound waves is the novel idea presented by Dr. H. This type of technology would be applicable to cell phone cameras and other devices requiring a small camera.

Two companies have already developed products that use a liquid lens concept. Company X has invested in this technology for the past 10 years. As a result, it is able to manufacture liquid lenses for commercial use. In collaboration with company Y, half a million lenses per month are being produced since September 2013. The main clients presently for liquid lenses are miniature camera phones, in which power consumption is a major concern. The process using their technology requires between 10 and 100 volts. By comparison, Dr. H's invention requires a couple of millivolts.

Dr. H has already proven the invention's camera capabilities. In tests, his camera was able to take 250 images per second at varying focal lengths. He envisions a camera that could instantly capture tens of images with different focal lengths, and then use simple image-analysis software to determine the sharpest image.

In short, the novelty of this technique lies in the creation of a high-speed, adjustable lens using a liquid lens and an oscillating device. Its main advantages over its existing competing technologies are higher speed and lower power requirements.

Appendix 1B: Technology's Key Features and Associated Comments

Cost	Inexpensive as the lens is made up of a drop of liquid
Robustness	More resistant to accidental damage as there is nothing to break
Power Consumption	High efficiency, requires a few millivolts
Speed	Very fast, claimed to be able to capture 100,000 frames per second

Appendix 1C: Claims of the Invention and Who Might Value Them

Claims of Invention and Key Attributes of the Technology	Reasons for Claims/Attributes Being Benefits and Who Would Value This Benefit
A key feature of this new technique is that the water stays in constant, unchanging contact with the surface, thus requiring less energy to manipulate.	Presently, cell phones consume a lot of power while shooting a video or clicking a picture. Most of the power is consumed during focusing the object.
There is no need for high voltages or other exotic activation mechanisms. This means that this new lens may be used and integrated into any number of different applications and devices, making many applications feasible.	Low voltage requirement is attributed to the method of creating oscillating through sound. Potential applications including cell phones, web cams, and satellite imaging will be the ones the most benefited.
The great benefit of this new device is that you can create a new optical system from a liquid lens and a small speaker, which along with its driving circuit can be easily manufactured in a small and lightweight package.	Presently most of the cell phone camera packaging is primitive and creates a bulky look. The tiny camera can fit in a few square millimeter area on a cell phone.
With small enough apertures and properly selected liquid volumes, it is able to create a lens that oscillates as fast as 100,000 times per second—and still be able to effectively capture those images.	Fast focusing lens is very important in shooting different frames and then integrating them to make a video or panoramic picture. For example: in the movie The Matrix some of the shots were shot at 108-frames/sec speed.

Claims of Invention and Key Attributes of the Technology	Reasons for Claims/Attributes Being Benefits and Who Would Value This Benefit
The liquid lens that captures 250 pictures per second and requires considerably less energy to operate than competing technologies.	The contraction and expansion of the liquid take considerably less energy than moving a mechanical lens. Cell phone users can greatly benefit from it.
The lens is simpler than earlier liquid lens designs that use a combination of water (or some other fluid capable of conducting electricity) and oil as well as an electric charge by using water, sound, and surface tension to adjust the focus.	The technology enables the lens to be packaged in a tiny space, takes only fraction of energy needed in competitive lens, and simple mechanism will benefit cell phone users and manufacturers the most.

Appendix 2: Technology Market Mind Maps

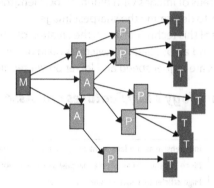

Figure 13A.2a: Start with a market need.

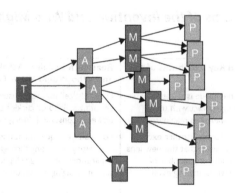

Figure 13A.2b: Start with a novel technology.

References

Bjelland, O. M., & Wood, R C. (2008). An inside view of IBM's "innovation jam." *MIT Sloan Management Review*, 50(1), 32–40.

Buchanan, R. (1992). Wicked problems in design thinking. *Design Issues*, 8(2), 5-21.

Christensen, C. M. (2000). *The innovator's dilemma*. New York, NY: Harper.

Cooper, R. H. (2001). *Winning at new products: Accelerating the process from idea to launch* (3rd ed). New York, Perseus Book Group.

Jaruzelski, B., Loehr, J. & Holman, R. (2013). The Global Innovation 1000: Navigating the digital future. *Strategy + Business,* 73(Winter). Retrieved July 11, 2014, from www.strategy-business.com/article/00221

Kelley, T., & Kelley, D. (2013). *Creative confidence*. New York, Crown Business.

Leonard-Barton, D. (1995). *Wellsprings of knowledge*. Boston, MA: Harvard Business School Press.

O'Connor, G. C., & DeMartino, R. (2006). Organizing for radical innovation: An exploratory study of the structural aspects of RI management systems in large established firms. *Journal of Product Innovation Management*, 23(6), 475–497.

O'Connor, G. C., Leifer, R., Paulson, A., & Peters, L. (2008). *Grabbing lightning: Building a capability for breakthrough innovation*. San Francisco, CA: Jossey-Bass.

Ruggles, R. L. (1997). *Knowledge management tools*. Boston, MA: Butterworth Heinemann.

Wenger, E. (1988). *Communities of practice: Learning, meaning and Identity*. Cambridge, England: Cambridge University Press.

About the Authors

V. K. NARAYANAN is the Associate Dean for Research, Director of the Center for Research Excellence, and the Deloitte Touche Stubbs Professor of Strategy and Entrepreneurship in Drexel University, Philadelphia, Pennsylvania. His research focuses on three themes: innovation, corporate entrepreneurship, and strategy during industry emergence; political and cognitive process in strategy formulation; and the epistemological foundations of strategy. His work has spanned biopharmaceutical, aerospace, and information industries, and educational and innovation-focused governmental agencies. His consulting assignments have been with large pharmaceutical and high-technology companies primarily in strategy implementation and corporate innovation. Direct correspondence related to this article to Le Bow College of Business, Drexel University, vkn22@drexel.edu.

GINA COLARELLI O'CONNOR is Professor of Marketing & Innovation Management and Associate Dean for Academic Affairs at Rensselaer Polytechnic Institute's Lally School of Management. Her research examines how established companies link advanced technology development to market opportunities and how they build capabilities for breakthrough innovation. She has published numerous articles in refereed journals including the *Journal of Product Innovation Management*, *Journal of Marketing*, *Organization Science*, and *R&D Management*, and is co-author of several books about breakthrough innovation in mature industrial firms. Gina teaches and consults on the topic of organizational change for breakthrough innovation. Direct correspondence related to this article to Gina Colarelli O'Connor Lally, School of Management, Rensselaer Polytechnic Institute or oconng@rpi.edu.

14

STRATEGICALLY EMBEDDING DESIGN THINKING IN THE FIRM

Pietro Micheli

Warwick Business School, Coventry, United Kingdom

Helen Perks

Nottingham University Business School, United Kingdom

Introduction

We already know that organizations are starting to use design strategically—to do something different from the competition, to launch new brands or strengthen existing ones, and to inform strategic choices. Yet, while success stories abound—from Apple to Dyson, from BMW to Alessi—little is known about *how* organizations actually embed design thinking at a strategic level. What do organizations really do to make a design perspective part and parcel of an organization's culture? This chapter goes to the heart of this issue and sets out clear guidelines on how to strategically embed design thinking within the firm. The chapter is studded with examples and lessons drawn from a large-scale research program. Getting design thinking into strategy is not a straightforward process, but we show how to use three fundamental levers to really make a difference. These are the roles of key personnel, organizational practices, and organizational climate and culture. Essentially, in this chapter we highlight the main enablers and barriers to making design thinking a strategic driver of success.

The chapter is structured as follows: We first provide a brief introduction to what research tells us about strategy and design thinking. We then talk about how our research was carried out and present new insights around the three themes, discussing them first in general and then through real company and product examples, both positive and negative. We conclude by providing practical suggestions on how to embed design thinking at strategic level.

Research studies show us that design can have a positive impact on companies' financial performance. For example, Hertenstein, Platt, and Veryzer (2005) demonstrated that investments in product design lead to enhanced corporate financial and stock market performance. Yet this design investment-performance relationship is not clear-cut. Scholars have looked into this further, and we now know that design's impact on performance is contingent on several factors, including level of product/service innovativeness (is the product radical or incremental?), industrial sector (such as service vs. product), and designers' roles and skills. For example, in their study of midsize to large U.K. manufacturing companies, Perks, Cooper, and Jones (2005) found that designers can undertake three different roles in new product development (NPD): lone players as functional specialists, members of cross-functional teams, or strategic/process leaders. Each of these roles can have different effects on performance. So what research is starting to tell us is that design's contribution depends on whether it is used strategically, utilized as a means to differentiate from competition, and whether design thinking is embedded in organizational processes.

So far, researchers have focused on whether and how investments in design lead to enhanced financial performance, rather than on understanding *how* and *why* organizations succeed, or indeed fail, to introduce design thinking strategically. This chapter fills this gap. Drawing from an extensive study of a wide range of organizations, we show that the strategic embodiment of design thinking is determined by three main factors:

1. Roles of key personnel
2. Organizational practices
3. Organizational climate and culture

Each of these factors is important, but all three are interconnected: new initiatives, based on each factor, should not be considered or implemented in isolation. For example, CEO and senior management support for design is fundamental if you want to instill design thinking strategically, but it is not enough. As our examples will show, support from senior management needs to be coupled with an iterative approach to NPD and with a tolerant approach to failure. You need combinations of things in place to elevate design thinking as a strategic imperative within the firm and, more broadly, as a way of thinking and doing things.

In the rest of the chapter, we discuss each of these areas and highlight how these factors can both inhibit and facilitate how design thinking is embedded strategically in organizations.

How the Research Was Done

A large-scale research program was undertaken with companies where design was either established or being newly introduced. For most of these firms, design was undergoing change and shifting its role and influence either positively or negatively. Companies participating in the research program were both product- and service-based organizations. The program included large multinationals and small to medium-sized enterprises (SMEs). To gather different perspectives on the roles of design, we interviewed CEOs and design and marketing executives, as well as senior representatives of other functions (e.g., product development, finance, and operations). We investigated both broad approaches to design and NPD, and also focused on specific projects, examining reasons for success or failure, the roles design had played, and principal enablers and barriers.

14.1 Role of Key Personnel

People matter, and our research shows that investment in the right roles is the first building-block to embed design thinking at strategic level. Table 14.1 shows how these roles can pan out and influence the strategic embodiment of design thinking.

Support from top management and, especially in SMEs, from the CEO or chairman is a necessary step in introducing and eventually embedding design thinking. How does this work in practice and over time? Well, early support can be fundamental when appointing a new design director and forming a new design team. But the key issue here is sustaining this effort, through active support and internal promotion of the design team by the board, for instance. In our research, we found that where this does not

Table 14.1: How Different Roles of Key Personnel Can Influence the Strategic Embodiment of Design Thinking

Role of Key Personnel	Enablers	Barriers	Outcomes
CEO/Senior management support for design	Establishment and introduction of a design function, continuous support demonstrated	Establishment of a function, but poor management of internal dynamics/tensions	**Positive:** Design becomes more institutionalized and part of the organizational structure and culture. **Negative:** If not accompanied by further efforts, design may not become embedded or be relegated to an operational role.
CEO/Senior management engagement in design	Direct link between design and decision making	Continuous redesign—needing to amend or change products/services according to senior management wishes	**Positive:** Design can be elevated at a strategic level in the organization; senior management can provide "air cover" to design initiatives. **Negative:** Lengthy product development cycles and wasted effort on failing products/services.
Design Director	Influencer: Lobbies for design to play a more strategic role	Technical specialist: Fails to connect with other functions	**Positive:** The design director can enhance awareness and understanding of design's role and capability across the organization. **Negative:** Does not elevate design's role beyond that of a service.

happen, design teams are often relegated to more operational roles. But watch out—at the same time, endorsement from the top and having a direct link with the CEO/board can create problems for the design team. Jealousy can set in. For example, once the new corporate design unit was set up at one of our case companies, other non-design departments started to act negatively toward it. The shiny new unit was seen to be taking away some of their autonomy and share of the local budget. At another firm, the design-oriented company CEO had a particular liking for specific services or products and hung onto them at all costs. This meant dictating their redesign multiple times, despite continuously disappointing results.

We also need to pay attention to the nontechnical roles of the design leader, often called design director or chief design officer. These are what matter when introducing and shifting design thinking within the firm. A persuasive personality and skill set within this role—someone able and willing to promote design inside the company—can emerge as critical. Let's see how such roles are played out in practice in the following two vignettes.[1]

Introducing and Embedding Design Thinking at Diageo

Diageo is the world's leading premium drinks business, operating in 180 markets in 80 countries, with over 28,000 employees. Its portfolio of brands includes Guinness, Johnnie Walker, Smirnoff, and Baileys. In 2006 the Diageo executive committee worked out that design was playing too marginal a role in the company, so it decided to appoint a global design director and build an internal design function. As the design director recalls, "I think where the intervention as a senior board level came was things like the [brand name] redesign which was really badly done when there was no internal team, people didn't understand how to manage design." This early appointment investment by the Diageo executive led to enduring support of the design director and his team. Indeed, the design director emerged as an important figure, not in terms of skills (the actual product design is mainly undertaken by external agencies), but as the person capable of promoting design's value to the various business units inside the company. This made a big difference between successful and unsuccessful introductions of design at a strategic level in the different business units. In the words of the design director:

> We have got a big influencing job here to do, to find our sponsor, to persuade, to help people understand. And if we only talk to people on the financial level, then we're completely missing the boat. Of course, we have to talk the language of business, and finance, and return on investment … but it is also our job as leaders of design within organizations to help people understand the thinking behind it.

These early investments and ongoing attention to the nature of senior management roles have borne fruit. Currently, design is used to inform strategic choices concerning branding and positioning of products. A good example of this lies with the redesign of the Johnnie Walker label, as part of the wider relaunch of the brand. This being a very iconic brand and a luxury product, the project was quite risky. However, the project was

[1] For reasons of confidentiality, company names included in negative examples have been anonymized.

eventually very successful thanks to strong leadership, which drove through the new design. According to the company, "The unveiling of the new Johnnie Walker Blue Label bottle was the relaunch of an icon. The design speaks to the rarity and authentic luxury credentials behind the ultimate expression of Johnnie Walker, fueling global net sales growth of 27 percent." (Diageo Annual Review, 2012).

Too Much CEO Control Can Hamper the Strategic Embodiment of Design Thinking

The CEO of company C, a medium-sized enterprise, attended a design conference and mused over how the showcased SME design-led brands had become so successful, with loyal customers and large revenues. How could he propel design's role within his own company, get the product design more customer-focused, and see that link directly to financial performance? This event pushed him into action: he recruited a number of new designers and rapidly involved them in several projects. However, while his support was always solid, his engagement with the design team was often too direct. He constantly looked over the shoulders of his design team, checking and telling them what to do. In the end, this hindered the design team not only in getting their job done, but in developing a capacity to take greater responsibility and influence strategic decisions. Coupled with this was the firm's overreliance on customer feedback. Indeed, this reflected a lack of trust in design's capacity to propose something new to the market. This is summed up neatly in a quote from the commercial director. Talking about an unprofitable product, he said, "The internal designers have reworked it, probably eight or nine times, and customers have always come back with: no, we don't want to change it.... So we have tried redesigning it lots of times and none of them have come to fruition, so it's a complete waste of time." Heavy interference from the CEO, twinned with an overresponsiveness to customer feedback, really hampers what designers can do and what design thinking can deliver.

We can see complexities and contrasts in the way roles influence the embodiment of design thinking in the firm over time. At Diageo (as well as at Virgin Atlantic and Herman Miller—see further examples), senior managers not only support design but trust what design can achieve and what designers could do. Such trust might come from the company history (as is the case with Herman Miller), from the leader's belief that design and innovation can make a positive difference (such as Virgin Atlantic), or from the design director's influencing efforts (see Diageo). In other cases, the CEO himself can simply espouse the embodiment of design thinking. Such is the case with Trunki, a small but leading company in children's travel equipment. Here, the founder of the company, whose background is in industrial design, played a crucial and positive role to embed design and design thinking: "Being the CEO with design training, it's all about problem solving and looking at things differently. Why do we do things that way? Is there a better way of doing it? So that thinking has been applied across the business rather than just in the product." Compare this with those companies where support is ad hoc and lacks continuity, or where the design director acts as a mere technical specialist, or where support and control from the CEO ends up stifling the creative potential of the design function.

14.2 Organizational Practices

Defining the design brief, involving customers in the NPD process, collaborating across diverse functional units, measuring and evaluating design: all these are practices that can either support or hinder a firm in embedding design thinking at a strategic level. Table 14.2 shows how these practices can have both positive and negative outcomes, depending on the prevalence of either barriers or enablers. The two following vignettes exemplify different situations where some of such practices are at play.

Virgin Atlantic's Upper-Class Suite

Airline Virgin Atlantic is well known for its innovativeness and use of design. But what is not so well known is the way its organizational practices shape and influence design

Organizational Practices	Enablers	Barriers	Main Outcomes
Table 14.2: How Organizational Practices Can Influence the Strategic Embodiment of Design Thinking			
Definition and communication of the design brief	The brief is developed in collaboration between different functions	The design function receives a narrowly defined brief to which it has to respond	**Positive:** Design elements and perspectives are included in the brief; a more strategic design perspective is likely to be adopted. **Negative:** Design is kept out of strategic decisions and performs mainly a "service" role.
Systems to promote collaboration across functions	Formal and informal systems are used to create a shared perspective among functional groups	Strong demarcation of functional silos	**Positive:** More effective collaboration and greater sharing of information lead to higher project success rate and reduction of time to market. **Negative:** Design is not embodied over the long term.
Customer involvement	Design plays an important role in identifying and addressing customers' problems	Customers are either over- or underinvolved	**Positive:** Design can uncover and help address "hidden needs" and generate successful radically new products/services. **Negative:** Overinvolvement leads to constant changes and "design by committee"; underinvolvement (often triggered by overconfident designers) means that customer voice is not heard and radical designs may fail to heed customers needs.
Measurement and evaluation of design	Evaluation undertaken at the end of each project	Thorough justification of design involvement from the beginning of each project; constant request of detailed financial information	**Positive:** Greater freedom awarded to design and feeling of trust across functions, particularly at the beginning of projects; greater potential for innovation. **Negative:** Conservative designs, longer time to market, and lower proportion of product/service launched.

projects and, ultimately, how design thinking permeates the organization. The design of the upper-class suite is a good example of this. In the early stages of the project, the company developed a concept briefing—how to create a flat bed to provide customers with a better flight experience. However, rather than focusing down on this clear, albeit narrow, brief, designers and marketers collaborated to open up and broaden the initial idea. This led to a more customer- and benefit-oriented concept: how do we create a space for customers and all their needs? Designers refined the brief through customer ethnographic practices, directly observing customers' in-flight behaviors. The head of Design stated, "It wasn't necessarily asking people what they want; it's about looking at someone and seeing how they behave, seeing the things that irritate them, not necessarily asking them, because often they don't know what they want, until you show them what they could have."

As the director of Brand and Customer Experience recalls, "perhaps some of our competitors have concentrated on: how do you get a seat to go flat so the customer can sleep? This was more about: how do we create the space for the customer for all of their needs?" The decision to focus on such a broader question enabled the company to formulate a much broader brief, which then led to the involvement of key internal people and customers, and to questioning existing offerings. In so doing, Virgin Atlantic adopted a more holistic view of service design, and connected organizational practices—in this case, how stakeholders are involved in the design and development process—with design thinking. Eventually, the project was very successful, as it created a suite that other airlines have since copied.

Design as a Service

Like many other companies, the CEO of company C and his senior management team realized that design thinking seemed to be driving higher levels of brand recognition and customer loyalty of many of their competitors. Therefore, they decided to make some changes and proudly established a new separate corporate design function. However, despite leadership support and good intentions, the design team really struggled to be appreciated within the company. Designers' work was often treated in a dismissive way, and designers were given orders by other functions. As the design director explained, "Other units say, 'I tell you what I want and you need to organize it for me.'" Design managers felt frustrated and saw themselves fulfill an almost passive role vis-à-vis the marketing function, unable to influence tightly defined briefs. One design manager stated, "In many projects, we are a service industry. In many cases, I don't think our projects are breakthrough or really exciting consumers because they've been ... restricted at the very beginning." So while internal staff perceived the company as adopting a "cowboy culture," its approach to design and product development was actually risk averse, heavily relying on consumer feedback obtained through surveys and focus groups.

The two vignettes presented above show how practices that support collaboration across functions, particularly allowing questioning and shaping of the design brief, are fundamental in introducing design thinking. We also learn that the way organizations engage with customers can influence design's strategic role. Organizations, such as

Virgin Atlantic, observe and shadow customers, rather than involve them in formal surveys or focus groups.

Finally, how do you measure and evaluate design thinking and do such practices influence the strategic embodiment of design? Again, we see contrasting practices with diverse outcomes. At company A, for example, great emphasis was placed on quantitative data to formally measure the input of design, particularly at the beginning of the NPD process. Also, various development stages were tightly defined up front in terms of both functions' involvement and budgets. Financial rewards were given to project leaders depending on how many ideas passed selection gates. Requests for formal evidence and the introduction of performance measures happened much later at Virgin Atlantic, Gripple (see below), and Herman Miller. For example, at Herman Miller performance is measured only on the completion of projects during learning reviews. Indeed, our research shows that if practices are in place to constantly assess and challenge the value of design, its impact is likely to be negligible. A design consultant said, "The worst thing is when that trust disappears or is not there, and so you become constantly questioning the values, the expertise, the knowledge, and the drivers of the other people, and that's hell, you know." Practices that seek to constantly measure and justify design seem to marginalize design's strategic contribution. Putting practices in place to catch design out and devalue its worth can end up as a self-fulfilling prophecy.

14.3 Organizational Climate and Culture

Successfully embedding design thinking at a strategic level can mean making brave and risky decisions that threaten the status quo. This is where organizations often really struggle. Insights from our research tell us that the nature of organizational climate and culture is crucial in facilitating or hampering brave decisions. Core to this are attitudes toward risk, use of formal processes, and the work environment (see Table 14.3). Real cases of how companies addressed such challenges are given in the following vignettes.

Herman Miller's SAYL® chair

As a well-established leading producer of interior furnishings, Herman Miller has created several iconic designs, such as the Eames lounge chair, Action Office, and the Aeron chair. Such a record, coupled with strong organizational values (quality, ergonomics, and environmental sustainability) and a clearly recognizable brand may suggest strong investments in internal design units. On the contrary, when it comes to design, Herman Miller relies almost entirely on collaboration with external agencies. Also, while most companies in the industry commission design through marketing departments in response to what they consider market opportunities, Herman Miller is more exploratory: The director of finance stated, "Find a few really good designers ... then trust them .... We kind of go: 'Well, here's a problem to solve. Send it out to the designers and see what they bring back.'" In addition, many ideas arrive unsolicited from design consultancies. External designers are even given the freedom to challenge briefs—say they don't like them, suggest alternatives. The company's

Table 14.3: How Organizational Climate and Culture Can Influence the Strategic Embodiment of Design Thinking			
Organizational Climate and Culture	Enablers	Barriers	Outcomes
Risk aversion/ appetite	Failure is not stigmatized, and it is treated as an opportunity for learning	Fear of failure; success is rewarded, failure punished	**Positive:** Greater exploration for opportunities and potential for innovation. **Negative:** Design is incremental/marginal and not used to differentiate the firm. Fear of doing anything too ambitious.
Reliance on formal processes	Processes are structured, but not too rigid, especially in the more exploratory phases	Processes are sometimes too rigid as they are intended as control mechanisms; sometimes they are virtually not existent, as they are perceived as constraining individuals' entrepreneurial spirit	**Positive:** Consistency of process enhances innovation outcomes. Especially in service-based organizations, designers were often involved in developing processes themselves, in order to create consistency, while not reducing creativity. **Negative:** Too rigid processes may stifle innovation and create perverse consequences (process is followed, but results are disappointing). Lack of a formalized process tends to lead to lack of clarity, and to create conflict across functions and inefficiency.
Work environment and physical space	Creation of a collaborative space, often also to reflect brand values internally	Physical divisions among individuals and functions	**Positive:** Greater collaboration potential and sense of belonging. **Negative:** Functional divisions are reinforced.

success does not rely on tight specifications and controls, but on an open and deep appreciation of design thinking, positive attitude toward risk, and capacity to select and work with external designers, and to execute projects.

At the same time, a structured and transparent process is in place. The director of Insight and Exploration stated: "There's a level of thoughtfulness that is going into both product design as well as the application of the brand presence. So we aren't doing things randomly; we're doing things very, very deliberately." Three distinguishing features characterize the design and development process at Herman Miller:

1. Senior management goes through an extensive vetting procedure before engaging with an external design firm. Successful collaborators become trusted partners and are often kept on the roster for a long time.
2. While the company has a documented design process, from insight and exploration through to launch, iteration and experimentation are defining aspects of Hermann Miller's design thinking. As the senior VP of Global Marketing stated, "A fair amount of variation is tolerated given the difference in things that we might be developing or designing."

3. Business cases and financials are developed organically over the duration of the project, rather than determined at the beginning of the process. As the finance director argued:

> [Let's say that] we want a designer to go do a chair that does this. We could probably give him some financial information, like, hey, the material cost needs to come in around here, and the volume is ... [but] now you're starting to constrain the designer too much and put him in a box.... What [we] are really doing is: [we] start out with kind of a theory and guesses and estimates, and [we] are trying to eliminate all of those that you can as you go through the process, toward a business proposal.

The SAYL® chair, designed by Yves Béhar of the studio Fuseproject, is a good example of such design practices. Initially, Herman Miller identified the need to enter a low-price market, aiming to create an office chair that could retail at $300 to $400. This was a substantially lower price than its other products in that category. The challenge was to design a low-cost chair while not compromising on the company values. Herman Miller sent out a brief to several design agencies expecting good but pretty standard responses. Yves Béhar reacted. He had worked with the company before but had never designed a chair. He came back with a response to the brief with a visually provocative and higher-performance product at that same price point. Herman Miller's openness and trust in its external designers meant that this latter option was eventually chosen, despite the external designer's lack of experience with the product category. The result was the SAYL® chair, a product that is performing well, hitting margin targets, and seeing soaring sales in Asia and the United States.

The Challenge of Overcoming Cultural Barriers

Company B adopted a typical design process, which goes as follows: Developers work on a concept from definition of requirements to implementation; there is little interaction with customers; the new product is launched. Following disappointing results and poor customer feedback, fresh thinking about innovation finally emerged and a new design function was established. This sounds straightforward enough, but while positive results were achieved, cultural barriers have proved to be a major headache. As a design officer lamented: "It's where you have designers and design working with people who are not design literate, who are uncomfortable with the design process. That's where things go wrong." Similarly to company A, lack of awareness of the role and value of design and difficulties in interactions among functional groups proved to be major obstacles to embedding a design way of thinking at this company too. A design officer stated: "[The challenge] is getting an organization that is not historically design-led to understand: what are the benefits of that? And that is a massive challenge." Indeed, in most of its projects, the company preferred to use known and tested processes rather than developing new ones that involved the design unit. In this case, stubbornness around maintaining the status quo through existing processes prevented the introduction of new ways of working.

Things have progressed, but this has meant hard work in developing a different attitude toward failure and a more iterative way of working. As the digital director argued:

> I think there's a growing acceptance within the company that if we are going to take this kind of design-led approach, we need to get used to failing sometimes .... And if there is a failure, typically it's only remembered for a few weeks, because what designers do very, very quickly is move on from it; either learn from it quite quickly, or get the next thing out there and iterate it very quickly.

Here, changes in the physical work environment were also important in the way design worked. Traditionally, offices were structured as cubicles or with people sitting next to each other along pretty standard, long tables. According to the head of Design:

> Very noncreative, noninnovative, noncollaborative, no wall space to write and draft and create and design—there's none of that. We have ripped out flooring and designed and developed a new floor—a new way of working .... It's just very different than the rest of the floors. And so, that in a lot of ways is driving the cultural change.

Changes in the work environment parallel changes to the structure and dynamics of the project team:

> We [now] have an agile work environment .... If we're going to start a new project, we'll assign a designer, a business analyst, an operations person or a technology person, and a program manager or account manager, and put all of them at one table ... and they're solely responsible for that product. Instead of multiple meetings throughout the building, they all work very closely together.

Embedding design thinking as a strategic perspective relies on a culture that takes risks, trusts its designers (whether internal or external), and, importantly, tolerates failure. Another good example is midsize manufacturing company Gripple, where we found substantial freedom to explore ideas and learn from failure. The special product manager said: "We don't see failure on the path of a project as a bad thing ... [In fact] fear of failure actually means that you're not actually going to take a design to the extreme."

Our cases have shown that risk aversion and fear of failure can easily relegate design to a simple "service" role, and therefore inhibit the introduction of design thinking. How you formalize processes is also tricky: while Virgin Atlantic, Herman Miller, and Diageo all have established and structured routines and processes in place, they worked out how to develop sufficient flexibility to accommodate specific aspects of a project. In company B, subtle but key changes in processes and the work environment were necessary to start embedding a new way of working around design thinking.

14.4 Embedding Design Thinking

So far, we have shown how firms need to think very carefully about the way they initiate and develop design roles and new organizational practices and shape the climate if they want to strategically embed design thinking. We have shown how these approaches and techniques can act as both enablers and barriers. However, these techniques can-not be applied in isolation. They are interrelated. Let's take Herman Miller, for example. The collaborative way the brief is defined is linked both to the company's positive atti-tude toward failure and senior management's deep understanding of design's role and contribution. Conversely, at company A, we see that the tight definition of briefs by mar-keting, the upfront focus on financial data, and fear of failure led to the demeaning of design's role, conflict across functions, and disappointing outcomes.

Different structures and ways of operating can trigger virtuous or vicious cycles, which can lead either to the successful embodiment of design thinking in strategy or to the marginalization of design to a service function. This is illustrated in Figure 14.1. On the left-hand side we see a situation where the value of design is unclear. Exces-sive scrutiny and measurement of design at the beginning and during the develop-ment process leads to practices which overly constrain design, exacerbated by fear of failure. This, in turn, means that design plays a narrow, operational role, and its value will be marginal. On the contrary, the right-hand side of Figure 14.1 shows how, with a similar starting point, firms allow design to play a more strategic role (as in Diageo's case and, partly, at company B), and analyze projects at the end with the aim of learn-ing, rather than sanctioning failure. This increases trust toward designers and leads to a higher appreciation and evaluation of design. Finally, key attributes of design think-ing emerged in several companies, from a clear focus on problem solving at Trunki, to observation of users and adoption of a human-centered approach at Herman Miller, to the capacity to redefine the brief at Virgin Atlantic, and to a more iterative way of working at company B.

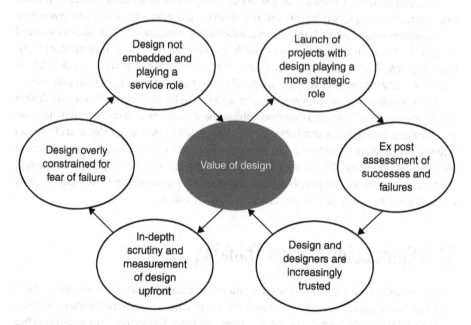

Figure 14.1: Value of design thinking.

Practical Implications

It is not easy to embed design thinking in an organization. Stubborn adherence to existing processes and routines, conflicts among organizational functions, overreliance on customer feedback, fear of failure, and lack of senior management appreciation of design can often form barriers hard to overcome. However, there are several actions that can be taken to strategically embed design thinking:

1. Design can help spark the first connection with the customer, and hence strengthen the brand. To do this, a deep understanding of customers' often hidden needs, rather than a response to narrowly defined briefs, should initiate the design process. Such understanding can be gained through observation, shadowing, creating "personas," and adopting other design approaches and tools. These are more effective than investigating explicit customer needs. The process should be user focused (i.e., creating value for customers), rather than user driven (constantly reacting to customer preference changes). A clear example is the Virgin Atlantic's creation of the upper-class suite.

2. Briefs should be developed through collaboration between multiple functional groups (e.g., marketing, engineering, design), with room left for making changes over time. As we have seen in both Virgin Atlantic and Herman Miller, briefs were reworked to broaden their scope and address wider problems: How can we enhance in-flight customer experience? And how can we create an affordable but comfortable and environmentally sustainable chair? Importantly, having wider briefs does not mean lack of discipline and overly autonomous designers coming up with unsellable products. Instead, they enable NPD teams to question existing offerings and develop innovative solutions.

3. Design investments should not be limited to simply recruiting good designers or partnering with successful design agencies. It is important to create an environment for design to flourish and to establish real trust between top management and designers. Both aims require strong senior management sponsorship and sufficient appreciation of what design can do. We can learn from novel practices at Diageo, for example. The design director introduced an annual event where senior management reviews all the design work of the year, and to which external design consultants are invited. This event has two positive effects: further promoting the appreciation of design in the company and spurring competition among design agencies.

4. The work environment may need altering if design thinking is to be successfully embedded. For example, as at company B, novel and unusual office layouts can kick-start greater interaction among functions and support the work of project teams.

5. Managers should gather evidence of design successes (and failures). This triggers learning but also enhances design's position in an organization. Such evidence may be internal or external, as in the case of benchmarking exercises. However, be warned that while performance measurement can play a very positive role, it

should not be undertaken too early in the development process as innovation may be stifled. This can be a real differentiating factor between those companies that successfully elevate design thinking to a strategic role and those that do not.

6. Managers should identify the diversity of roles that designers can fulfill. Designers often start their careers as technical specialists with functional expertise. But for strategic design thinking to work, they have to be able to join cross-functional teams and act as influencers who champion design. They have to be capable of using and understanding different languages and perspectives and be fully aware of commercial considerations. Having gained a high level of trust and positioning, they can also play leading roles in which they get increasingly involved in articulating concepts and future scenarios.

Finally, how far are you prepared to go? Some organizations are capable of deeply embedding and elevating design to a way of thinking and operating, rather than keeping it simply as a perspective in development. Fundamentally, what sets some companies apart from others is their holistic approach toward design and their culture more broadly, rather than the skill set of individual designers or the position of the design unit in the company hierarchy. As Virgin Atlantic's head of design stated:

> What you are embedding is the important bit. Are you embedding design and the design team? Or are you embedding a point of view? ... You're only as good as the culture in which you sit.... So it's not the quality necessarily of the designers, the design team, it's the quality of ... the organization as a whole.

References

Hertenstein, J. H., Platt, M. B., & Veryzer, R. W. (2005). The impact of industrial design effectiveness on corporate financial performance. *Journal of Product Innovation Management*, 22(1), 3–21.

Perks, H., Cooper, R., & Jones, C. (2005). Characterizing the role of design in new product development: An empirically derived taxonomy. *Journal of Product Innovation Management*, 22(2), 111–127.

About the Authors

DR. PIETRO MICHELI, MSc, MRes, PhD, is Associate Professor in Organizational Performance at Warwick Business School, UK. He was awarded a PhD and a master of research in management from Cranfield University, and a master of management and production engineering from the Politecnico di Milano, Italy. His main areas of expertise are strategy execution and performance management, and innovation management, particularly in relation to the role of design in the development of products and services. As a practitioner and consultant, Pietro has worked with over 30 organizations, including Royal Dutch Shell, BP, KLM, British American Tobacco, British Energy,

Wartsila, USAID, the United Nations, and the British, Italian, and U.A.E. governments. Pietro's research papers have been published in leading academic journals such as the *Journal of Product Innovation Management, Long Range Planning, International Journal of Operations and Production Management,* and *Research Technology Management.* Pietro speaks fluent English, Italian, French, and Spanish.

DR. HELEN PARKS is Professor of Marketing at Nottingham University Business School and Honorary Visiting Professor of Marketing and Product Innovation at Manchester Business School, United Kingdom. Prior to academia, she held senior positions with multinational groups throughout Europe, including Olivetti (based in Italy), the PA Consulting Group (London), and the European Commission (Brussels). She is Associate Editor (Europe) of the *Journal of Product Innovation Management* (JPIM), academic chair of PDMA UK and Ireland, and serves on the editorial board of a number of leading journals. Professor Perks is a member of the EIASM International Product Development Management (IPDM) Conference Board and was conference chair in 2012. She has specific research interests in service innovation, design in innovation, innovation networks/collaboration, international NPD, and qualitative methodologies, publishing her research in journals such as JPIM, *Industrial Marketing Management, R&D Management, International Journal of Innovation Management, International Small Business Journal, Services Industries Journal, Journal of Services Marketing,* and *International Marketing Review.*

Part III
DESIGN THINKING FOR SPECIFIC CONTEXTS

15

DESIGNING SERVICES THAT SING AND DANCE[1]

Marina Candi

Reykjavik University Center for Research on Innovation and Entrepreneurship

Ahmad Beltagui

University of Wolverhampton

Introduction

Design in the context of services continues to baffle practitioners and academics alike. What can be designed when there is no *thing* to design? Design is usually understood as the activity of giving form to manufactured products. So when it comes to intangible services, it may seem like there is not much that designers can apply their skills to. Nevertheless, as the importance of service industries has grown and jobs in manufacturing industries have declined, there is increased recognition that service design may in fact offer a means for the scope of the design profession to grow (Candi, 2007). Examples such as the Apple iPhone, Bang & Olufsen sound systems, and the Volkswagen Beetle have imprinted the clear message that product design can make a huge difference. None of these examples represents the most advanced technological innovations, but each can command premium prices and persistent market recognition based on outstanding design. Similarly, design—unlikely as it may seem to sound—can be used to create competitive differentiation in services (Candi, 2010).

The key to understanding service design is to recognize that design need not only be about giving form to physical objects. Design thinking involves reframing problems in a manner that allows novel solutions to be developed (see Chapter 1 of this book). Whether or not these solutions take the form of tangible artifacts, the focus is on those

[1]This work was funded in part by the European Union under the FP7 Marie Curie Industry-Academia Partnerships and Pathways Programme, project number 324448.

who will benefit from them. In other words, the ultimate aim is to influence customers' experiences, to form lasting impressions in their minds, and encourage them to return. Interaction designers recognize that the customer experience is beyond their direct control, but they design with the aim of influencing the emotional connection between customers and products. This means the focus of design has moved from the product or service that customers interact with to the behavior of the customers themselves (Redström, 2006). As Herbert Simon's definition in Chapter 1 suggests, design is the act of changing a current situation into a preferred one. So good service design should have a transforming effect on customers, offering a positive emotional outcome and improving their lives in some way.

Theater can be a useful metaphor for understanding services. Like theater, services involve one group of people performing for the benefit of another group of people. The processes that have been developed over thousands of years of theater can, therefore, be used in managing service processes (Grove, Fisk, & Bitner, 1992). Like theater, services require preparation and planning (or what we might call *design*) with the aim of creating a particular reaction and a lasting memory in the customer's mind. As in script writing for the theater or the screen, the result of design thinking in services should be to produce something with a powerful narrative, to engage the audience, and to offer something unexpected but ultimately memorable.

In this chapter, we start by describing the differences between designing products and services. We introduce the theater metaphor to show how this offers a useful starting point for service design. Next, we discuss the challenges that make experiences elusive before offering three principles that can be applied to service design: *narrative, participation,* and *surprise.* To illustrate these principles, we describe two examples of companies that engage customers by designing for "singing and dancing" and thus create compelling and memorable service experiences.

15.1 Products, Services, and Experiences

When we think about outstanding services, design is probably not the first element that comes to mind. Ritz-Carlton is well-known for empowering its staff to spend money to solve customers' expressed, or even unexpressed, problems. Southwest Airlines is well-known for its pioneering efforts to make flying fun. And many of the stories of Nordstrom's liberal return policies are mythical in proportion, being told and retold and passed between generations. But are these examples the result of design? Certainly not in the traditional sense of the industrial design of objects. But if we take the object of service design to be the customer and its success to be a memorable and compelling experience, then employee roles, service environments, and return policies are all aspects that can be used to set the stage. By broadening our view of design, we can extend the theater metaphor to encompass all the areas that are managed in the typical service business, see Figure 15.1. Indeed, the notion that all business is a stage has been espoused for a long time (Grove et al., 1992; Pine & Gilmore, 1998). Customers can be regarded as the audience, and the business is responsible for staging a performance that engages and involves this audience in order to make the experience memorable.

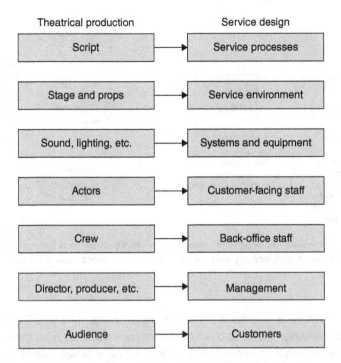

Figure 15.1: **Analogies between theatrical production and service design.**
Adapted from Pine and Gilmore (1998).

Managers and owners play the roles of directors, producers, and backers, who help to shape the production. The performance is created and delivered by employees who can be thought of as the actors appearing on stage and the technical crew, working backstage to make the show happen. The environment in which the performance takes place can be thought of as a stage, requiring aesthetic design to capture the audience's imagination, while the information systems and equipment of a business are akin to the sound and lighting equipment on the stage. Finally, the script consists of the procedures put in place to ensure a consistent performance by the actors.

Building on the theater metaphor, the task of a financial service firm is to choreograph the performance of financial services to create a desired experience for customers. This entails a focus on the psychological and emotional impact service encounters have on customers in addition to the technical core of the value proposition. Likewise, health care providers transform customers physically, but by deliberately staging an experience, they can capitalize on the importance of the psychological element of physical health and leave positive memories even in the context of crisis or tragedy.

Experiences Are Elusive

Designers often express broad, idealistic aims like making the world a better place through the things that they design. However, the things that are designed are usually judged in terms of their own characteristics, rather than on how well they achieve their

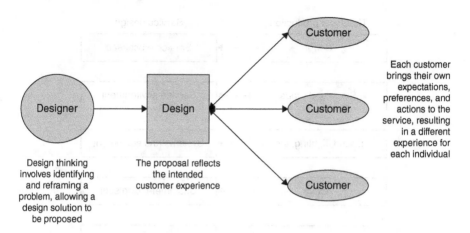

Figure 15.2: Companies can design the conditions (or prerequisites) for a service experience. But it is only when customers interact with the service that the service experience itself comes into existence.

ultimate aim. A famous statement, attributed to the management guru Theodore Levitt, states that people don't want to buy a one-quarter-inch drill bit; what they really want is a one-quarter-inch hole. In other words, the quality of a design cannot be judged by looking at the thing that is designed, but at its impact on customers' lives. If we take this perspective, *we can see product design and service design as one and the same,* even when there is no *thing* to design. When the aim is to create a positive experience—as it always should be—products and services are the tools that are used to achieve the aim. The experience only comes into being within an individual's mind as a result of their interaction with what has been designed. The challenge that designers face is twofold: they do not directly control the experience, and no two experiences are identical even when the same props and script are used. This dilemma is illustrated in Figure 15.2.

Compare this challenge of designing an experience to that of writing a play. The script writer seeks to make a lasting impression on the audience, to educate, entertain, or challenge. The writer does this by appealing to thoughts and emotions, but can do so only indirectly through the words that are written, but will then be interpreted by directors and delivered by actors. Exactly what each individual takes from the final performance depends on factors that cannot be fully controlled. These include what the individual knows and has experienced in the past as well as the physical and social environment in which the performance is experienced. Just as two people can have entirely different views on how good a performance they watched together was, two customers can have entirely different experiences as a result of the same service. The objective of design thinking becomes identifying a customer problem, reframing it in a manner that enables a solution, and then being able to adapt that solution for specific customers. The customer's experience is undeniably ephemeral, changeable, and difficult to control, making the challenge equivalent to that of creating classic theater. What companies can do is control and design the prerequisites for an intended service experience. In the following sections, we offer some guidelines that can help in doing so.

15.2 How to Design for Compelling Service Experiences

Businesses normally have three main levers at their disposal for putting in place the conditions under which a desired service experience can be created. They can control the *environment* in which the service is offered, including some or all of its tangible and intangible aspects; and they can design *service processes*, including the behaviors that are expected of *employees*. In a theater, the environment is the stage, but also the auditorium, the foyer, and the backstage area. All of these exert an influence on the mood of the audience, and designers should be aware that anything can help or hinder the creation of a positive experience. The service processes include the script followed by actors, but also the procedures followed by sound and light technicians and even the process of checking tickets on entry. We can see that there are at least two layers here—the performance itself and the supporting activities. All of the components need to be closely aligned and follow a clear plan for it all to make sense. If not, the audience will be distracted from the emotional impact of the script and the acting.

Sometimes acting looks so natural that it appears to be spontaneous—as if those on stage are speaking the words of the script for the first time. But watch a performance on two consecutive nights and it becomes clear that improvisation happens and takes an incredible amount of preparation as well as the ability to read the audience and adapt the performance accordingly. Successful service businesses are those that plan their service processes meticulously and work toward flawless delivery. But success also lies in allowing employees the freedom to improvise within these service processes. This doesn't mean throwing service process design out the window. Planning to do things by the seat of your pants is seldom a good idea. What is called for is designing robust service processes that anticipate multiple permutations of possible service journeys and include a great deal of built-in flexibility. Service processes should empower employees to be helpful, responsive, and flexible, while still ensuring that service delivery is streamlined and economical.

But what should be included in a service process—or script, if we continue to use the theater analogy—to create a compelling and memorable experience that customers will want to repeat and recommend to others? There are many things that could be placed at the core of a service value proposition to achieve this, but we emphasize three that are particularly promising, while not necessarily obvious. These are *narrative, participation,* and *surprise.*

Narrative

An experience can be seen as a sequence of events that take place over time, involving a number of different actors and seen from several different perspectives. To describe an experience is therefore to tell a story, or narrative—one that is unique to the person telling it. For a designer, the story ends the moment ideas are put into practice and sales begin. For the customer, however, the narrative is just about to begin. Designing an experience requires the designer to consider the customer's perspective and see the delivery of a service as part of the customer's narrative.

Successful service businesses typically plan their processes meticulously, making use of mapping tools, such as the *service blueprint* (Bitner, Ostrom, & Morgan, 2008). The service blueprint identifies *touch points*, for example, the point when a waiter takes a customer's order in a restaurant, as well as processes that happen in the background, such as passing the order on to the kitchen staff who prepare a meal. It identifies possible points of failure where things could go wrong and which designers should pay particular attention to. Design is typically a visual activity, in part because the creativity it requires is likely to involve the right side of the brain, which processes visual rather than verbal information. Mapping tools that allow the visualization of otherwise intangible interactions are therefore extremely useful in the service design process. However, such tools tend to focus primarily on business processes. This means they capture only the business's side of the story while neglecting the customer's side.

A subtle but important distinction can be made between touch points (seen from the business's perspective) and moments in a customer's journey (seen from the customer's perspective). Jan Carlzon, credited with taking the airline SAS from the brink of ruin to the top of its industry, famously claimed his company had 50,000 *moments of truth* every day. Every time an employee and customer interact, there is an opportunity to shape the customer's experience, for better or for worse. Unfortunately, what happens in each of these moments is not fully controlled by the company. For example, an airline could do everything correctly but find its customers unhappy because of delays in traffic or at airport security or simply because of a bad night's sleep. While these things cannot be predicted fully, taking them into account calls for considering the emotional impact of service interactions and how these contribute to customers' narratives.

All the processes, environments, tangibles, and interactions that come into play in a customer's typical journey through a service can be identified and mapped. A crucial task is to examine the emotional impact that each of these can have on the customer. The key here is to look not only at what happens but at how the customer feels about it. This exercise should help reveal inconsistencies in the narrative of the customer experience or identify opportunities for improvement. Looking for failure points, as service blueprinting does, is an important task, but when mapping the customer journey, the failure points are not in the efficiency of the process, but rather a failure to contribute to a positive experience.

Part of the appeal of design thinking is its ability to address so-called *wicked problems*, which are complex and consist of many interconnected issues. The application of design thinking can develop an understanding of problems from the customers' perspective, considering all the details, while also retaining the bigger picture. What seem like unrelated processes that are the responsibility of different people in an organization all come together in creating a customer's narrative of the service. So, in addition to considering the details of each process, the designer needs to see how they fit together and ensure that they convey a consistent message.

In a bank, employees are likely to adhere to a formal dress code to convey a message that this is a reputable establishment and that these are serious, hardworking people who can be trusted with your money. Employees would not dress in T-shirts or jeans because this would be inconsistent with the expected narrative. Similarly, a bank that wants to be perceived as a high-class establishment and wishes its customers to feel respected and valued would not present its customers with the gift of a cheap, mass-produced pen, since this would be inconsistent with the desired customer narrative.

Participation

Although we tend to think of theater audiences as passive consumers, there is a long history of active audience participation. At the very least, actors expect to hear a response from the audience, and this helps encourage and inspire their performance. Yet there are also many examples of attempts to break the *fourth wall* by engaging the audience directly in the story. Stuart and Tax (2004) draw attention to the traditional British pantomime, in which active audience participation is absolutely essential to the show. Well-known children's stories such as Peter Pan or Jack and the Beanstalk always feature a villain, whose every appearance is to be booed by the audience, and a hero, who enlists the help of the audience. The experience relies on an auditorium of children shouting, "He's behind you," to warn the characters of danger. Another example is the cult film *Rocky Horror Picture Show*, which people show up for repeatedly and usually in full costume, not so much to watch the film as to be part of a collective experience by engaging with it and with others in the audience. Likewise, participants in comic book or science fiction conventions, music festivals, and sporting events add color, humor, and sometimes noise by arriving in costume and providing an atmosphere. Such events would be nonevents without their participation.

Leaving the theater for a moment, services always involve customer participation. Indeed, the defining feature of any service is not its level of tangibility or lack thereof, but the presence of the customer in the production process. Customers always supply an input, which may be their possessions (e.g., when a customer ships a package), information (e.g., when insurance companies process customer details to produce a quote), or it is their mind or body (e.g., education, health care, or theater). Rather than simply processing these inputs, however, services can be designed to facilitate more active involvement from individual customers and, increasingly, from communities of connected customers.

A well-known type of customer participation is self-service, which is in many cases simply an ill-disguised ruse pretending to improve customer convenience while actually being a way to cut costs. However, two approaches can be seen when we consider self-service.

One does indeed involve identifying processes that can be passed on to customers in an attempt to reduce costs. The airline sector has been revolutionized by budget airlines that require booking, ticket printing, and even check-in to be conducted by customers online. The savings are passed on to the customer, allowing these airlines to offer cheaper travel options. What they generally fail to do, however, is to offer a pleasant experience. It may be a memorable experience, but when it is, it may be remembered for the wrong reasons and leave customers liable to switch to other airlines.

The Irish airline Ryanair has been as successful as it has been controversial, with its deliberate attempts to shake up the established European market leaders. It has recently acknowledged the negative experience it has offered through its focus on efficiency and cost while treating customers (and employees) with disdain. Restrictions on hand luggage and other small details that irritate passengers unnecessarily have been reexamined by the company in an attempt to change its image. These actions came about suddenly after years of intense criticism from customers. The catalyst? A conversation with customers via Twitter in which the CEO recognized the level of annoyance and vowed to shed the company's macho image while trying not to upset people unnecessarily. The old-fashioned approach of managers taking to the phones to listen to customer complaints or manning the checkout register still work in the Internet age.

The second approach to self-service is examplified by the Swedish furniture giant IKEA. Like Ryanair, it passes many processes, such as transportation and assembly, to customers and shares the savings by offering cheaper products. IKEA's popularity can partly be explained by its low prices and the convenience of buying everything required for a home from one store. Another factor, however, is what a group of behavioral scientists have referred to as "the IKEA effect" (Norton, Mochon, & Ariely, 2012). Their experiments demonstrated that people value products more when they have contributed some effort to their creation. Deliberately or inadvertently, IKEA stumbled upon a formula for creating emotional attachment between customers and products, which leads to positive experiences of the IKEA brand. By combining design for functionality, along with design for manufacture and assembly, with perhaps a touch of emotional design, IKEA arguably generates customer loyalty based on customer participation in the form of self-service.

The biggest driver of self-service approaches in recent years has undoubtedly been the technology that enables e-commerce. From bookstores to banking, entire segments of commerce have moved or are moving to the Internet. In most instances, this simply removes the human touch and replaces it with a depersonalized and frequently daunting and stressful experience. However, there are many examples of customers forming communities based on support and collaboration—communities that in turn help power innovation. The open source software movement, in which global communities of like-minded individuals use their spare time to develop software is a case in point. The objective of most participants is recognition or challenge rather than financial gain, and the results can be seen, for example, in Wikipedia and OpenOffice.org.

Humans have a strong need to belong, and a sense of community among service customers can provide a powerful way to fulfill this need. The rise and impressive popularity of social network sites speaks volumes about the leverage that businesses can gain by connecting customers. A good example is American Express's Open forum, which is an online forum created by American Express for its small business owner cardholders. On this forum, small business owners get to know other small business owners and exchange ideas, tips, and stories about running small businesses. The benefit to American Express is decreased customer churn due to the strength of the community of customers.

Online customer communities or offline user groups can provide a wealth of information that companies can leverage for service design. In the online realm, this activity is sometimes referred to as *netnography* or the analysis of *big data*, in which businesses track customers' conversations or analyze statistics on their behaviors to try to discern what they want or might want in the future. Although such market research may be a reason for companies to try to encourage the creation of customer communities, these communities, if they take life, can also build loyalty among customers. Such loyalty will, in many cases, be to the community rather than to the company. But if companies keep the community as a whole satisfied, an individual customer's loyalty to the community will make them reluctant to leave it and, by extension, reluctant to stop using the service.

Customer communities can become powerful forces for change, and there have been instances of communities rising up against a business. For example, a Facebook group created to protest a change of tea blend by the English brand Twinings forced the company to reintroduce its original Earl Grey blend alongside its new product. The historical tradition of protests over tea is clearly alive and well in cyberspace.

Thus, service designers should think about how they can design to create a sense of camaraderie, affiliation, belonging, or kinship with other customers. Building in a way for customers to converse and interact with each other is a possible way to do this, and leveraging existing social network sites is a promising means to do so (Roberts & Candi, 2014).

Surprise

For businesses, predictability is traditionally highly desired. Standardized processes are well suited to measurement and improvement as businesses strive for efficiency and profitability. For customers, predictability is desired when it means they know what to expect when making a purchase. When it comes to experiences, however, an element of surprise is a great way to keep things exciting and keep customers coming back for more. Everyone who flies regularly on a commercial airline knows what to expect from the in-flight safety briefing. Southwest Airlines has been known to surprise its passengers with lines of poetry inserted into the otherwise serious presentation. Air New Zealand, knowing many of its passengers have made a *Lord of the Rings*–related vacation decision, has begun producing themed in-flight safety videos, featuring Hobbits, Orcs, and all manner of Middle-Earth creatures. The stories about outstanding service providers such as Nordstrom and the Ritz-Carlton are full of elements of surprise. The Nordstrom store that accepted a returned set of car tires even though no Nordstrom store sells tires is one such legendary example. Surprise (aka variability) often originates from the customer, but outstanding service providers are able to cope by being flexible. The Ritz-Carlton capitalizes on its policy of delighting customers and publishes its own portfolio of stories on its website. This may be taking things just a step too far. If it is expected, it is no longer a surprise, is it? There is a fine line here.

Four Seasons employees follow what the company refers to as its Golden Rule: "Do to others (guests and staff) as you would wish others to do to you." This is made possible because employees are empowered to provide guests with a personalized service. Employees seem to approve (Four Seasons repeatedly appears on *Fortune* magazine's list of 100 Best Companies to Work For) and customers feel the benefit. There are two important lessons we can take from this example. First, employees are crucial to any service because customers are unpredictable, so it is not possible to plan for every eventuality when designing a service. Second, since employees are so important, it becomes vital that they also have a good experience. Employee empowerment is a very important condition needed for surprise in services delivered by people or facilitated by people. Employees need to be adept at "reading" each service encounter and deciding when a surprise might be welcomed.

Surprises are surprises only if they are surprising. This means customers should not be prepared for them. They are probably more appreciated when seeming to be genuine one-off, rather than carefully choreographed, events. That being said, surprises can, of course, be planned, for example, when a small art agency enrolled a *flash mob* of singers to surprise guests at its opening.

15.3 Services that Sing and Dance

We next provide two examples of services that "sing and dance," which we hope will inspire.

A Service that "Sings"

A web development company struggling with limited office space capitalized on this challenge by allowing prospective customers to have a glimpse of developers while they work through a glass wall separating the customer meeting room from the work area. Thus, the glass wall provided a window onto the company's backstage, similar to the way some restaurants bring their kitchens into open view. Being a young company with a relatively young staff, everyone works to a random mix of contemporary music. No isolating headphones here—everyone listens to the same music, and every once in a while, someone starts singing out loud and a magic moment happens when everyone joins in. When current or prospective customers experience this, they get the message that the employees love their work and that this love will go into the services delivered. The singing is not part of a defined service process, and it's not part of anyone's employment contract. Rather, the conditions for spontaneous bursts of singing have been built into the company's culture, which emphasizes creating memorable experiences for customers and, indeed, for employees as well. This brings up a potential added benefit worth mentioning. Research by Candi, Beltagui, and Riedel (2013) has shown that businesses that emphasize the creation of compelling experiences for customers are typically more successful in attracting and retaining great employees than are businesses that neglect experience design. Thus, customers are not the only beneficiaries of a company's experience creation; employees also benefit, and, by extension, the company benefits.

This company has incorporated an element of real, rather than premeditated, surprise into its service. The possibility of surprise is supported by the company's culture and by the sense of community among employees. The element of surprise reinforces the desired customer narrative of a happy company that will provide a happy service with happy results.

A Service that "Dances"

Our example of a service designed for dancing also centers on a window and comes from a small shop selling one-of-a-kind art and design. This shop struggles with the configuration of its retail space. There is a window at the pavement level, but to enter the shop itself, customers have to cross the psychological hurdle of venturing down a steep flight of steps into a windowless basement. This company decided to capitalize on the window, and rather than use it to try to display as comprehensive a sample of the wares available at the subterranean level as possible, to use it as a stage for "dancing." One day, passersby see an artist hard at work at his easel in the window. On another day, they might see musicians performing or models showing the latest from a fashion designer. The performances are not necessarily tightly scripted or choreographed, but certainly attract attention from people who pass by and might be compelled to venture into the retail space and part with their cash. This is a good example of creative and ever-changing design of the service environment to create experiences of surprise for customers. An additional benefit is the interactions and communication that can form among the audience members outside the window. These might be transient communities, but crossing the psychological hurdle of the staircase is probably easier with a new friend than on one's own.

A word of warning is in order here. While businesses should strive to create compelling experiences with their services and even go so far as to place the experience at the center of the service business model, service delivery is no less important. Beltagui, Candi, and Riedel (2012) refer to the trap of the *hollow core,* meaning a service that is all "song and dance" and no substance.

15.4 Designing a Service Experience Is Never Finished

The design of a service experience is an ongoing process. Experiences need to be constantly refreshed lest they become stale and uninteresting. This means one should never view a service as fully designed. It also means that in many cases there is scope for experimenting with different types and variants of experiences. By trying out various experience staging methods and initiatives, companies can simultaneously keep their customers engaged and interested and continually refine the experiences staged.

Designing a service experience is an iterative process. A business should in most cases start with the service processes and service environment, keeping the experience narrative firmly in mind, but can then add other elements such as customer participation or surprise. The design is never complete, and companies must continue to rethink, re-create, and re-design their service experience while attention to the core service delivery should never be neglected. Table 15.1 provides a worksheet, made up of a list of questions for transforming a service into a compelling experience based on narrative, participation, and surprise.

This chapter suggests only three ways to create a compelling and memorable service experience, namely, narrative, participation, and surprise. However, there are more options, such as incorporating memorabilia (souvenirs), focusing on ergonomics, offering customization options, and others. Each business will need to carefully consider the nature of its service and discern what ways to create a desired service experience are most likely to result in competitive advantage.

Table 15.1: Service Experience Worksheet
What is the narrative (or story) the service should deliver?
❑ How do the service processes need to be modified to support this narrative?
❑ How does the service environment need to be modified to support this narrative?
❑ What do customer-facing employees need to do to support this narrative?
Is there scope for leveraging positive self-service?
❑ How do the service processes need to be modified to support positive self-service?
❑ How does the service environment need to be modified to support positive self-service?
❑ How can customer-facing employees support and encourage self-service?
Is there scope for including active customer engagement in service delivery?
❑ How do the service processes need to be modified to support active customer engagement in service delivery?
❑ How does the service environment need to be modified to support active customer engagement in service delivery?
❑ How can customer-facing employees support and encourage active customer engagement?
Is there scope for the creation of a community of customers of the service?
❑ How do the service processes need to be modified to support a customer community?
❑ How does the service environment need to be modified to support a customer community?
❑ How can customer facing employees encourage or be part of a customer community?
Are there elements of surprise that could be introduced into service delivery?
❑ How do the service processes need to be modified to create surprises?
❑ How does the service environment need to be modified to create surprises?
❑ How can customer-facing employees improvise to create surprises?

15.5 Conclusion

Creating compelling service experiences can lead to improved business success by improving profitability, enhancing a company's reputation, its attractiveness to employees, and its ability to enter new markets and attract new customers. We posit that designing a service should focus on creating a service that literally or metaphorically "sings and dances" and that the outcome will be a compelling and memorable service experience.

So how can a company design a service that sings and dances? Assuming the core functionality is fit for purpose and is at least as effective as competing services—by making use of narrative, participation, and surprise. Environments and processes should be designed with an understanding of how customers will perceive them and how they will help customers to build a *narrative* of the service that has a happy ending and encourages repeat visits. Customer inputs are necessary for any service, but rather than a disruption to smoothly running processes, they should be taken as an opportunity to enhance the experience. To this end, processes and employee roles can be designed around customer *participation* and to encourage the formation of customer communities. Last, but certainly not least, design thinking—reframing problems and developing novel solutions—can benefit service design by continuously redefining the service in order to deliver an element of *surprise* to customers. The company should start with solid service processes with lots of built-in flexibility and a service environment designed to support the desired experience. Also, the company should put the service experience squarely at the core of its service value proposition. The rest depends on continuous creativity, which may be fully premeditated or result from flexible service processes supported by empowered employees.

References

Beltagui, A., Candi, M., & Riedel J. (2012). Designing in the experience economy. In S. Zou & S. Swan (Eds.), *Interdisciplinary approaches to international marketing: Creative research on branding, product design/innovation, and strategic thought/social entrepreneurship.* Advances in International Marketing Series, Vol. 23, 111–135. Emerald Group Publishing Limited.

Bitner, M. J., Ostrom, A. L., & Morgan, F. N. (2008). Service blueprinting: A practical technique for service innovation. *California Management Review*, 50(3), 66.

Candi, M. (2007). The role of design in the development of technology-based services. *Design Studies*, 28(6), 559–583.

Candi, M. (2010). Benefits of aesthetic design as an element of new service development. *Journal of Product Innovation Management*, 27(7), 1047–1064.

Candi, M., Beltagui, A., & Riedel, J. (2013). Innovation through experience staging: Motives and outcomes. *Journal of Product Innovation Management*, 30(2), 279–297.

Grove, S. J., Fisk, R. P., & Bitner, M. J. (1992). Dramatizing the service experience: A managerial approach. *Advances in Services Marketing and Management*, 1(1), 91–121.

Norton, M. I., Mochon, D., & Ariely, D. (2012). The IKEA effect: When labor leads to love. *Journal of Consumer Psychology* 22(3), 453–460.

Pine, B. J., & Gilmore, J. H. (1998). Welcome to the experience economy. *Harvard Business Review*, 86(4), 97–105.

Redström, J. (2006). Towards user design? On the shift from object to user as the subject of design. *Design Studies*, 27, 123–139.

Roberts, D. L., & Candi, M. (2014). Leveraging social network sites in new product development: Opportunity or hype? *Journal of Product Innovation Management*, 31(S1), 105–117.

Stuart, F. I., & Tax, S. (2004). Toward an integrative approach to designing service experiences: Lessons learned from the theatre. *Journal of Operations Management*, 22(6), 609–627.

About the Authors

Marina Candi is Associate Professor at Reykjavik University's School of Business and Director of the Reykjavik University Center for Research on Innovation and Entrepreneurship. She received her PhD in business from Copenhagen Business School. Prior to entering academia, she spent over 20 years working in the IT sector as a software engineer and project manager and, during the latter half of her industry career, held positions in executive-level management as well as sitting on the boards of directors of IT firms. Her research interests include design-driven innovation, experience-based innovation, business model innovation, and interactive marketing. Her research has been published in the *Journal of Product Innovation Management, Technovation, Design Studies,* and the *International Journal of Design.* For more information, please see www.ru.is/staff/marina. Dr. Candi may be reached at marina@ru.is.

AHMAD BELTAGUI is a Lecturer in Operations Management at the University of Wolverhampton Business School. He holds a doctorate from Nottingham University Business School as well as a master of engineering (in product design engineering) from the University of Strathclyde. His research interests concern the role of design in business, particularly in service management and innovation, as well as product development. His research has been published in the *Journal of Product Innovation Management, International Journal of Operations and Production Management,* and *Design Management Journal.* Dr. Beltagui may be reached at a.beltagui@wlv.ac.uk.

16

CAPTURING CONTEXT THROUGH SERVICE DESIGN STORIES

Katarina Wetter-Edman
Karlstad University

Peter R. Magnusson
Karlstad University

Introduction

There is an ongoing trend of gradually increasing the amount of services product manufacturers offer. This is often referred to as *servitization*. Compared to products, services are considered to have longer life cycles and larger margins and to be more resistant to the business cycle. From a marketing perspective, services can be used to differentiate and increase value to the customer. Furthermore, services can also be a means of strengthening customer relations as they entail interacting with and understanding the customer. It is, however, often troublesome for product-oriented companies to handle services, partly due to lack of methods and tools. In this chapter, we present a narrative design-based method, CTN (Context Through Narratives) that captures users' current practices, experiences, situations, contexts, and expectations and integrates these in service innovation.

Servitization has been described as a multistaged change process. Common to different servitization models is that a supplier starts with offering services supporting its existing products in order to improve accessibility, for example, providing spare parts and maintenance. At the other end of the servitization spectrum is support for (at least parts of) the customer's operations, providing total solutions. Slightly simplified, servitization can be synthesized into three steps (Table 16.1): (1) services supporting the product; (2) services supporting product usage; (3) services supporting the customer's operations (processes). Servitization thus demands an increased understanding of the

Table 16.1: Steps of Servitization

Step	Description	Example
1.	Services supporting the product	Spare parts provision, product maintenance
2.	Services supporting product usage	Optimizing and customizing the robot by means of, for instance, customization, training, and programming
3.	Services supporting the customer's operations (processes)	Optimizing customer processes of which robots form one part

end user and/or the customer's operations. Providing spare parts and maintenance requires little understanding of the customer's operation.

In the last step, the supplier offers services that can also be independent of the physical robot, where the aim is to support the customer's operations by offering a total solution. For instance, if the customer is using a robot to spray-paint a specific component, the supplier can offer to design the whole process and assume overall responsibility for painting, that is, the service: "painting a component" according to customer requirements. The robot is only a means of performing the service.

There is a demarcation line between steps 2 and 3, where the supplier needs to gain a thorough understanding of the customer's processes, that is, what the product is used for, and also the customer's applications and context. This can thus be defined as the borderline to become a solution provider. The further a supplier goes through the servitization steps, the stronger the need for new methods and tools that innovate and develop the services since fresh knowledge regarding the use-side needs to be obtained by the producer.

Successful innovation requires two types of knowledge: *technology knowledge* and *use knowledge*. Technology knowledge concerns aspects related to implementing an innovation, including knowledge of the mechanics of materials, chemistry, thermodynamics, and so on. Technology also includes non-product-specific technology, such as service-supporting technology, and organizational routines. Accordingly, technology includes all the enabling organizational resources necessary to make products and services.

However, moving beyond providing only products but also solutions to customer problems implies the need to have "use knowledge," also referred to as use experience, or application domain knowledge. Use knowledge involves understanding the use of an innovation/technology from a user/customer perspective. In other words, what the technology should do for its intended users, requiring a deeper knowledge of the customer's processes and needs, often called taking a customer, or service, perspective. The important thing is no longer what a company's products *are*, but what they *do* for the customer. Products are perceived as a means of creating value when used by the customer.

Use knowledge is more abstract than technology knowledge and is normally disconnected from innovation activities, at least as regards physical products. The nature of services is, however, quite different from that of physical products. Services are intangible and can be described as a series of activities where the customer is a co-producer. The active interaction of users during services, and the fact that these often take place

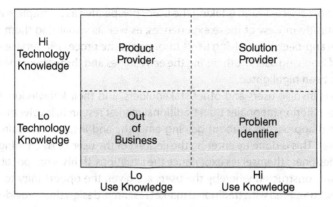

Figure 16.1: Two types of knowledge and company positions.
Adapted from Magnusson (2013).

in a context foreign to the supplier, puts greater demands on the supplier to understand the use side. Accordingly, use knowledge is much more important for services than for products, but can also be expected to lack when product manufacturers aiming to provide services.

A product-oriented company has, by definition, a lot of technology knowledge, but not the necessary use knowledge, depicted as a "Product Provider" in Figure 16.1. The added skills necessary to become a solution provider—and thus move to the next step of servitization—thus include use knowledge. Seeing use knowledge as a resource vital to competitiveness has implications for the company's innovation process. Gaining a deeper understanding of the use side is a necessity for providing more advanced services in the sense that these support not only the suppliers' products but also their processes and businesses.

Design has a long tradition of studying users in terms of understanding, interpreting, and translating their needs/problems into satisfying solutions—in other words, understanding the use side's problems and how proposing solutions can solve these. More recently, Service Design has emerged whereby the specific interest lies in designing solutions in terms of services.

16.1 Service Design

As discussed in previous chapters of this book, design has increasingly been seen and understood, over the past 15 years, as a possible driver of innovation. Design's contribution to the innovative capabilities of organizations can be ascribed to the iterative and user-centric approaches, the use of multidisciplinary teams, and the ability to externalize ideas and patterns using aesthetic skills such as visualization and prototyping. Furthermore, the designer's ability to interpret and reshape sociocultural relationships and configurations is even more prominent in service design than in traditional product design. Services are always co-produced together with the end user.

The core of service design is taking the user's perspective as a complete experience, giving an outside-in view of these experiences, as well as visualizing them and taking the iterative approach of involving users throughout the process. In service design, the approach of involving users both during the early stages and throughout the innovation process has been highlighted.

Designers involve users and other stakeholders, and their knowledge, in different ways and for differing rationales than traditional market research. Involvement through user-centered approaches is about gaining empathy and inspiration during the early design phases. This is done by entering the context of the user, or by creating situations where the designers themselves experience the situations. If this is not possible, staging situations are constructed whereby the users are given the opportunity to share their experiences. Compared with traditional market research, design thus provides distinctly different approaches to how users/customers should be involved.

16.2 Context, Stories, and Designers as Interpreters

The role of contextual understanding is to widen the focus to understand what broader role a product plays in the user's life. To understand the user's context, designers often move into that context to gain empathy via a deep understanding of latent needs, dreams, and expectations. Staging different workshop settings in which users, company (client) representatives, and designers interact is becoming an increasingly important part of the design profession. Approaches based on theories of play are gaining the interest of service design practitioners. In these approaches, users and other stakeholders are engaged in and encouraged to share their experiences, in addition to being a part of the co-construction of possible future solutions. The role of talk, stories, and dialogue is emphasized in order to understand the user's context and perspective.

Stories are one of the basic means by which we, as humans, communicate. We retell previous memories and experiences; we inform about our intentions, wishes, and dreams. In short, stories are one of the more dominant means by which we make sense of the world around us. In a story, actors, time, and contexts are often very efficiently captured and become beneficial for design purposes. Ethnographic stories can even serve as a bridge between the realm of the customer and the firm. Allowing customers to tell their own stories about their own experiences makes it possible to understand more about them than can be retrieved from an interview or a survey containing predefined questions. In the case (and model) offered here, designers are the interpreters of users' accounts. In design-driven innovation, designers are positioned as the interpreters of users' sociotechnical contexts and as brokers of knowledge across branches and organizations.

This intermediary role between the firm and its surrounding networks has been described as the brokering of knowledge, suggesting that designers, when moving between different companies, use and reuse known technologies in new areas. In the following section, we offer a method and a case whereby the means of involvement is stories rather than visualizations. We will point out the crucial role of the designer as an interpreter of these stories in order to achieve the intended outside-in focus.

16.3 Context Through Narratives—The CTN Method

A largely simplified description of the service design process can be illustrated using a double-diamond process containing four phases: (1) discover, (2) define, (3) develop, and (4) deliver. This process has a strong affinity with the design thinking framework presented in Chapter 1 of this book. The CTN method lies within the first diamond and represents the exploring and defining of needs and problem spaces for further innovation work. *Discover* largely focuses on extensive user and contextual research but is also aimed at exploring organizational prerequisites and strategies and potentially new or adjacent technologies that are suitable. It starts out from perceived problems/questions as defined by an organization; however, it attempts to go beyond specifics and explore a larger context.

The CTN method includes four steps: (1) preparation, (2) action, (3) processing, and (4) closure (Figure 16.2). We illustrate it using a service design pilot case. The aim of the service design pilot was to achieve a more extensive understanding of users' perspectives, situations, and experiences, rather than develop service innovations per se, which will come later on in the process. The service design pilot includes a repeatable workshop format whereby the facilitation experiences gained are both reused by the design company over time and, more importantly, transferred to the client company for future internal use.

The aim of the client company, IndComp (described in the box below), was to better understand one of its user groups, namely, farmers using automatic milking machines (AMMs). For this purpose, it had hired a design company, Veryday Agency; for a full description of this collaboration, see Wetter-Edman (2014).[1]

16.4 Case Illustration of the CTN Method

We try to get to meet the customers but it is not that easy. We need to ask permission, and we need to have a good reason to ask permission. Because there are good reasons and bad reasons. But that we just should hang around with a farmer is not popular."

Business Developer at IndComp, 2009.

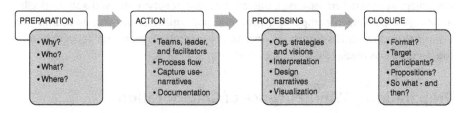

Figure I6.2: The CTN method in four steps.

[1] The case is real, as is Veryday. IndComp and all the personal names are fictitious.

Company Descriptions (Exhibit 1)

IndComp is a full-service supplier to farmers that develops, manufactures, and distributes equipment along with complete systems for milk production and animal husbandry. The milk production equipment ranges from vacuum-operated milking machines up to systems that can handle massive herd sizes of 5,000 to 10,000 head. They also make systems for herd management involving reproduction efficiency, health, and feeding. These systems need regular maintenance and occasionally require emergency services. IndComp also develops and sells services. Services are specifically handled by a central service division, but provided to the customer via local service organizations. The central service division manages the general characteristics of the service and, in addition, spare-parts management and service protocol preparation.

In the servitization model presented earlier, IndComp is in step 1, mainly providing services to support products. However, initial steps have been taken to move toward a more customer-oriented position. In February 2009, the organization launched its new service concept (NSC). It was communicated externally as a single feature but internally consisted of three parts: Connect—planned maintenance services, OnTime—emergency services, and Expertise—knowledge-based services such as consultancy. Expertise was the least developed and thus became the focus of the future design workshop. Additionally, Expertise is the concept with the strongest "service supporting the customer" flavor; however, IndComp had very little knowledge of how to move in this direction.

Veryday design agency (veryday.com) was established in the 1960s and has its basis in industrial design, ergonomics, and deep knowledge of user research. Today, this firm spans more than 10 design disciplines and has around 70 employees on three continents. The firm's philosophy is to take the user's perspective just as seriously as the aesthetic competence integrated into the design profession.

As described in Exhibit 1, IndComp is an industrial company aiming for increased service offerings beyond traditional maintenance service. This reorientation was supported by a biannual customer survey. Overall, the ratings were good, as usual, and the customers were satisfied. However, the latest customer survey showed a slight decline in ratings, especially in comparison with IndComp's main competitor. IndComp thus wanted to understand its customers' needs and expectations more qualitatively; hereby, the decision was made to employ service design.

Preparation—Why? Purpose of Collaboration

The purpose of the preparation step is to agree on the aim, purpose, and expectations regarding the project to be conducted, the "Why." It develops a shared understanding of the client company's resources but also those available for future solutions; the client's internal relations vis-à-vis their users; and how mature the client is as regards the use of design methods and tools. The latter is important when selecting methods for the project; in effect, what should be done?

As is often the case at companies with strong and sometimes independent sales organizations, the customers' voices are only implicitly present by way of a "whispering game": for instance, a customer said something to a salesperson, who took this further to his regional manager, who then continued to the person responsible for the local sales organization, via a designated forum, and eventually it reached someone responsible for service development. In a whispering game, the message is virtually always distorted.

Instead of direct user/customer involvement, the main input was a customer survey, mentioned above, conducted every other year, resulting in a "hard number." Even if satisfied, the number told little about *why* they were satisfied. One important issue when conducting a service design case is establishing a link between the service and business development units of the client company and its customers.

During the preparation phase, IndComp's business developer, Walter, met with the two designers from the design firm, Anna and Victor, on several occasions to set the scope of the workshop. These meetings involved:

1. Why—set the aim and focus of the project, select the methods to use.
2. What—decide on themes to be discussed, image selection, format of the workshop.
3. Who—
   a. Decide on participants, responsibilities for selection, and the invitation of customers (in this case farmers and company representatives).
   b. Define and invite the "right people" for the final delivery meeting and the presentation meeting.
4. Where—venue and time frame of the workshop.

In this case, IndComp was responsible for inviting its customers (farmers), and for the practicalities, for example, the venue, refreshments, and so on. A regional sales manager who had close relationships with the farmers was responsible for farmer invitations. The workshop included seven farmers, ranging in age from 45 to 65, who had AMMs and herds of between 160 and 200 cattle.

Veryday developed the workshop format, including content, processes, and materials. It was decided to use a method called *Landscaping* for the generative session, based on design dialogues (Brandt, Messeter, & Binder, 2008). The design company and the client chose the photos; they were vaguely connected with life on the farm, and were thus open to rather free interpretation and association. These initiated the discussions, whereby the participants are triggered into remembering various events or situations, and the dialogues occur around these.

The workshop was arranged around six themes exploring different aspects of these farmers' lives, as well as their interactions with IndComp: (1) preservice, (2) service, (3) invoicing, (4) emergency services, (5) purchasing automatic milking machines, and, finally, (6) working on the farm. The themes were prepared using preformatted sheets of paper and the farmers were asked to remember and tell about a particularly good experience as well as another bad one relating to these situations.

Action

The action step captures users' experiences through stories and images. It is important to have a framework for how to document the stories and for starting to make sense of these together with the users. The workshop participants included seven farmers, two company representatives who had made the preparations, and three designers.

After a presentation of the participants and the activity, the workshop started. The format was to work with the six themes in two parallel groups whereby each group worked with three themes (Table 16.2). A designer in each group took notes, documented the accounts using keywords and short sentences, and added an image to the prepared sheet of paper together with the farmers' narratives. Each theme session lasted for 30 minutes; in between these, the teams gathered and the situations were mapped out in a landscaped style. The aim was to make a collaboratively constructed landscape of the farmers' stories and experiences by the end of the day.

The facilitating designer moved between the teams, kept up the pace and motivation, and directed the construction of the joint landscape (Figure 16.3), positioning the sheets of paper containing descriptions in accordance with themes and experience.

The introductory question for each theme was framed along the lines of:

Service is about activities during and surrounding a service.
How do you experience that the service of IndComp works?
Describe some situations:
    Two typical situations where the service encounter works fine.
    Two typical situations where the service encounter doesn't work.

This phrasing focuses the discussion on experiences and situations rather than on discussing the extent to which the AMMs were functioning, or not. The designer on each respective team first noted down, on Post-its, both keywords and snippets from the stories while the farmers discussed the situation, then decided, in conversation with the farmers, on a set of important situations to report on, after which a note of them was made on the prepared sheets of paper together with an image. In Exhibit 2, we present some edited stories that were told by the farmers. Although these describe direct interactions with the AMMs, the emphasis in the stories is on the surrounding situation and context, that is, the role the AMMs and IndComp's actions play in the farmers' lives. Using the terminology of Chapter 1, these stories form the basis of finding issues, big and small, for "placing small bets" around customer insights.

Table 16.2: Organization of the Workshop		
	Group 1	**Group 2**
Introduction 15 min	Presentation of format	
Session 1, 30 min	Theme 1	Theme 2
Approx. 30 min	Collaborative Mapping	
Session 2, 30 min	Theme 3	Theme 4
Approx. 30 min	Collaborative Mapping	
Session 3, 30 min	Theme 5	Theme 6
Approx. 45 min	Collaborative Mapping and summary	

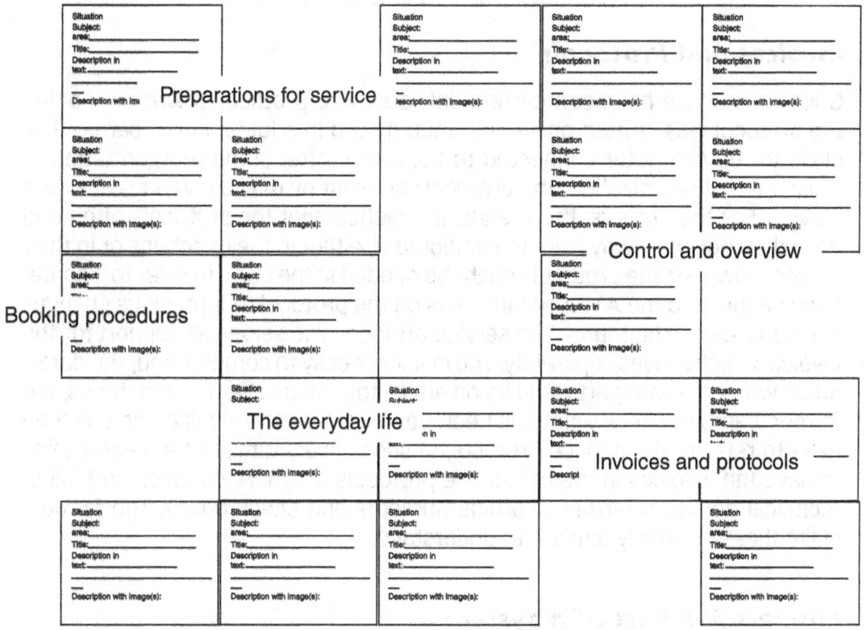

Figure 16.3: Schematic illustration of a constructed landscape of situations.

Examples of Use Narratives Told in the Workshop during the Action Phase (Exhibit 2)

Preparatory Milking

When maintenance is being carried out on the AMMs, the milking procedures and routines are disrupted. During certain operations, the AMMs cannot simultaneously milk any of the cows, for approximately an hour at a time. Accordingly, the milking procedure needs to be done either before the maintenance or carried out manually in parallel with it. Regular maintenance takes approximately four to six hours per AMM. Often, this preparatory milking is taken care of by the farmer before the service technician arrives. If the service technician is expected at 8 A.M., the farmer will probably have started milking by 5 A.M. in order to be finished in time for the servicing. Some service technicians assist with the milking, shortening the preparation time, but also lengthening their time onsite. Occasionally, service technicians receive emergency calls during the night and have to cancel scheduled maintenance. Due to nightly work hours, the service technician calls the farmer just before his/her scheduled arrival. However, by that time, the farmer will already have spent three hours milking in preparation. As one farmer said, "He could just have sent a text message, and then I would've known not to go out to the barn."

(continued)

Invoices and Protocols

Once the service has been carried out, a service protocol is written. Earlier, the protocol was written onsite and then placed in a folder in the barn office, close to the computer connected to the AMM. After some reorganization of the industrial organization, the protocols are sent by mail a few days after the service. For the farmers, this means, in practice, that the protocols often end up in their houses, away from the equipment, either in their kitchens or in their offices. However, the protocols might be needed in the barn, to refer to if something happens to the AMM because it is on the protocol that previous changes and services can be found. The service protocol also serves as support for the invoice, which arrives separately. The invoice is easy to comprehend; the dates, hours worked, spare parts, and so on are all totaled up clearly. Sometimes, the farmer wants to check what has been done vis-à-vis the invoice and will then have to refer to the service protocol; in such cases, it might be a good thing to have the invoices in the office. The protocols are more complex and full of technical details, referring to article numbers and check boxes. The farmers claim these are really difficult to understand.

Farmers Are Part of a System

The AMM includes several units that are not owned by IndComp but are crucial in order for the AMM to function. The experience of one farmer was that when his computer crashed, a service technician arrived on an emergency call-out. Since he could not fix the computer, and it is not part of IndComp's own technology, he changed it for a new one from emergency stock. A bit later on, when things had calmed down, the farmer called the computer company directly and was surprised to hear how easily and quickly the computer could be fixed. The very next day, the computer was fixed and returned to the farm again. If the computer company and IndComp had been collaborating, this would have saved the farmer approximately €1,000.

Processing

The aim of processing is to make sense of the outcomes from the workshop. What do these stories really mean for a client company's service business? This is done through organizing the stories in relevant themes and formatting the insights in ways that are actionable for the client company. Combining and interpreting the input from the users and the company creates design narratives, later combined in scenarios, as will be presented later in the chapter.

Directly after the workshop, the designers and the business developer started debriefing the workshop experience. The main issues that came up included ongoing projects that touched on issues that had been discussed, the service strategy of the company, and new ideas that could be related to this. Three weeks later, the first formal interpretation meeting was held at Veryday at which only the designers took part in structuring, analyzing, and discussing the outcomes. As input they used the situations captured on paper, where the use narratives had been noted in the workshop. The short descriptions reminded them of the full story told at the workshop by the farmers. It was a full-day meeting at which the designers, Anna and Victor, told and retold accounts that they had heard, referred to the documentation, and also acted out short scenarios to understand their meaning. The whiteboard was initially used to note down interesting aspects. Soon they moved over to the computer to create a digital mind map. In this, topics were organized in relation to opportunities and tentative themes.

The following day, a meeting was held with Walter (business developer at IndComp) at which a draft of the insights and opportunities was discussed, the mind map being the main input. From the slides in the draft presentations, the focus points argued by Veryday included primarily understanding more about IndComp's service vision and its potential relationship with this particular project. They also discussed what value meant to IndComp and its customers, and where value is or could be created. The company's aims and visions were thus integrated with the farmers' experiences, situations, and expectations as a basis for constructing design narratives. After the meeting, the insights were shared and updated between the designers and Walter. The findings and insights from this early stage research were presented using two scenarios. The first scenario represents the farmers' situation today (2010), and the other a future scenario (2015). Needs, problems, possibilities, and everyday life practices were integrated into these scenarios.

Closure

The final step of the CTN method is called closure, where the project in its current state comes to an end and is summarized and communicated to stakeholders. The preparations encompass arranging the meeting format and inviting the relevant participants.

People were invited from different parts of the organization who held positions from which they could actually act on some of the results of the project. In total, eight key members of staff from IndComp participated. In addition to the service division manager and Walter, the participants were managers from national and northern European markets including the global service coach.

Veryday presented the two scenarios that described present and future situations, thus revealing the gap between them and IndComp's need to act with purpose in order to move its positions forward. The scenarios consisted of five scenes, each focusing on one specific situation discussed during the workshop. The scenes were the same in both scenarios. The representation of the Service Day used in both scenarios is presented below.

The first scenario talks about the service technician in positive terms; however, it seems as if he does not really have the support and tools to do a proper job. The main focus is on attending to the equipment itself, changing spare parts, and the like (Figure 16.4).

In the corresponding scene in the 2015 scenario, many of the identified needs and problems are addressed via new service offerings. The farmer had ordered an additional wash, the service technician called to confirm his visit, the service protocols and other documents are accessible from different places. Thus, how the farmer wants to receive the service is in focus, rather than the way IndComp is capable of providing it today.

Figure 16.4: Scene Service Day 2010.

These scenarios, specifically the 2010 scenario, gave rise to a lot of discussion and involvement among the participants. The way in which the stories were retold and reframed by the designers made the farmers' situations and experiences urgent and important to act on for the organization. The design narratives helped them to empathize with the farmers, and to understand their perspectives.

16.5 Conclusion and Recommendations

This chapter has presented the increased interest in and need for servitization and the three basic steps that companies go through during the servitization process, and has argued that increased understanding of the customer is a prerequisite for this development. Further, service design is argued to be a design mind-set suitable for integrating a customer perspective. We presented the Context Through Narratives method exemplified using a case for moving toward an increased focus on customers' lives, contexts, and needs. In the following sections, we summarize this method's contribution to servitization and provide some hands-on advice about how to implement the CTN method.

Conclusion: CTN's Contribution to Servitization

Services demand a true understanding of the service from a user perspective, that is, the customer's processes and context. As illustrated by the case application of CTN, narratives widen the scope of information. From the workshop, the client company learned that it could not see the AMM as an "island." It was a part of a system where other suppliers' components and products interact and result in the farmer's final experience with his/her AMM. Take, for instance, the broken computer, which the farmer regarded to be an integral part of the AMM, since it was used to control it. Nevertheless, IndComp took no responsibility for any hardware malfunctions and did not even have any spare computers in stock. This is just one example of how knowledge is gained that can be transformed later on into a new service whereby the supplier takes full responsibility for the farmer's system.

From a knowledge perspective, the narratives capture the use side of technology, including the often complex context wherein the products and services are intended to be used. Design narratives have proven to be a fruitful method of gaining vital knowledge when moving to the more advanced steps of servitization.

Prerequisites and Recommendations for Successful CTN Usage

The presented CTN method suggests an approach to understanding the user's perspective through the use of narratives. The method is composed of four steps, all located in the first diamond of the Service Design process: preparation, action, processing, and closure. Each step puts the focus on different issues that need to be solved, negotiated, and attended to, as illustrated in the case. We strongly recommend that client representatives in service development or staff in customer responsibility positions are present at the workshop. Their role is to listen to the users' stories directly and not to defend the company. Below, we present checklists used to facilitate the workshop, and the analyses and interpretation steps:

Checklists for Implementing the CTN Method

Ahead of the workshop:
- ❏ Invite knowledgeable and experienced participants: users and employees in service development positions.
- ❏ Define subject areas covering the entire user experience and surrounding events.
- ❏ Formulate experience-oriented questions for initiating discussions.
- ❏ Prepare documentation templates for noting the users' stories.
- ❏ Prepare images that show different situations from the users' day-to-day lives for triggering discussions.

During the workshop:
- ❏ Be open and attentive during the workshop; the aim is to obtain and understand the users' stories.
- ❏ Let the users present and describe their situations in their own words when constructing the landscape of use narratives.
- ❏ Make notes on sticky notes, then on the situation template, and add images selected by the group.

❑ Construct the landscape collaboratively with the participants.
❑ Document the full landscape using photos.

Analyses and interpretation:

❑ Revisit company visions and strategies for identifying where the use narratives can complement, strengthen, and/or threaten existing ideas.
❑ Read and retell the use narratives with the purpose of remembering the nuances and aspects of the original story; pose the question "What are these situations 'really' about?"
❑ Involve company representatives in aligning with company visions and facilitating implementation.
❑ Cluster the use narratives according to affinity, good and poor, pros and cons.
❑ Identify problematic and good situations, as well as possible solutions; allow for iterations with next checkpoint.
❑ Combine use narratives and company visions.
❑ Create design narratives, based on present and potential futures, for example, scenarios identifying gaps.

As described above, the users' stories (use narratives) are to be noted down on preformatted sheets of paper; sticky notes may be added containing keywords or adjacent stories; images that bring life to the story are to be added (Figure 16.5); and the paper is to be labeled "good" or "poor" to depict the quality of the experience.

CTN explicitly puts the focus on the end consumer's situations and priorities. Companies using this method will thus need to be open to fresh input and prepared to change

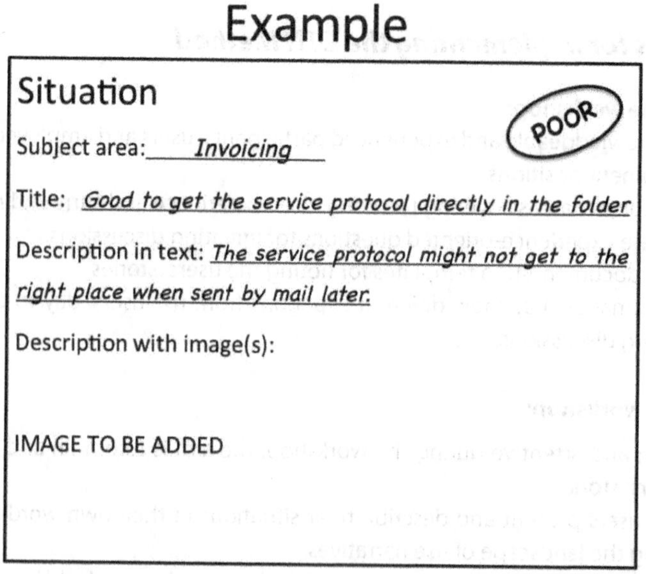

Example

> **Situation**
>
> Subject area:___ *Invoicing*___
>
> *POOR*
>
> Title: *Good to get the service protocol directly in the folder*
>
> Description in text: *The service protocol might not get to the right place when sent by mail later.*
>
> Description with image(s):
>
> IMAGE TO BE ADDED

Figure 16.5: Example of a documented use narrative for a given situation.

their own perspectives accordingly. Although stories are told and collected in various forums, we see, in this case, that design competence is vital when it comes to interpreting and transforming the users' stories into actionable knowledge for the company. Design competence is also important when it comes to framing the situation and posing questions aimed at experience rather than functionality.

However, if the firm's strategy is to move along the servitization continuum and to fundamentally change how customers' experiences and contexts are taken into account regarding service innovation, then the company will need to develop internal design capacity. Servitization will also affect the company's strategy; product features will no longer be the focus of selling, but offering solutions to the customer will be. The link with strategic reframing turns design work into something that cannot be handled solely by external consultants, but by in-house staff. This can be done in two supplementary ways—first, through the internal learning of design methods and tools, building up internal service design knowledge and facilitating skills across different functions, for example, by using the CTN method.

Second, because we have presented the role of the designer in the above case, we firmly believe that professional design knowledge plays an important role in interpreting and articulating design narratives. Thus, we additionally suggest developing internal design capacity by employing professional service designers as a resource spanning across the marketing and research and development functions. Thus, the user's perspective will be integrated strategically and professionally within the organization.

The CTN method is a well-tailored method whose purpose is to explore, expose, and articulate what the service provider's offerings will actually achieve in the customer's life, for better or worse. Additionally, the method includes tools for designing and proposing new offerings.

References

Brandt, E., Messeter, J., & Binder, T. (2008). Formatting design dialogues—games and participation. *Co-Design*, 4(1), 51–64.

Magnusson, P. R. (2013). Service innovation in manufacturing. *Managing industrial service in dynamic landscapes—A flexibility perspective:* MTC., Fredrik Nordin, editor. Östertälje tryckeri, Sweden.

Wetter-Edman, K. (2014). *Design for service: A framework for articulating designers' contribution as interpreter of users' experience* (PhD dissertation). University of Gothenburg, Sweden. (No. 45)

About the Authors

DR. KATARINA WETTER-EDMAN is Senior Lecturer in Design at Karlstad University and holds a master of fine arts in industrial design from HDK-School of Design and Crafts, Gothenburg University. Dr. Wetter-Edman has 10 years practical experience in industrial design and design management. Her research focuses on articulation of the emerging field of service design, Design for Service, the potential contribution of design practice and user involvement through design. She has an increasing interest for the role of service design in the public sector.

DR. PETER R. MAGNUSSON is Associate Professor at the Service Research Center at Karlstad University, Sweden. He holds an MSc in electrical engineering from Chalmers University and a PhD from the Stockholm School of Economics. Dr. Magnusson has 20 years' practical experience in research and development in the computing and telecommunications industries, working for companies such as Ericsson and Telia Sonera.

His research focuses on new product/service innovation, focusing on open innovation and user involvement. He has received several nominations and rewards for his research and has been published in leading refereed journals, including the *Journal of Product Innovation Management, Journal of the Academy of Marketing Science, Journal of Service Research and Creativity,* and *Innovation Management.*

17

OPTIMAL DESIGN FOR RADICALLY NEW PRODUCTS

Steve Hoeffler
Vanderbilt University

Michal Herzenstein
University of Delaware

Tamar Ginzburg
Vanderbilt University

Introduction

In this chapter, we provide prescriptive advice for new product development professionals who are interested in designing and creating radically new products. Radically new products are defined as products that allow us to do something we could not have done before (Hoeffler, 2003). Examples include 3D printers, TiVo, and more recently, Google Glass. While the topic of design and the domain of radically new products are both relatively new and emerging topics in marketing, crossing the desire to create radically new products with emphasizing the role of design in the success of these novel products requires altering many of the approaches that are currently used.

This chapter contains a series of processes design professionals should use in order to create an independent process that results in novel ideas. We offer six processes for firms that are seeking to improve their ability to come up with breakthrough products:

1. Communicate the challenge goal toward radically new products.
2. Shift time frames to future and past.
3. Promote emerging technology focus across the consumption chain.

4. Promote the use of analogical thinking.
5. Look for novel ways to solve simple problems.
6. Leverage more ideators via crowdsourcing.

Table 17.1 gives specific examples of how these processes translate into action and supplies specific examples of companies and products that have employed these processes.

17.1 Communicate the Challenge Goal toward Radically New Products

When trying to break out of existing new product development (NPD) patterns and create completely novel designs, it is important that upper management explicitly communicates those goals to both internal new product development personnel (when the task is in-house) and to any external ideators. We suggest that periodic attempts to create radical designs are added to the traditional NPD process and highlighted as unique so that they can receive special attention. To facilitate these attempts, the organization's top management should take three general steps.

First, management has to create a culture in which failure is acceptable; otherwise, the development team will be less willing to take big risks. It has been shown that if the manager directly in charge of the new product development teams is supportive and has an open-minded response to questions and challenges, then the NPD team is more likely to feel psychologically safe and take more risks in its work. A culture that allows NPD teams to make mistakes along the way promotes more radical ideas and designs. The measurements of performance of the NPD team can also influence how radical the team's ideas are. To be effective in that way, the performance measurement should not emphasize short-term gains, but rather focus on the long term. Further, it should emphasize the process rather than the outcome. For example, did the team learn from its experience, even if the outcome was ill-fated?

Second, management can help the innovation development team by viewing this process as having two stages, the first including open behaviors and the second closed behaviors. Open behaviors include taking risks, drawing on different domains, being radical. Closed behaviors, on the other hand, focus on what is possible—on implementation. The goal should be to start with the radical and audacious ideas, and only much later on in the process to think about implementation. The tension that implementation creates should be avoided in the early stages, but it is helpful in later stages because it can push innovators to come up with creative solutions.

Third, management has to allow the team to avoid routine organizational procedures. It other words, management should allow the development team to be autonomous and free of bureaucracy. This means that the development team should have more access to resources (such as adding other employees to the team, adding materials, purchasing and training with new technologies and equipment, etc.), and more freedom in terms of reporting on their progress. After communicating the

Table 17.1: Implementation of the Six Ideas

Area for Novelty Audit	Implementation Ideas	Examples
Communicating the Challenge Goal ■ Add periodic attempts to create radical designs to the traditional NPD process. ■ Leaders must create a culture in which failure is acceptable. ■ Do not let the tension between open and closed behaviors affect the open innovation team.	**On the individual level:** People in the NPD team should have X% of their time assigned to develop radical ideas they are personally interested in.	Google allows its employees one day a week to work on whatever they wish to develop.
	On the group level: The organization should create groups that include people with different backgrounds. The group would have high degree of autonomy and would work secretly.	Lockheed Martin Skunk Works allow the freedom to create without restraints; Google X-Lab is a semi-secret facility, tasked with developing technological advancements, like the self-driving car.
Shift Time Frame ■ Ask for ideas that will only be relevant in the distant future. ■ Avoid prescribed imagination and remove mental boundaries.	**Questions to ask the design team:** ■ Think about something that is impossible (today, in 1 year, in 5 years, in 10 years) ■ Alternatively: Think about something you want but hasn't been invented yet.	President JFK announcing on May 25, 1961, that the United States will put a man on the moon by the end of the decade.
Promote Emerging Technology Focus ■ How can a particular emerging technology be used? ■ How can we hack together two technologies to create something new?	**A brain storming activity for a skunkworks group:** Each person brings to the meeting one technology they want to discuss, then the group brainstorms: What can this technology combination do? What are the outcomes? Are they relevant in our domain?	Hybrid ways to deliver products to consumers: Fuse together home 3D printers (new tech) with same day delivery (new challenge). Amazon can e-mail a file to be printed at home and thus allow just-in-time delivery.
Promote the Use of Analogical Thinking ■ Making an analogy between your industry and a distant domain. Once the obvious similarities are clear, think about what else the distant domain can add to your idea.	**A brain storming activity for a skunkworks group:** Half of the group brings in a problem, and the other half brings some unrelated ideas/things that have superior qualities. Then the group brainstorms how the problems and the "superior things" are similar and how they are different.	Sungard, a cyber-security cloud computing company, developed the idea that moving to the cloud is like surviving a zombie attack. In both cases the attacker wishes to "eat your brain" and the only way to survive the attack is by being prepared.
Do Not Write Off Simple Problems ■ What are the nuisances or challenges in your daily routine or in your industry's processes? Can you remove them?	**A two-part exercise:** (1) Ask the design team to make a grocery list of the nuisances/ challenges they encounter daily. Then, after a month of keeping such records, (2) hold a skunkworks group meeting.	Coravin allows the pouring of a single wine glass without removing the cork. It was invented because the inventor wished to enjoy wine while his wife was pregnant.
Leverage more Ideators via Crowdsourcing ■ What does the crowd want and think is needed? Crowdsourcing helps in creating a steady stream of new innovations.	Acquire information about users and engage them in ideation activity. Use current communication technologies and virtual platforms to attract worldwide crowd to share ideas. ■ Let the crowd be seen in their creativity, offer rewards, employment opportunities and invitations to co-create with the brands they love.	Quirky.com lets everyone submit a product idea (from a little doodle to a chemical formula). Next, people who frequent the website vote whether they like the idea. Finally, the company designs and manufactures the chosen idea of the week.

challenge goal of developing a radically new product, the following steps should be taken to implement the change:

1. **Develop new criteria.** First, senior management should develop and communicate new evaluation criteria for challenges geared toward radically new products. Thus, the focus of the idea generation phase should be on uniqueness rather than on feasibility. This may be difficult for managers to allow because they are often evaluated on concrete outcomes, thus this directive has to come from top management. For example, feasibility (which is often a new product review criterion) should be eliminated (or lessened in importance) when the focus is to create radically new product ideas.

2. **Set target goals.** Specific target goals should be set for the number of radically new ideas that are put forth. These targets could be in the form of a percentage of ideas that are more radically new and/or a percentage of time that each employee should devote to generating radical ideas.

3. **Perform systematic idea audits.** In order to measure the success of the program to create radically new ideas, the firm can bring in outsiders to rate the newness or novelty of the ideas. The novelty ratings should be tracked over time and shared with the NPD team, on both a team and an individual basis.

4. **Measure the effectiveness of the team.** Teams that develop radically new products should be able to be autonomous. This needs to be confirmed from time to time in order for the team to function well. Occasionally, senior management should question the team as to (adopted from Edmondson, 1999):

    a. Does your team get all the information it needs to do your work and plan your schedule?

    b. Is it easy for your team to obtain expert assistance when you do not know how to handle something that came up?

    c. Does your team have access to useful training on the job? This way management makes sure that the developing team stays effective.

Moreover, management stays in the loop and learns about issues before they become too complicated to solve.

17.2 Shift Time Frames to Future and Past

One method to free up ideators from the bounds of today's traditional designs is to specifically ask for ideas that will not be relevant until some future date (i.e., 5 or 10 years into the future), or as Google X Lab defines it, "science fiction–sounding solutions." This farther-into-the-future thinking may relax the normal constraints associated with developing a novel design. Moreover, this shift in time focus may eliminate one's common self-imposed restriction on new ideas to merely those where the ideator can envision a path to fulfillment. Innovators' prescribed imagination is the killer of radical ideas. Therefore, top management should find ways to remove mental boundaries by allowing the innovation team to generate ideas that will not be feasible in the near future. The following steps should be taken when implementing this change in time frames:

1. **Specific future time frames.** Instead of just having the NPD teams think about products that will be developed in the future, specific challenges should be created for unique time frames (i.e., unique challenges for products that could be available in 5 years, 10 years, 25 years, etc.). These challenges can be technological, such as amputees who control a fully functional hand that can handle delicate objects with their brains, or clothes made of fiber-based nanogenerators that provide a flexible, foldable, and wearable power source that allows people to generate their own electrical current while walking, and harvest/storing the power generated for later use. These challenges also can be related to social acceptance, as we see happening now with Google Glass, which raises concerns over privacy and safety; or related to the law, as is the case of the driverless car and the legal questions that arise.

2. **Look into the past.** In order to get some sense of where new technologies might come from, an interesting approach is to examine where current technologies came from. So much of what is included in today's products is taken for granted. If one examines the history behind these technologies to identify the original purpose for which the technology was developed, one can gain insights into the seemingly idiosyncratic development of current technologies. Many of these technologies were likely intended for industries other than the ones they ended up being used in.

3. **Create a pictorial archive.** One method that can be used to examine changes over time is to document the history of those changes in an easy to display manner. The idea is to capture every form of all competing products in the industry, going back to the introduction of the product category. These pictures should be captured and displayed in such a way that the history and development of the category can be easily examined.

17.3 Promote an Emerging Technology Focus across the Consumption Chain

One way to promote uniqueness is to fuse the process with a particular emerging technology and ask how that emerging technology could be used in the focal domain. Another way is to attempt to hack together two emerging technologies and then place the outcome in a specific domain (related to the company's core industry). In addition, technological road maps can be created for all potential emerging technologies that may impact the industry, consequently promoting combinations of several emerging technologies in a novel design challenge.

The idea to focus on new technologies is similar to the blue ocean concept in which innovators create new industries rather than compete with current players. Nintendo's Wii console is a perfect example. The company did not wish to directly compete with

Sony's PlayStation or Microsoft's Xbox in terms of the resolution and animation of the games. Instead, Nintendo created a new control and thus appealed to segments new to the video game industry (women and older folks). By promoting a focus on emerging technologies, firms can reconstruct market boundaries and introduce novelty. New product developers can use cross-conventional technologies to create new demand in a new, unknown space, rather than compete over existing markets. This emerging technology focus in the design process will lead to innovative developments. The following steps should be taken when implementing the emerging technology focus:

1. **Identify steps in consumer consumption chain.** The first step here is to identify the steps associated with consumers' consumption chain (MacMillan & McGrath, 1997). Then, for each step within the consumption chain (Table 17.2), the NPD team can identify emerging technologies that are likely to have an impact on how consumers make decisions at those different stages.

   Strategically thinking about the various steps consumers go through when interacting with a product enables the development team to creatively enhance the consumer experience every step of the way with unconventional technological solutions.

2. **Mix and match technologies.** Those emerging technologies that the team has identified then become the focal technologies. Next, the team should examine and list implications for mixing and matching different technologies together to come up with radically new product ideas. As a simple example we may consider two "hot" ideas: 3D printers and same-day delivery. How can we combine these to create a meaningful radical new product? One idea is that for simple products the designer can send the client a file to be printed at the client's home with the 3D printer—delivery is immediate. The beauty of this example is that it hacks together two seemingly unrelated products/services to create a potentially meaningful offering.

Table 17.2: Steps in the Consumer Consumption Chain
How do people become aware of their need for your product or service?
How do consumers find your offering?
How do consumers make their final selection?
How do consumers order and purchase your product or service?
How is your product or service delivered?
What happens when your product or service is delivered?
How is your product installed?
How is your product or service paid for?
How is your product stored?
How is your product moved around?
What is the consumer really using your product for?
What do consumers need help with when they use your product?
What about returns or exchanges?
How is your product repaired or serviced?
What happens when your product is disposed of or no longer used?

17.4 Promote the Use of Analogical Thinking

The design team should find ideas from analogous domains in extremely different industries that have dealt with some of the same abstract problems (Kalogerakis, Luthje, & Herstatt, 2010). Analogy use is a common and vital technique in creative problem solving and complex innovation tasks. The goal of this exercise is to find both similarities and, more importantly, differences between the problem at hand and the base of the analogy, in order to infuse the solution to the problem with attributes that otherwise would be unthinkable.

The story behind the invention of the Nest Learning Thermostat is a good example for analogical thinking. The team that designed the thermostat aimed to reinvent something that has not changed in the past 20 years. The head of this team was the person who invented the iPod. His thought was, how can we make a thermostat that is more like an iPod? Indeed, if you look inside the Nest, you will basically see a smartphone—a very high-powered processor with memory, flash, wireless radios, and antennas. By using analogical transfer, the team was able to reinvent this "unloved but important home product."

In addition to analogical thinking, scenario thinking can be helpful in mentally simulating future consumption and experiential usage scenarios. One aspect that may help ideators think through the potential usage scenarios is to think through the specific types of consumer uncertainties that are thought to impact the usefulness of the new product or service. Hoeffler (2003) demonstrated that consumers have greater uncertainty when estimating the perceived usefulness of RNPs, and proceeded to partition the sources of uncertainty into:

- *Benefit uncertainty*—an estimate of the perceived benefits provided by the new product or service;
- *Learning cost uncertainty*—how much work the consumer will need to do in order to fully utilize the new product; and
- *Symbolic (or affective) uncertainty*—which is the more gestalt affective reaction to the adoption of a radically new product.

Thinking through these uncertainties may help identify particular areas where completely novel opportunities exist. Specific tasks to elicit novel design ideas may include sketching, prototyping, and storytelling. In addition, visual examples may be shown to ideators to promote novelty, as long as the timing of the exposure to examples is strategically designed to avoid design fixation. The following steps should be taken when implementing this emerging technology focus:

1. **Identify important consumer uncertainties.** The first step is to identify the uncertainties that consumers have with respect to the existing products available. This can be done by examining the different types of uncertainties presented in Table 17.3. After listing the associated uncertainties, the development team can better focus on identifying creative solutions that can ease the consumer's experience and lower the degree of uncertainty. In the following three subsections, we show examples of how these uncertainties have been dealt with in the past. Development teams can learn from others' experiences and solutions.

Table 17.3: Types of Uncertainty

Prediction	Uncertainty	Examples
Predicting benefits	Consumption/usage uncertainty	▪ Replaces something you are currently doing ▪ Provides new benefits ▪ Unknown consumption constraints
	Performance uncertainty	▪ New features (additional or improved) ▪ Combined functionality of several products
	Network externalities uncertainty	▪ Evolving architecture/technology (standardization) ▪ Availability of supporting products ▪ Evolving functionality
Predicting drawbacks	Switching costs uncertainty	▪ Physical (remodeling) ▪ Psychological (threatens or enhances existing knowledge)
	Learning curve uncertainty	▪ Learning how to purchase ▪ How to use (Extract benefits)
	Price change uncertainty	▪ Price will decrease after the consumer buys
Predicting social implications	Symbolic uncertainty	▪ Consumer (How do I feel about the image it portrays?) ▪ Peers (What image does it portray to my peers?)
	Affective reaction uncertainty	▪ Seductive or repulsive

2. **Reducing benefit uncertainty:** When Apple introduced the first iPod in 2001, the company faced a lot of consumer uncertainty—consumers did not understand the product, what it does, and how. The commercials said "A thousand songs in your pocket," and "Say hello to iPod. Say goodbye to your hard drive." But consumers still resisted and sales were relatively slow, until Apple introduced the iTunes music store in mid-2003 (Apple announced the sale of one million iPods on June 23, 2003, and the sale of two million iPods on January 6, 2004). Once iTunes was available, consumers benefited from the iPod substantially more and their uncertainties regarding usage, performance, and network diminished.

3. **Reducing switching costs and learning curve uncertainty.** As another example, in 2007 the U.S. Mint announced it wished to replace the $1 bill with coins (because bills last only 18 months and coins last much longer). To attract people to the new $1 coin, the U.S. Mint announced it would issue a series of dollar coins that bear portraits of past presidents. But consumers experienced elevated uncertainty regarding switching costs and learning curve, and thus continued to use the one dollar bills. A few years later, the U.S. Mint discontinued the aforementioned series of coins. In striking contrast, in 1987 the Royal Canadian Mint introduced a one dollar coin and in 1996 it introduced the two dollar coin. Both coins are widely used. What is the cause for this difference? The Canadian Mint removed the $1 and $2 bills from circulation when the new coins were introduced. While it is true that most companies do not have the power to eliminate used products or competitors' products from the market, they may be able to economically incentivize their customers to switch.

4. **Reducing social acceptance uncertainty.** Nokia's introduction of the first Bluetooth earpiece for mobile phones in 2001 demonstrates how effectively

the company has dealt with symbolic uncertainty. Consumers were uncertain about the Bluetooth earpiece's social acceptance—being seen wearing a strange device and seemingly talking to themselves. As more companies joined Nokia in producing these earpieces (like Motorola, LG, and Samsung), they lobbied to change the laws in many countries around the world such that drivers would not be allowed to use their phones while driving. Once consumers had to find hands-free solutions for their phones, the use of the Bluetooth earpiece took off worldwide.

17.5 Look for Novel Ways to Solve Simple Problems

Often, companies focus on solving "big" problems in an effort to create extremely novel designs. We suggest that firms can alternatively focus on breakthrough ideas for solving common everyday nuisances. The goal here would be to start with the focus on common "small" problems, and then to think about designing extremely novel approaches to these simple problems.

For identifying day-to-day problems that need to be solved, a company might choose to expand its ideators group and invite input from a variety of factions. Involving customers and potential customers in the brainstorming stage can broaden the pool of ideas and thus promote novelty (and we provide suggestions for "crowdsourcing" to get novel ideas in the next section). By carefully defining the various problems brought up and identifying their core, the design thinking process can result in creative and previously unidentified solutions.

As an example, Leonard Bosack and Richard Troiano managed the computers at two different departments at Stanford, and wished to have their computers communicate with each other. They ended up developing the first router that allowed computers in different networks to be connected, and named their company Cisco. A simple problem triggered a solution that created one of the largest computing companies in the world.

17.6 Leverage More Ideators via Crowdsourcing

Toward a goal of coming up with a continual stream of new products, firms have traditionally relied on an internal staff of professional innovators to generate ideas and to evaluate those ideas. Recently, there has been a push to seek out novel ideas from any avenue available. One method that firms are employing is to outsource their ideation efforts in an attempt to get fresh ideas into their innovation process. The idea behind crowdsourcing is that firms can tap a dispersed "crowd" of nonexperts (e.g., consumers, employees). The steps to leverage more ideators include:

1. **Find more diverse pools of ideators.** A key aspect that may be important when implementing a crowdsourcing effort to general novelty in design is associated with the makeup of the participants of the crowd. Parjanen, Hennala, and Konsti-Laakso (2012) examine the use of a virtual idea generation platform and

advocate the use of people with diverse experience, expertise, and perspectives to enhance the chances of success. One of the reasons why firms should use ideation teams that are diverse in knowledge and skills is to mitigate the effect of a competence trap that results in more incremental ideas rather than novel ones. An example was Nespresso's 2005 Design Contest, where the idea was to imagine the future of coffee rituals. One of the ideas that users from around the world came up with was the Nespresso Chipcard. This card stores your personal coffee preferences, and when inserted into a vending machine it brews your personalized cup of coffee. The open contest resulted in a completely new idea, and not an improvement to an existing Nespresso product/service.

2. **Run comparative challenges against company ideators.** One aspect that may help to spur on creativity of a "crowdsourced" idea challenge is knowing that the ideas that they come up with are going to be compared to the ideas found from within the company. This call for ideas can challenge outsiders for a "let the best idea win" competition, knowing their idea might be the one the company ends up developing. Who does not like to show off his innovativeness and creative thinking? Moreover, if coming up with a creative idea might result in an employment opportunity, even better. Co-creating with the development team can serve as a strong motivation for users to share their creative skills.

3. **Highlight a goal of unique approaches.** As we mentioned in the first process idea of communicating the challenge goal, the idea here is to make sure that there is special recognition for the most novel approaches created by the crowd. A company may want to consider offering a big reward or even a profit sharing for the best idea that will move the company forward into the future.

A summary of the aforementioned six ideas to improve product development is presented in Figure 17.1.

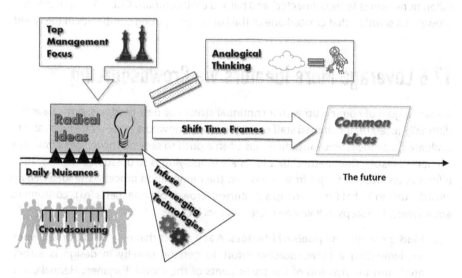

Figure 17.1: Six ideas to improve the development of a radical new product.

17.7 Conclusion

In order to achieve greater innovativeness, product development professionals interested in enhancing the novelty of their new product designs should first understand the unique differences between designing for improvements in existing products and designing for creating radically new products. As Figure 17.1 shows, a combination of effective communication of the challenge to create novelty, together with shift in temporal focus, inclusion of emerging technologies, analogical thinking exercises, focus on day-to-day issues, and crowdsourcing can effectively enhance novelty in design. With the aid of our six processes we hope that new product design teams will be more likely to find that breakthrough product to help redefine their industries going forward.

References

Edmondson, A. (1999). Psychological safety and learning behavior in work teams. *Administrative Science Quarterly*, 44(2), 350–383.

Hoeffler, S. (2003, November). Measuring preferences for really new products. *Journal of Marketing Research*, 40, 406–420.

Kalogerakis K., Lüthje C., and Herstatt C. (2010, May). Developing Innovations Based on Analogies: Experience from Design and Engineering Consultants. *Journal of Product Innovation Management*, 27(3), 418–436.

MacMillan, I., & McGrath, R. (1997, July). Discovering new points of differentiations. *Harvard Business Review*, 133–145.

Parjanen, S., Hennala, L., & Konsti-Laakso, S. (2012). Brokerage functions in a virtual idea generation platform: Possibilities for collective creativity? *Innovation: Management, Policy, and Practice*, 14(3), 363–374.

About the Authors

STEVE HOEFFLER is a professor at the Owen Graduate School of Management at Vanderbilt University; previously Steve was an assistant professor of marketing at the Kenan-Flagler Business School at the University of North Carolina at Chapel Hill. Steve's research on such topics as positioning multiple category products, marketing radically new products, and the advantages of strong brands have appeared in such publications as the *Journal of Consumer Psychology, Journal of Product Innovation Management,* and *Journal of Marketing Research*. He has worked in marketing for NCR/AT&T and consulted for Procter & Gamble, IBM, and Fujitsu.

MICHAL HERZENSTEIN is an Associate Professor of Marketing at the University of Delaware's Lerner College of Business and Economics. She has been a member of the Lerner College since 2006, upon completion of her doctoral studies at the University of Rochester. Michal's research focuses on consumer decision making and specifically on financial decision making, crowdfunding, and prosocial behaviors. Her papers were published in the *Journal of Marketing Research, Journal of Consumer Psychology,* and *Organizational Behavior and Human Decision Processes,* among other outlets. Prior to her doctoral studies, Michal worked as a marketing consultant at a prominent consulting firm in Israel, and was a lieutenant in the Israeli Defense Forces. She holds a BS in economics, statistics, and operations research, and an MBA in marketing, both from Tel Aviv University.

Tamar Ginzburg holds an MBA in marketing and strategy from the Owen Graduate School of Management at Vanderbilt University, and a BA in economics and Arabic language and literature from Tel Aviv University. Currently, she is an at-large member of the Nashville Opera Executive Committee and serves on the Opera board. She is a board member at Akiva School in Nashville, Tennessee, as well as the PTFA co-chair of the school and a member of the school's recruitment committee. She has worked as a consultant and a marketing analytics manager at an advertising company, as a marketing analytics intern at Xerox, and prior to that in investment banking.

18

BUSINESS MODEL DESIGN

John Aceti

Analogy partners LLC

Tony Singarayar

Analogy partners LLC

Introduction

In 2006, Henry Chesbrough wrote in the opening lines of his now classic book, *Open Business Models*, that "knowing that innovation is a core business necessity is not news, but that innovating a business model is news."(Preface xiii) Today, thinking about innovating your business model has become more common and less news worthy. What is news, though, is the currently available methodologies by which business models may be analyzed, designed, and implemented. For this reason, business model design is appropriate in a volume dedicated to design and design thinking.

Business model design will be of greatest interest to: executives responsible for finding better ways to compete, product managers who look to launch a new product, research and development (R&D) managers who need to prove their innovation has market value, or marketing managers who look to increase market share. Many of these professionals have already heard of business models, but in our experience they lack an ability to articulate exactly what their model is or how it works.

Although innovative business model thinking is becoming more prevalent, there are still lingering questions as to what it is, what is its value, and where does it fit with design thinking and strategy. This chapter addresses these questions and more. We will show how business model design can insightfully provide bold new opportunities for competing more effectively, launching a new business more successfully, or growing your business and market share.

18.1 What Is a Business Model?

The first use of the term *business model* came at the advent of personal computing and the invention of spreadsheets to create what-if financial models. These models were helpful for exploring sensitivities of a business's financials to numerous variables before ever launching the business. The ability to model and simulate a business's financials allowed questioning and exploration, which provided management with insight and increasing confidence. It stands to reason then that if the financials of a business can be modeled and *in charta* stress-tested why not other aspects of the business.

In 2002, Joan Magretta noted in her seminal article "Why Business Models Matter" that business modeling is becoming much more than just the financial modeling. She suggested that business modeling is the combination of two elements: the first was a *logical story*, which she called the narrative, and second, an *economic model* based on the narrative. The story, she added, was explaining who your customers are, what they value, and how you will make money. The economic model is then based on, and dependent on, the assumptions introduced in the narrative. She concluded that fault with either element would be fatal to the business. Although modest in detail, for the first time a method for business modeling beyond the financials was proposed.

In 2008 Mark Johnson, Clay Christensen, and Henning Kagermann wrote "Reinventing Your Business Model," in which they defined the business model as the sum of three elements: the Value Proposition, the Profit Formula, and Key Resources and Processes. They demonstrated that business model innovation can substantially create new value in an industry or category and they defined the conditions warranting a business model review. For the first time, a criterion for triggering a business model evaluation was proposed, including a framework for creating a business model. Further, they boldly offered that the value of business model design was that it has the potential to help revolutionize a market or even to create new markets.

However, these critical thinking experts stopped short of delivering a framework that management and product managers could use to analyze, design, but also implement a business model. Figure 18.1 shows the evolution of business model thinking, from spreadsheets to the current thinking in business model design.

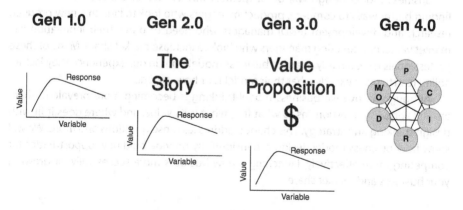

Figure 18.1: Evolution of business model thinking has progressed from Gen 1.0, a pure financial model, to Gen 2.0, the logical story supporting the financial model, to Gen 3.0, combining value proposition, resources, and the financial model, to today's Gen 4.0, a model defined herein with six key and interrelated aspects of business.

Regardless of the framework you may consider using, keep in mind the guiding principle of all business model design concepts, and that is, that the implemented model must both optimize *value to the customer* and *value to the organization* with a risk profile that is acceptable to the business. Identifying a business model that can deliver on this principle is the objective of all business model design methods. But this simple principle does not allay the confusion about business models and herein we will address four key questions:

1. When do I need to think about my business model?
2. What value should I expect from a business model design?
3. What method can I use to design a business model?
4. How do I implement my new or revised business model?

18.2 When Do I Need to Think about My Business Model?

In our experience, we have found two key triggers for management to consider a business model design: either the business is under obvious or perceived threat, or there is intent to launch a new business or new product. Here, we can refer to the first trigger as the Black Cloud scenario and the second, of course, as the White Cloud scenario.

In the Black Cloud scenario, management considers the business's potential to sustain itself into the future as they face a changing environment and evolving customer needs. As the dynamic unfolds, prescient managers recognize that they need more than a bandage to reinvigorate their position. They see need for an integrated analysis that will question if the company's business model is keeping pace, anticipating change, and positioning it ahead of market needs. Anticipatory Black Cloud thinkers see their markets changing and act—some successfully, like IBM, Netflix, and HP; and some not so successfully, like Kodak, Research-in-Motion (Blackberry), or Blockbuster. Jim Collins (2009), in "How the Mighty Fall," characterizes this condition superbly:

> I've come to see institutional decline like a staged disease: harder to detect but easier to cure in the early stages, easier to detect but harder to cure in the later stages. An institution can look strong on the outside but already be sick on the inside, dangerously on the cusp of a precipitous fall. (page 5)

In the White Cloud Scenario, entrepreneurs see the world in a new light with a new technology, new product, or new processes or service. Their choice is to introduce this new idea in a traditional business model or strengthen the offering by wrapping it in a more creative and potent model. There is a vast list of entrepreneurial companies that have wrapped their innovation in a new business model—like Google introducing its search engine in an advertising model, Fresh Direct introducing its fresh meals in a home delivery model, Zip Car introducing its automotive card swipe technology in a rent-anywhere-anytime model, or Apple wrapping its iPod in the iTunes music delivery model.

In both Black Scenarios and White Scenarios, there is that moment when management knows they must to do something different. And rather than jumping into reevaluating their strategic plan, which feels good because it is action oriented, they first consider how a business model analysis can inform their strategic planning process. So the time to think about your business model is when you have a Black Cloud or White Cloud trigger but before you conduct your strategic planning.

In our experience, it benefits to start with a critical self-evaluation of how your business stacks up against the competition. One way of doing so is to complete the Business Model Strength Survey (found at www.analogypartners.com/methodology.html). The survey provides a means of directly scoring your business against that of your competition and reveals where possible weaknesses exist.

18.3 What Value Should I Expect from a Business Model Design?

Michael Porter (1985) in his classic book *Competitive Advantage* defined a business's potential as always being relative to its competitors: "Competition is at the core of the success or failure of firms" (page 1). He suggested that the main purpose of a business model is to create competitive advantage by creating superior value for buyers. Thus, any business model design method must include comparison to your competition.

Improving on your business model will depend on the type of competition and the strength of competition. Are you competing with a well-defined competitor or more nebulous competition such as with your customer's time or attention (there is always competition whether it is implicit or explicit)? If your competition is weak, lacks growth, or has razor-thin margins and yet the market's need remains, then there is less of a challenge in developing a competitive business model. However, if your competition is like Apple, Starbucks, or Nike, your challenge is, of course, greater. Nevertheless, to win in your market is to understand your competition, and improving your business model so that it delivers to the needs of your consumer in the multiple ways that are important to them.

Michael Porter further defined a business's competitive advantage as either being Cost Advantage or Non–Cost Differentiation Advantage. Some contemporary experts suggest that companies must drive both to stay competitive in our global market. Regardless, a business model design can help identify multiple and perhaps unexpected means of achieving both. For example, Target achieves cost and differentiation advantage relative to Wal-Mart, with its comparable low costs but with superior product offerings and a better shopping experience. Olay achieves relatively low cost compared to prestige cosmetics but does so, and differentiates itself in its channel. Their model is *masstige,* a prestige product sold through a mass channel. Toyota achieves differentiation through its smart product design and low cost through its relentless productivity improvements in manufacturing. To the consumer, this translates into highly reliable vehicles at very affordable prices. A business model design can and should define elements that both deliver cost and non–cost differentiation.

Finally, there is a common misconception that achieving value from a business model design is difficult or arduous to implement. Of course, some changes can be difficult and risky, but in many cases, extreme makeovers are unnecessary to achieve tangible results. You have options—some will deliver significant business value but are high risk; others are low risk but of limited value. Other options are the appropriate choice, but the business lacks the resources or know-how. Homing in on the appropriate options, in a methodical way, is the process of business model design. You should therefore expect that the business model design process will answer the questions of (1) how to consider the "many different options" and (2) how to "select the option with an effective change that offers an acceptable risk of failure."

18.4 What Method Can I Use to Design a Business Model?

Business model design is still an evolving discipline with multiple design methods available, but when considering which to choose, assess the method's ability to:

1. Allow teams to easily understand and work fluidly with the method.
2. Connect competitive realities with the model.
3. Facilitate innovative low-cost and differentiation advantages across all aspects of the business.
4. Facilitate transition of innovative ideas into the business's strategy and execution processes.

While business model design methodologies generally satisfy the first three objectives, we have not seen any that facilitates the fourth. Methodologies that use more abstract building blocks, such as Value Proposition, Key Partners, and Key Activities, while essential to the business model discussion are nonetheless less rigorously connected to the organization's defined roles and responsibilities and therefore do not fluidly translate into, or enable, the execution process. We suggest that the six-cornerstone methodology (below) uniquely achieves all four objectives.

The six-cornerstone methodology uses a framework depicting the six fundamental cornerstones or building blocks of a business as seen in Figure 18.2. The six cornerstones are titled: (1) Product (or Service), (2) Customer, (3) Influencer, (4) Revenue, (5) Channel, and (6) Manufacturing or Operations. Moreover, the framework is applicable to any organization: for-profit or non-profit, large corporations or start-ups.

Let us first define what each business cornerstone represents; here as a guide are some characterizations:

Product defines the physical product or service you intend to provide and the means by which you develop and protect it. It also includes all ancillary effects that create uniqueness for the offering including packaging, accessories, training or education, support or complimentary services, or defines defensive assets such as intellectual property (e.g., patents, trademarks, trade secrets, and copyrights). If you offer a

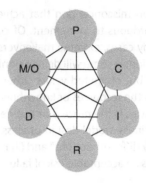

Figure 18.2: Six-cornerstone business model framework where P = Product/Service, C = Consumer, I = Influencer, R = Revenue and Profit, D = Channel of Distribution, and M/O = Manufacturing or Operations

service, this can include customer experience, customer service, presentation, or the environment in which you deliver the service.

Customer defines the ultimate user or decision maker. They can be a consumer buying shampoo or an intermediary who selects and purchases a product for an end user such as a surgeon selecting an implantable pacemaker for a patient. Marketing is a key contributor to this cornerstone; consider, then, how effectively the Functional (physical, utilitarian), Experiential (emotional), and Symbolic (spiritual or self-actualization) aspects motivate the customer to purchase your product (for details see Park et al. "Strategic Brand Concept-Image Management").

Influencer defines anyone in the product's ecosystem that has credentials to influence the awareness, trial, purchase, or product loyalty of your product. This can be an individual, media, associations, buying groups, key opinion leaders, recognized experts, or what Malcolm Gladwell called "Mavens" in his book *The Tipping Point*. The Influencer helps the customer make a product selection, or to help realize the maximum benefit after buying a product or service.

Revenue defines the manner in which the business will generate revenues and profits. Even a "free" business model, where the product is given away free, must have some aspect, such as sponsor-paid advertising, that helps secure compensation to sustain the business.

Channel defines the means by which product is delivered to the ultimate consumer. There are three kinds: channels that require the consumer to go to a location (bricks-and-mortar stores), channels that deliver product to the consumer (by mail, by Internet, by drone, or service in the home), or channels that are virtual (television and online sales). Increasingly, the channel must offer the consumer alternatives and an excellent shopping experience to provide a competitive advantage.

Manufacturing or Operations defines the means by which you produce your product or a service. It can be a tangible production process such as that used to make ball bearings, automotive engines, or electric toothbrushes. Or it can be less tangible production such as software. Or it can be production of an experience such as Disneyland, a Broadway play, or Viking Cruises.

The six-cornerstones are the foundational elements of business model design, but there is more. You will note in Figure 18.2 that each cornerstone has a link to each other cornerstone. These links can have meaning and importance. Whenever the definition of one cornerstone is inextricably linked to another, then the link has unique meaning and is of critical import. For example, if the product is defined or designed by the customer, then both the Product and Consumer cornerstones become linked in a manner different than if it were a mass-produced product. A good example of this is the Threadless Tee Shirt Company that produces T-shirts designed by consumers. The link between the Product cornerstone and the Consumer cornerstone becomes an essential and differentiable element of their business model. For simplicity, we will not expand on the nuances of the links in this chapter, but mention it only to show that the framework embodies depth beyond the six cornerstones.

18.5 Process of Designing a Business Model

Given this framework for defining a business model, we will now show how to use it for designing your model. This process will help your team explore all options that are plausible to outdo the competition and will prioritize which to include in your strategy. The six steps are:

1. **Define the business model of competitors.** Since this is a comparative effort, it is necessary to first define the business model of your competitors by evaluating their strengths and weaknesses using the cornerstone model. For consistency, assume the point-of-view of the consumer when doing the comparative analysis—how would the consumer rate the value they derive from each cornerstone. If consumers express need for improvement from even the most successful company in your category then even that company does not earn best-in-class scores. In some markets, an "Influencer" or recommender's point-of-view of competitive strength is more relevant to the evaluation. For example, a surgeon influences which type of pacemaker a patient will receive even though the patient is the consumer of the pacemaker.

   We suggest you do your first analysis by color-coding the six cornerstones of your competitor. Use green to highlight your competitor's cornerstones where consumers consider them the best in your market. Use red for your competitor's cornerstones where they are clearly subpar, and use yellow for any remaining cornerstones.

2. **Define your own business model** by giving fair-minded thought to how you measure up against the competition. If starting a new business, this would be your prediction of how customers will perceive your offering relative to the competition. Similar to the competitive analysis, use green where you believe consumers will perceive your business to be equivalent to, or better than, the best in the market; red for your weak cornerstones; and the remainder yellow. (You can conduct a more thorough comparative analysis, cornerstone by cornerstone, by using the comparative questionnaire "Three Steps to Assessing Your Current Business Model

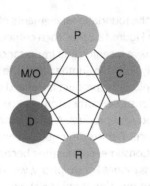

Figure 18.3: A self-evaluation of your business model. Here, for illustration, M/O (manufacturing/operations) and C (branding) are strengths, and D (channel of distribution) is weak.

Strength" found at www.analogypartners.com/methodology.html.) Your business model may look like that in Figure 18.3.

3. **Decide which cornerstones to improve.** Now compare your own business model with your competitors. Using a table such as Table 18.1 you can visualize how your self-evaluation stacks up against the competition. In this example, competitor #1 has a product inferior to your own but is doing a better job at leveraging influencers in the market, and this is delivering better revenue performance. Comparing your business to competitor #2, they have superior product strength but because their customer and influencer strengths are weak, they are doing no better in revenue generation. Clearly, competitor #1 should be the focal point in your competitive strategy to gain market share and growth.

4. **Develop a rich set of Value Accelerators™** to creatively decide how to improve any cornerstones that appear weak in your analysis. We refer to these ideas as Value Accelerators, which are assets that deliver value or help extract value from the market. Value Accelerators are the ideas you have (we will use both terms interchangeably) that, if introduced, would help bring you to parity or superiority relative to your competition. (For more in-depth instructions, see www.analogypartners.com/methodology.html for a "Guide for Developing Powerful Value Accelerators.")

Table 18.1: Comparing Your Business Model Evaluation by Cornerstone against Your Evaluation of Your Two Most Important Competitors

	Self-Evaluation	Competitor #1	Competitor #2
Product			
Customer			
Influencer			
Revenue			
Channel			
Manufacturing/Operations			

This exercise is the most intense and soliciting input from managers, key opinion leaders, and customers is most helpful. Further, there are tools and methodologies for the do-it-yourselfer, and there are professional *innovation facilitators* available to accelerate the creative process. The ideas generated may appear more challenging than what you believe you have the ability or resources to introduce, but include them nevertheless. As business guru Ram Charan suggests in every Business is a Growth Business (Preface viii).

> The mindset of growth starts with an insatiable curiosity about the world's needs. It does not accept the limits of existing products and existing markets; it quests endlessly for new opportunities to expand beyond their artificial boundaries. In a phase we use, it's about broadening your pond—enlarging the scope of your business activities by defining and meeting the new needs that change is always generating.

Reviewing the example in Table 18.1, this company should focus on its weaknesses relative to the main competition and develop Value Accelerators to improve their Influencer and Channel cornerstones.

Why not only focus on fortifying the strengths of your company rather than fixing weaknesses? If you have scored a cornerstone of your business as "strong" and your competition equally strong, then this means two things: (1) if you improve on this strength, then customers may not value these improvements; and (2) customers already perceive you as equal in strength to your competition, so further improvement may not improve customer's perceived differentiation. Can you improve on the taste of Starbuck's highly rated coffee? Assuming you are equal in taste, will making yours even better create sufficient differentiation for consumers? If you put effort and resources into improving the strength of an already strong cornerstone, it will at best provide diminishing returns and at worst dilute resources needed for improving your weak cornerstones. If either of these is not true, then you should rate yourself inferior to the competition.

5. **Prioritize Value Accelerators based on impact.** At this point, you should have a rich set of ideas to enrich your business model. Some ideas will profoundly help your business while others may only provide short-term or nonsustainable improvement. To select the best ideas for your business, we suggest prioritizing them by evaluating them against two criteria: value to the business and level of risk. Those ideas that deliver high value to the business and are least risky, of course, are preferable to those of low value and high risk. In order to "keep score," we suggest using a score card that will help you evaluate each of the ideas. The score card has two major categories—Value and Risk—with subcriteria to provide a richer characterization of each. For example, subcriteria for Value may include: is the expected incremental revenue sufficient, is it aligned to the overall corporate strategy, and how defensible is the idea to prevent imitation? Then develop a number of subcriteria for Risk, such as: the level of investment required, do you have the internal resources, and will management accept the risk?

The score card may look like Table 18.2. Here, we use a scale of 1 to 10 for scoring, where 10 is for the most valuable ideas and the least risky ideas to introduce.

Table 18.2: Score Card for Evaluating Value Accelerators

| Value Accelerators | Business Value | | | | | Business Risk | | | | |
	Revenue Potential	Sustainable Advantage	Good Strategic Fit	Defensible	Value Sum	Resource Investment	Have Skills & Experience	Confidence of Execution	Can Mitigate Risk	Risk Sum
Product Value Accelerator	10	8	8	8	34	3	8	8	3	22
Consumer Value Accelerator	9	4	6	2	21	9	9	6	7	31
Influencer Value Accelerator	7	7	4	4	22	7	7	4	4	22
Revenue Value Accelerator	4	9	6	2	21	1	3	3	1	8
Channel Value Accelerator	8	7	1	4	20	8	7	7	4	26
Operations Value Accelerator	9	9	7	9	34	6	6	7	3	22
Value Accelerator #7	3	7	3	3	16	3	7	1	1	12
Value Accelerator #8	6	2	9	3	20	6	2	7	3	18
Value Accelerator #9	8	7	3	7	25	3	7	9	7	26
Value Accelerator #10	2	8	9	2	21	2	8	9	2	21
Value Accelerator #11	8	6	4	4	22	8	6	1	4	19

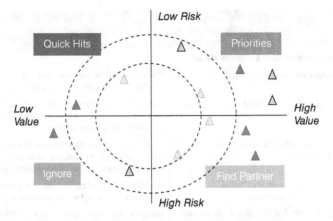

Figure 18.4: After scoring Value Accelerators, you can plot by coordinates to visualize which to pursue first.

The score card should sum the subcriteria for a total Value score and a total Risk score for each idea. For example, if you have four subcriteria for Value, sum them to get a total Value sum. Do the same for Risk. To visualize how the ideas prioritize we then suggest creating a simple plot of the results. Each idea has two coordinates, Value sum and Risk sum, with which you can plot each on a 2 × 2 matrix, as shown in Figure 18.4.

The matrix helps to categorize all the ideas. Ideas that fall into the upper right-hand quadrant are Priorities, as they are most potent for the business and relatively ease to implement. Ideas that fall into the lower right-hand quadrant are important but difficult to implement and you should find a partner to help execute these in a competent manner. Ideas in the upper left quadrant, Quick Hits, are easy to do but likely to be imitated by the competition, so choose these wisely. Of course, those ideas in the lower left, Ignore, are both difficult and of low value and should be ignored.

At this point, you have defined in quite some detail all the ideas by cornerstone that will make up your new or revised business model. You have considered each idea and have had your management team score each to the best of their prognostic capabilities. You now need to set a threshold for identifying those ideas you will implement and those you will not. This will largely depend on your resources and timing for enabling and launching each of the ideas.

6. **Leverage your top Value Accelerators into a business strategy**. This, the most crucial step, brings life to those Value Accelerators that will bring you to a superior position relative to your competition. We suggest using your plot of Value Accelerators (see Figure 18.4) in your strategy process. You may elect as many of the Value Accelerators from this process that meet your needs and consistent with available resources. If you are planning to drive revenues in the short term, look in the quadrant labeled "Quick Hits" and select those with the highest Business Value. Those ideas that are deemed important but difficult to execute for your firm will likely appear in the quadrant marked "Find Partner." Those ideas that you find

Table 18.3: Process to Design a Business Model Using the Six-Cornerstone Framework

Six Steps to Develop Your Business Model

1. Define the business model of competitors.	Use the six-cornerstone framework to define the relative strengths of your competitors.
2. Define your own business model as it is currently.	Use the six-cornerstone framework to define your own strengths or for a new business where you perceive your strengths to be.
3. Decide which cornerstones need to be improved to be more competitive.	If you find a cornerstone lacking in strength or differentiation, then consider what Value Accelerator you can incorporate to improve that relative situation.
4. Develop a rich set of options or Value Accelerators for each cornerstone.	Develop as many ideas or Value Accelerators as necessary to strengthen a cornerstone or a linkage.
5. Prioritize your Value Accelerators based on their impact to the organization and their level of difficulty.	Using the score card, prioritize your Value Accelerators to find those that are most important to the business and understand which you will tackle yourself and for which you will need a partner.
6. Leverage your top Value Accelerators into a business strategy.	Select those Value Accelerators you wish and integrate them into your business plan for execution.

in "Priorities" should be where you focus most of your resources, noting that the further away from the center of the chart the idea is located the more transformational the idea will be for your business. If many ideas are attractive but you lack resources, create a Value Accelerator road map showing how you intend to tackle multiple Value Accelerators over time.

Table 18.3 provides a summary of the Six Steps for Designing a Business Model.

18.6 How Do I Implement My New or Revised Business Model?

Strategy& (formerly Booz and Company) published a Strategy Execution Survey in 2014. In it, they reported that 55 percent of the leadership surveyed believed that their companies were not focused on executing their strategy. Their top three concerns were: the strategy is not bold enough, it is not coherent enough, and it is not clear enough. They also expressed concern that the organization was not aligned behind the strategy and that the work of the strategy creates conflicting priorities. Business model design mitigates these concerns and will significantly simplify and improve your strategic planning process.

Your strategic plan must define how the organization will implement its business model. With business model design, the key challenges and objectives are quantified, internal discussion of trade-offs completed, as is leadership buy-in by its scoring and prioritization. Strategic planning then becomes easier given a clear set of Value Accelerators. If there is any dissension, it will more likely be about the lesser, but still critical, issues of timing and resources. Critically, business model design will have addressed these stated concerns:

Being bold. One of the most potent aspects of business model design is the ability to identify weaknesses and to generate break out ideas—the Value Accelerators. However, ideas around product or branding alone do not necessarily create a highly competitive offering that comes from the persistent development of superior ideas in each cornerstone—where the integrated whole is significantly more bold and powerful. Business model design will deliver the bold steps that management wants to see.

Being coherent. A key advantage of using the six-cornerstone business model framework, and a weakness of other models, is its solid connection between its output and your organization's leadership. Each cornerstone can be coherently mapped to a specific leadership position, and this expedites the best Value Accelerators into action. For most organizations, it is clear that:

- Product is the responsibility of R&D.
- Customer, Influencer, and Channel cornerstones belong to marketing.
- Revenue is the responsibility of finance.
- Manufacturing or Operations owns the how-to-make aspects of the business.

Being clear. Business model frameworks that use more abstract building blocks, such as Value Proposition, Key Partners, and Key Activities, while essential to the design process, are nonetheless less clearly connected to the organization's defined roles and responsibilities. As the narrative of the six-cornerstones permeates the organization, key objectives and responsibilities become clear.

Being aligned. To complete the business strategy plan, management must be aligned and have their roles and responsibilities defined. The score card provides a perfect mechanism for testing alignment. If scores from management are consistent, then the organization understands the need and rationale to introduce the selected Value Accelerators. If scores are widely divergent, then there is either a lack of understanding with the Value or the Risk associated with the idea. In this case, more discussion, debate, or data is needed to resolve the issue.

18.7 Conclusion

Business model design is first and foremost a methodology to significantly enrich your offering and enhance your competitive advantage, and must be integratively developed with the design thinking of your product or service. Further, the six-cornerstone method allows your organization to build a common language, provides a framework for presentation and debate, delivers bold innovation, and enables a strategy for attaining superiority in your market. The framework provides a simple terminology to easily learn, yet does not hinder more complex thinking. It allows idea exchange and communication more easily, with fewer tendencies for misunderstanding. This framework stimulates innovation but is grounded in a reality of the competitive environment. Once business model design is embraced as a standard practice in your organization, all design thinking will have a consistency and power to communicate and to gain buy-in and alignment. Finally, the ideas, scoring, and recommendations are well documented and provide clear lines of responsibility for execution.

This framework has been introduced in many management teams, from start-ups to Fortune 100, from for-profit to non-profit, from business-to-consumer to business-to-business companies. What we find to be most helpful in the design of a business model is the engagement of the team, an embrace of the terminology, and free and creative thinking.

When it comes to creating a growth business, Ram Charan (1998) summed it up best in his book *Every Business Is a Growth Business:*

> For those that want to build profitable growth companies there is no single answer, no cotton candy high, no quick fix. What we are talking about is not easy. It takes a deep commitment to change on the part of leaders grounded in reality, with clear teachable points of view on growth. (Preface ix)

References

Charan, R. (1998). Every business is a growth business. New York, NY: Three Rivers Press.

Chesbrough, H. (2006). *Open business models: How to thrive in the new innovation land-scape.* Boston, MA: Harvard Business School Press.

Collins, J. (2009). How the Mighty Fall, HarperCollins, New York.

Gladwell, Malcolm (2000), The Tipping Point, Little Brown and Company, New York.

Johnson, M. W., Christensen, C. M., & Kagermann, H. (2008, December). Reinventing your business model. *Harvard Business Review, 86*(12), 51–59.

Magretta, J. (2002, May). Why business models matter. *Harvard Business Review* 80(5), 86–92.

Park, C. W., Jaworski, B. J., & MacInnis, D. J. (1986, October). Strategic brand concept-image management. *Journal of Marketing, 50*(4), 135–146.

Porter, M E. (1985). Competitive advantage: Creating and sustaining superior perfor-mance. New York, NY: Free Press.

Strategy & (2014). Strategy execution survey & key findings. Retrieved from www.strategyand.pwc.com/media/file/Strategyand_Slide-Pack-Strategy-execution-survey.pdf

About the Authors

JOHN ACETI is a technologist, inventor, business executive, and entrepreneur; he has worked for 30 years in three national US laboratories. In each, his responsibility has been to transition new and advanced technologies into commercially viable innovative products in the pharmaceutical, medical device, consumer health, and consumer markets. He is also the inventor and founder of three medical device companies in the hearing, diabetes, and health monitoring markets. He holds 30 US patents. Today, John's interest is in facilitating design thinking teams to create innovative, new-to-the-world products and business strategies and to help them secure early-stage investment.

TONY SINGARAYAR is a business model architect who advises senior business leaders on competitive advantage and revenue and margin growth. His expertise stems from five experiences: a 20-year career at Johnson & Johnson including roles in business model innovation, new categories and products, R&D, business development/L&A, supply chain, corporate social responsibility, emerging markets, finance, marketing, and information technology; he founded a team named one of the top 15 internal consulting units in the world in an independent study of 700 companies; he founded Analogy, a strategy consultancy that accelerates revenue and profit growth by using a proprietary database of hundreds of business models; and he is Managing Director of a 44-year-old, 400-employee, office technology business in Sri Lanka.

19

LEAN START-UP IN LARGE ENTERPRISES USING HUMAN-CENTERED DESIGN THINKING: A NEW APPROACH FOR DEVELOPING TRANSFORMATIONAL AND DISRUPTIVE INNOVATIONS

Peter Koen

Stevens Institute of Technology

Introduction

Delivering breakthrough innovations often requires companies to reach beyond the technology itself to rethink the business model using an iterative or probe and learn approach which represents a key tenet of design thinking. Corning's optical fiber program, General Electric's development of computerized axial tomography, Motorola's development of cellular phones, and Searle's development of NutraSweet (Lynn, Morone, & Paulson, 1996) created entirely new markets to achieve success. The technical innovation in each of these cases was accompanied by a new business model, as these new products required different operational competencies, vendors, and customer channels than the companies' existing offerings.

However, large enterprises, which are particularly adroit at exploiting their existing business models, often have considerable difficulty in developing new business models. For example, Sony developed the Walkman audio player, establishing the market for portable music devices. But Apple displaced it in the portable audio space with a new business model that included a new delivery channel—iTunes. Kodak, which dominated the film photography market, failed to embrace the business models needed to support digital photography and ultimately ceded the market to companies such as Canon and Nikon.

The lean start-up process, with its iterative learning cycles, is particularly suited to breakthrough innovations that require an iterative process and a new business model. Sustaining innovations, which represent the majority of product development activities in large companies, don't require a lean start-up process since customer needs are well understood and companies are able to exploit their current business model. Most large companies have a well-honed process and a formal Stage-Gate process that comprises a set of serial activities (i.e., stages) and decision points (i.e., gates). An iterative process, embraced by the lean start-up process, could be counterproductive to the sequential Stage-Gate process.

The lean start-up process is beginning to be used at enterprises (Blank, 2013a), such as GE and Intuit. The methodology has some unique features that are congruent with both the probe-and-learn process as well as design thinking, but it's most important contribution is its focus on the business model. This is an artifact of its origins in entrepreneurial start-ups, which all need to create a new business model. In contrast, enterprises already have business models for their sustaining business, but those sustaining business models may not be appropriate for breakthrough innovations. Thus, the lean start-up process provides a needed focus on business model development.

The objective of this chapter is to introduce the lean start-up process, integrate it with key concepts in human-centered design, and show how it can be used for developing breakthrough innovations. The chapter is broken into five sections. In the first section, the principles and methodology of the lean start-up approach is discussed. In the second section, breakthrough innovation is defined within the context of sustaining, transformational, and disruptive innovation. The third section provides a definition of what a business model is and demonstrates how the lean start-up approach makes the business model a key outcome. The fourth section discusses the lean start-up approach through the lens of human-centered design principles and evaluates the attributes of different business model canvases. The final section offers a discussion of lessons learned from implementing the lean start-up approach in enterprises.

19.1 Lean Start-up

The Lean Start-up Process

The lean start-up process, schematized in Figure 19.1, involves four parts. Three were described by Blank in his explication of the model: the business model, customer development, and agile development; the fourth element, the minimum viable

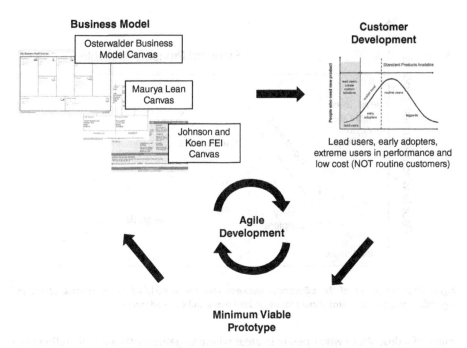

Figure 19.1: Schematic showing the four elements of the lean start-up approach: the business model, customer development, the minimum viable prototype (MVP), and agile development.

prototype (MVP),[1] is added here since it is the main experimental tool used by lean start-up teams to validate their hypotheses. The process involves continuous iterations of customer development, MVP, and business model changes, repeating until a scalable, repeatable business model emerges. The value of the lean start-up approach is that the business model, which is schematized using the business model canvases, is the principle convergence point of the process.

In the customer development stage, the team validates its business model through ethnographic studies of customers in relation to their environments. Visiting customers is a central theme of both the lean start-up approach and human-centered design. Start-ups often make the mistake of visiting "routine users" (Figure 19.2). These customers are often satisfied with the current solutions and product offerings, and thus provide limited insight. Lead users or early adopters who are not satisfied with current solutions offer far more potential for real insight and learning. Lead users (von Hippel, 1986) and early adopters are different from other customers because they are at the leading edge of an emerging product or process need and have a high incentive to find original solutions to meet their own needs. For example, a team developing new farm irrigation systems would benefit from spending time with farmers who are in the

[1] The lean start-up movement defines MVP as "minimum viable product." The author prefers "minimum viable prototype" because the term *product* implies something that can be sold. In contrast, a prototype incorporates only the feature set necessary to get a response from the customer and often is not a full, saleable product.

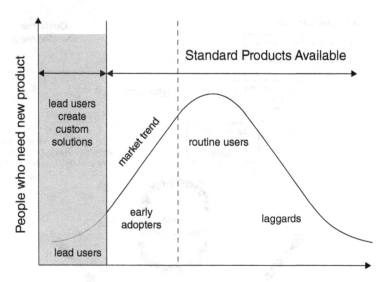

Figure 19.2: Schematic of the differences between lead users, early adopters, routine users, and laggards; lean start-up teams should focus on lead users and early adopters.

midst of a drought or who operate in areas where irrigation costs are high, rather than farmers who have access to sufficient affordable irrigation using current solutions.

The third part of the lean start-up approach is the development of an MVP. There is frequently confusion around what exactly constitutes an MVP. Most, when first confronted by the concept, believe that the MVP is actually a minimal-featured version of the final product. This is not the case. Rather, *the MVP incorporates the minimum set of features necessary to get early customer validation that the company's long-term vision makes sense*. MVPs may take many forms, depending on the stage of development and the information the prototype needs to yield. For example, the MVP shown in Figure 19.3a illustrates only the basic design features for a new nasal debrider; the final version of is shown in Figure 19.3b.

Figure 19.3: Example of an MVP. (a) A very rough prototype, constructed to demonstrate the minimum feature set in terms of look and design needed to get rapid, candid feedback from ENT surgeons). (b) Picture of the final Diego Gyrus ENT debrider.
Source: (a) Image courtesy of IDEO; (b) courtesy of Olympus.

Blank (2013b) offers an illustrative example demonstrating the need to focus the MVP on the customer needs. A California-based start-up planned to develop a series of unmanned serial drones to carry hyperspectral imaging cameras that could tell famers where their land required more fertilizer or water. The team envisioned the MVP as a drone equipped with a hyperspectral camera. Their business model was to build a fleet of drones with hyperspectral imaging cameras. The farmer didn't really care how the data was collected—just wanted the data. The team confused the MVP in trying to develop an early working prototype of their envisioned product as a drone with a hyperspectral camera. In fact, the farmers didn't really care if the data was collected with a drone, or a plane, the MVP, for this customer set, was the data. In the end, the team rented a hyperspectral camera and leased a crop duster single-engine plane that flew over the fields to collect data, which they then showed to the farmers in their target market.

The final component of the lean start-up process is the iterative cycle of developing and testing MVPs, which can be described either as agile development or as build-measure-learn feedback loop (Reis, 2011). A key metric for this process is how quickly the team loops through the process, developing successive MVPs.

19.2 Transformational and Disruptive Innovation: Defining the Domain Where the Lean Start-up Process Should Be Used

In order to see where the lean start-up approach can be most productively implemented in enterprises, it is important to develop a common framework and typology. Not every radical innovation will benefit from a lean start-up approach. For example, Intel's dual-core processer doubles performance while reducing power consumption. This is a radical innovation, but it doesn't require a new business model: Intel can leverage its current business model since the product is sold to its current customers using the company's existing channels. Technology project management tools designed for high-risk projects, such as Technology Stage-Gate (Ajamian & Koen, 2002), are more appropriate to manage these kinds of innovations. In contrast, Intel might have found the lean start-up methodology to be valuable in its failed attempt to get into the mobile phone market, with chips built using existing technology but sold through a new channel to new customers based on a new value proposition.

Innovating outside an existing business model has always been difficult for large companies. In a study of 154 companies, Bain and Company found that the odds of success dropped as low as 10 percent when large companies tried to develop products two steps from their core, where one step was a single change in the business model (Edwards, 2012).

The principle area that causes problems for large enterprises is innovating into a new value network. Many schematics of the innovation space map two dimensions, with newness of the market and the technology as the two critical axes. Christensen and Raynor (2003) and Koen, Bertels, and Elsum (2011) suggest a value network dimension that is more encompassing than the traditional market dimension, capturing the unique relationships enterprises build with both their upstream (supplier) and downstream (distributor and customer) channels.

Koen et al. (2011) suggest a three-dimensional innovation typology that captures value network, newness of the technology, and the financial hurdle rate; Figure 19.4a shows the value network and technology dimensions of this model. Within the technology dimension, incremental, architectural, and radical innovation are demarcated. Incremental innovation involves the refinement and improvement of existing technology. Architectural innovation involves new ways of integrating existing components into a system, but no new technology. The iPod, for instance, incorporated no new technology but provided an entirely new design. Finally, radical innovation, exemplified by Intel's dual-core processor, incorporates new core technology.

Procter & Gamble developed its own definitions for the different types of innovation: sustaining, transformational, and disruptive (Brown & Anthony, 2011); these are overlaid on Koen et al.'s model in Figure 19.4a.

Sustaining innovations bring incremental improvements to existing products; they may include radical technology innovations, as in the case of the dual-core microprocessor chip.

Transformational innovations, sometimes called *adjacencies*, bring a significant improvement to the existing product line and often direct the company into new value networks. An example is Nespresso, which engaged Nestlé's coffee business into a new value network focused on young urban professionals willing to pay a premium price for fine coffee.

Disruptive innovations establish an entirely new value network that involves nonconsumers—customers who have not entered the market. Sony's Walkman is an example of an architectural innovation focused on teenagers who had not previously owned audio playing devices.

Different combinations of innovation and value network require different project management tools, as shown in Figure 19.4b. Stage-Gate and Technology Stage-Gate should be used for projects in the sustaining space, as the company already has intimate knowledge of the value network and the iteration required by lean start-up will add costs and time to the process. In contrast, a lean start-up approach should be used for the transformational and disruptive innovation, where a probe-and-learn approach is required to glean needed customer insight.

19.3 Why Is a Business Model a Valuable Part of the Lean Start-up Process?

The concept of a business model was first mentioned in an academic article in 1957 (Bellman, Clark, Malcom, Craft, & Ricciardi, 1957) in the context of building business games for training purposes. The term continues to confuse academics and

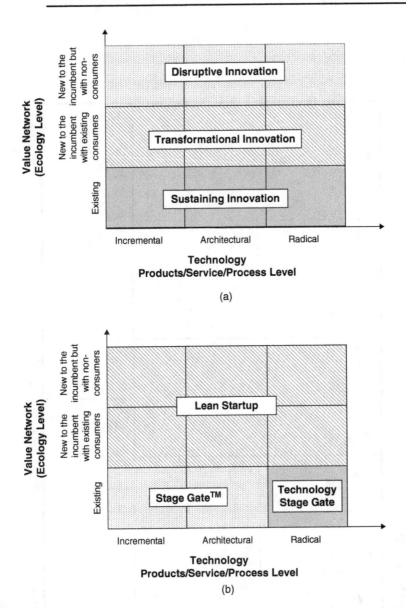

Figure 19.4: (a) Business model typology showing the relationship between sustaining, transformational, and disruptive innovation (b) and the areas where the lean start-up methodology may best be applied to.

practitioners alike. Wirtz (2011), reviewing the academic literature around business models, showed that there was little, if any, agreement in the academic literature to what constitutes a business model.

The business model canvas (Figure 19.5), introduced by Osterwalder and Pigneur (2010), addresses this confusion by providing a visual encapsulation of the business

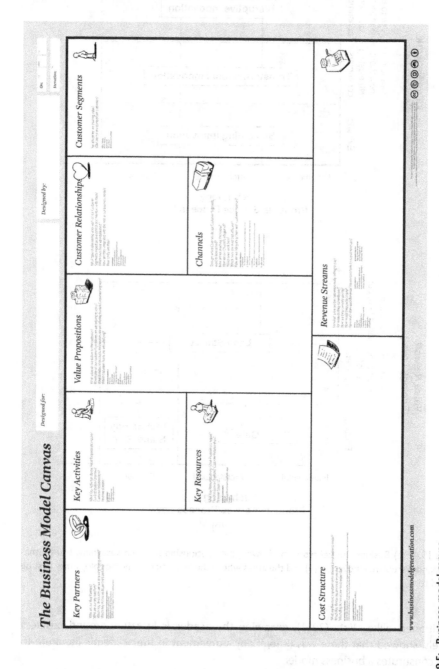

Figure 19.5: Business model canvas.
Source: Osterwalder & Pigneur (2010).

288

model and a clear vernacular, which facilitates discussion and debate without sacrificing the complexities of the business. The business model canvas allows the development team to evaluate the different parts first separately and then together, thereby facilitating new insights that would not have been possible. As part of a lean start-up approach, the business model canvas helps the team validate business model hypotheses until it finds one that is repeatable and scalable.

Edward Tufte (1997), a renowned scholar in the area of information design and visual literacy, encourages the use of data-rich illustrations and emphasizes the importance of being able to see all of the key data "in one common eye span."[2] Exploring the decisions leading up to the 1986 explosion of the space shuttle *Challenger,* in which seven astronauts died because of leaking O-rings, Tufte posits that the disaster could have been predicted had the critical information all been plotted in one descriptive illustration that could be surveyed within a single eye span (Tufte, 1997, p. 49). The business model canvas provides just such a layout for the team, capturing all the data needed to visualize the business within one easily viewable graphic.

Because the business model canvas functions as a convergence tool for the project team, it is a critical element of the lean start-up process. But most teams, in the beginning, fail to understand its value, feeling that the canvas contains no new insights and replicates what they already know. Teams quickly come to understand its value when they begin to use it as a tool to organize and test hypotheses while simultaneously accounting for the linkages that connect the different elements of the business model.

19.4 Lean Start-up through the Lens of Human-Centered Design

Lean start-up codifies many elements of the human-centered design process, which solves problems by matching people's needs with what is technologically feasible by developing simple prototypes and then iterating them until a viable business strategy emerges that can be converted into customer value.

To accomplish this goal, the human-centered design process always begins with a focus on the central question: what is the business problem? This approach helps teams avoid the typical error of focusing too quickly on the idea or solution. Many innovations fail not because of a fatal flaw in the solution, but because the company fails to understand what problem it is solving. The team developing Newton, Apple's PDA, was so enamored with the technology underlying the concept that they failed to consider the unique set of problems that the mobile user needed to solve. Segway failed because its development process was focused on transportation for everyone and not on particular jobs to be done for specific users; the company built a huge plant at the outset—based on the idea of transportation for everyone—and ended up with significant overcapacity.

---

[2] E-mail communication between Tufte and the author, January 10, 2014.

Getting to the right problem represents the pinnacle of the design process used by the iconic design firm IDEO. IDEO's methodology consists of three critical questions:

1. **What is the right problem?**

   As indicated in the preceding discussion, Apple's Newton and Segway failed since they did not understand the problem they were solving. A quote from Einstein further emphasizes the importance of understanding the problem:

   > If I had only one hour to save the world, I would spend 55 minutes defining the problem and only 5 minutes finding the solution.

2. **Who has the problem?**

   The heart of the human-centered design process is a focus on human values and a deep empathy with users. Thus, it is necessary to identify which customers the team plans to spend time with.

3. **What is the value to the user in solving the problem?**

   The value of a solution for the customer is determined by observing what people do, how they think, what they need, and what they want. These determine the attributes of the solution (as opposed to the solution itself).

The business model canvas allows teams to track the interactions between the various elements of the emerging business model. When the business model canvas is used in the context of a human-centered design method, it is extremely valuable to separately evaluate these three core questions and the solution, so that the solution attributes are not confused with the solution. Keeping the problem, the customer, the solution attributes, and the solution separate in the canvas allows the lean start-up team to build on the key tenets of the human-centered design process.

Unfortunately, Osterwalder and Pigneur's (2010) canvas does not allow for this to the extent that Maurya's (2012) lean canvas and the FEI canvas[3] do. The lean canvas, shown in Figure 19.6, was specifically developed for the start-up entrepreneur and is intended to better capture the uncertainty and risk of the start-up (Maurya, 2011). The FEI canvas, shown in Figure 19.7, was developed to support the front end of innovation in large enterprises.

The attributes of the three canvases are compared in Table 19.1. The lean and FEI canvases share five attributes with the Osterwalder and Pigneur canvas, but also encompass a number of other attributes. These differences reflect the different intents of the three canvases. For example, the lean start-up canvas does not have a box for external resources, as Maurya (2011) believes that entrepreneurial start-ups should focus on customers before looking at developing partnerships. In a similar vein, the FEI canvas includes additional boxes intended to capture the particular context of front-end innovation in a large corporation. Osterwalder and colleagues (2014) recently published the value proposition canvas (Figure 19.8), which fill many of the gaps in the original version.

[3] http://www.frontendinnovation.com/media/default/pdfs/fei-canvas.pdf

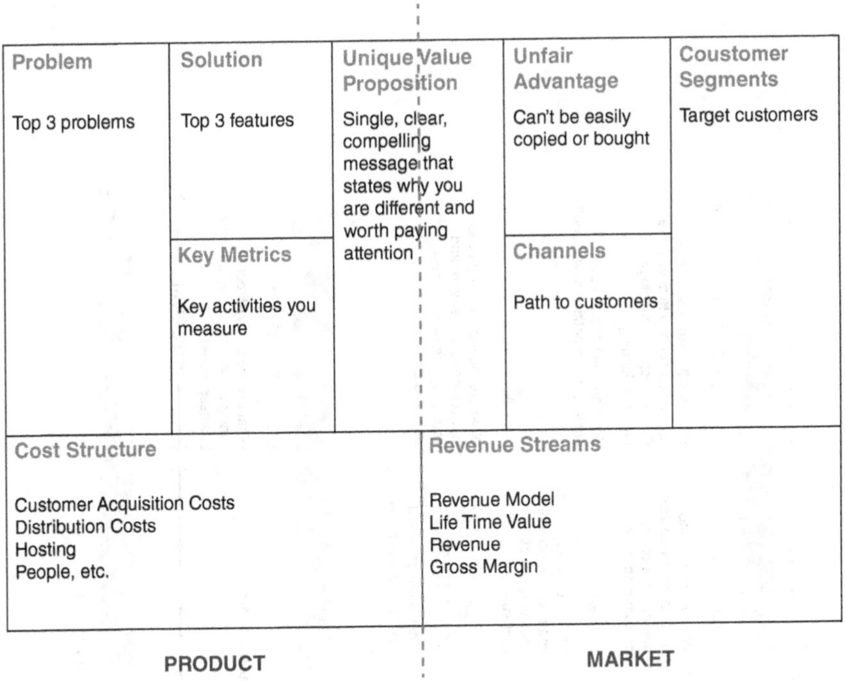

Problem	Solution	Unique Value Proposition	Unfair Advantage	Coustomer Segments
Top 3 problems	Top 3 features	Single, clear, compelling message that states why you are different and worth paying attention	Can't be easily copied or bought	Target customers
	Key Metrics Key activities you measure		**Channels** Path to customers	

Cost Structure	Revenue Streams
Customer Acquisition Costs Distribution Costs Hosting People, etc.	Revenue Model Life Time Value Revenue Gross Margin

PRODUCT   MARKET

Figure 19.6: The lean canvas.
Source: Maurya (2012).

Each of the three canvases aligns with the human-centered design approach to varying degrees, as illustrated in Table 19.2. In the original business model canvas, three of the four building blocks of human-centered design are not accounted for, although the value proposition canvas addresses all of these shortcomings. For instance, the problem definition is included in the customer segment portion of the value proposition canvas using "jobs to done" language and the value to users in solving the problem, captured only generically in the original business model canvas, is expanded with its own box in the value proposition canvas. The solution is also missing from the original canvas, but detailed in the value proposition canvas, although the need to pair the original business model canvas with the value proposition canvas violates Tufte's (1997) insistence that effective tools must capture all critical information in a single eye span.

The lean canvas separates the problem, which customers have the problem, and the solution into separate boxes. Solution attributes are not assigned to a particular box; presumably, they should be included in the value proposition box, which calls for a "single, clear, compelling message that states why you are different and worth paying for" (Maurya, 2012, p. 5). The FEI canvas, which was designed with the human-centered design perspective in mind, has separate boxes for all four of the core design principles.

In summary, the human-centered design approach evaluates the project through the lens of the problem, asking the development team to define the problem, identify

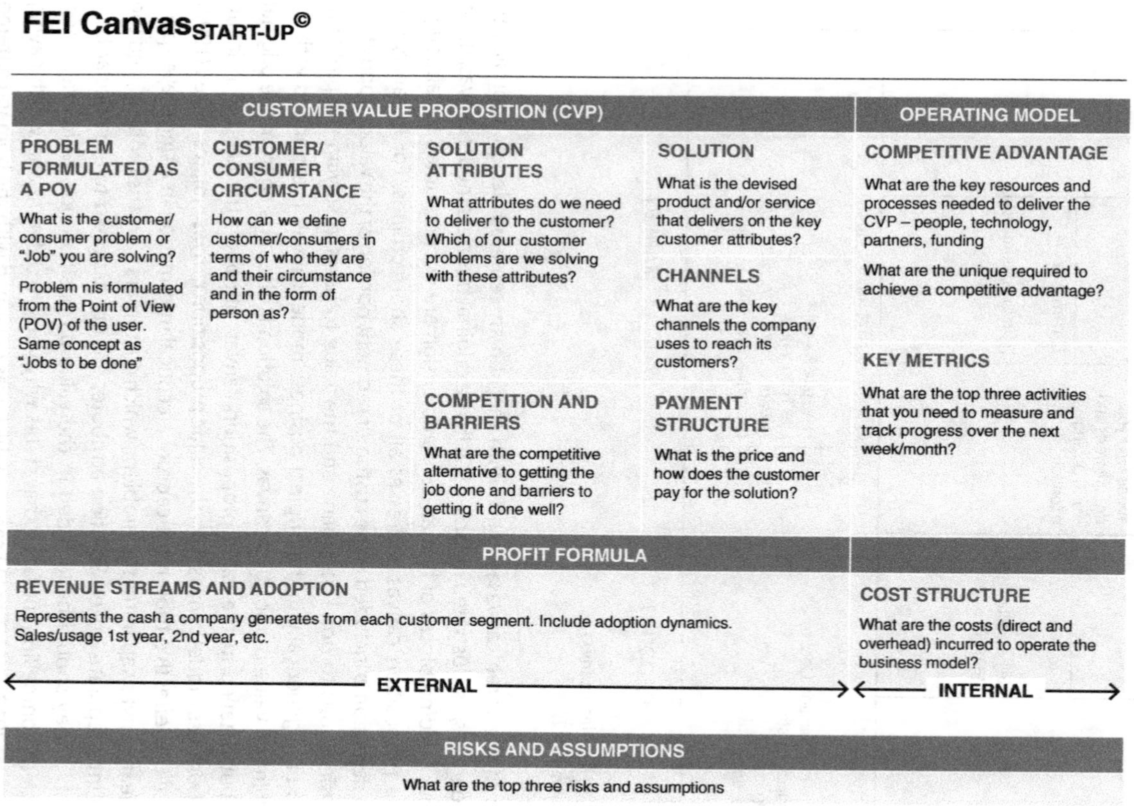

FEI Canvas_{START-UP}©

CUSTOMER VALUE PROPOSITION (CVP)			OPERATING MODEL

PROBLEM FORMULATED AS A POV

What is the customer/consumer problem or "Job" you are solving?

Problem n is formulated from the Point of View (POV) of the user. Same concept as "Jobs to be done"

CUSTOMER/ CONSUMER CIRCUMSTANCE

How can we define customer/consumers in terms of who they are and their circumstance and in the form of person as?

SOLUTION ATTRIBUTES

What attributes do we need to deliver to the customer? Which of our customer problems are we solving with these attributes?

COMPETITION AND BARRIERS

What are the competitive alternative to getting the job done and barriers to getting it done well?

SOLUTION

What is the devised product and/or service that delivers on the key customer attributes?

CHANNELS

What are the key channels the company uses to reach its customers?

PAYMENT STRUCTURE

What is the price and how does the customer pay for the solution?

COMPETITIVE ADVANTAGE

What are the key resources and processes needed to deliver the CVP – people, technology, partners, funding

What are the unique required to achieve a competitive advantage?

KEY METRICS

What are the top three activities that you need to measure and track progress over the next week/month?

PROFIT FORMULA	

REVENUE STREAMS AND ADOPTION

Represents the cash a company generates from each customer segment. Include adoption dynamics. Sales/usage 1st year, 2nd year, etc.

COST STRUCTURE

What are the costs (direct and overhead) incurred to operate the business model?

← EXTERNAL →   ← INTERNAL →

RISKS AND ASSUMPTIONS

What are the top three risks and assumptions

Figure 19.7: FEI canvas.

© 2014 Innosight LLC and Peter Koen.

Table 19.1: Attributes of Business Model Canvases

Attributes	Areas which are unique	Maurya's (2012) Lean Canvas	FEI Canvas
Major Focus	Sustaining projects	Start-ups	Transformational and disruptive innovation in large enterprises.
Key Partners	1. Who are the key partners, suppliers? What key resources and activities are we acquiring from the partners?	Missing since the start-up should first focus on customers rather than partners.	Partners are included as part of the redefined key processes box.
Key Activities	2. What are the key activities that our value proposition, distribution channels, customer relationships, and revenue streams require?	Missing since the key activities can be determined once you know the solutions.	Key activities required to accomplish the business model are embedded in the other elements of the canvas.
Key Resources	3. What resources do our value proposition, distribution channels, customer relationships, and revenue streams require?	Replaced by Unfair Advantage box since many key resources—but not all—create competitive advantage.	1. Key resources needed to deliver the customer value proposition (CVP).
Value Proposition	4. What customer value do we deliver? What problems are we solving? What solutions are we offering? What customer needs are we satisfying?	1. Value proposition: restated in terms of a compelling message that states why you are different and worth paying attention to.	The value proposition is the CVP, which is captured in elements 1 through 8.
Customer Relationships	5. What type of relationships do our customer segments expect?	Captured in the customer segment box.	
Channels	6. Through which channels do our customer segments want to be reached?	2. Channels	2. Channels
Customer Segments	7. Who are we creating value for, and who are our most important customers?	3. Customer segments	3. Formulated as customer circumstance
Cost Structure	8. What are the most important costs inherent in our business model?	4. Cost structure	4. Cost structure
Revenue Streams	9. What are our customers willing to pay?	5. Revenue streams	5. Revenue streams and adoption

(continued)

Table 19.1: (continued)

Attributes	Areas which are unique	Maurya's (2012) Lean Canvas	FEI Canvas
Unique to both Maurya Lean Canvas and FEI Canvas			
Problem	What is the problem you are solving?	6. Problem, separate box highlights fact that most start-ups fail because they fail to understand what problem they are solving.	6. Problem, formulated as either a POV or "job to be done" statement.
Solution	What is the solution?	7. Solution; broken out from the problem and value proposition boxes to help teams focus.	7. Solution
Key Metrics	Defines the key metrics that the start-up should be addressing.	8. Key metrics; encourages selection of three key metrics to foster focus.	Missing since this is not sufficiently important for enterprises.
Unfair Advantage	Competitive advantage or barriers to entry.	9. Unfair advantage: elements of advantage (or other firms' advantage) that can't be easily copied or bought.	8. Competition and barriers
Unique to FEI Canvas			
Key Processes	These are the key processes that a company uses to deliver its customer value proposition in a sustainable, repeatable, scalable, and manageable way.		9. Key processes— processes that are unique to the corporation and needed to deliver the value proposition and enable competitive advantage
Solution Attributes	What are the attributes which you need to deliver to the customer? Which problems are you solving with the attributes?		10. Customer attributes—separates solution attributes from the solution
Payment Structure	What is the price and how does the customer pay for the solution?		11. Payment structure
Risks and Assumptions	What are the top three risks and assumptions?		12. Risks and assumptions—All FEI projects have risks and assumptions that must be made explicit

who has the problem (i.e., who the customer is), and map the value proposition or the attributes required in the solution. Osterwalder and Pigneur's original business model canvas was designed to be used in a sustaining business, where it is less important to define the problem. This could limit its use as a brainstorming tool in transformational and disruptive innovations, where it is critical for teams to be able to work on problem, the customer, the solution attributes, and the solution separately. In contrast, the lean and FEI canvases separate out these four human-centered design attributes into separate areas.

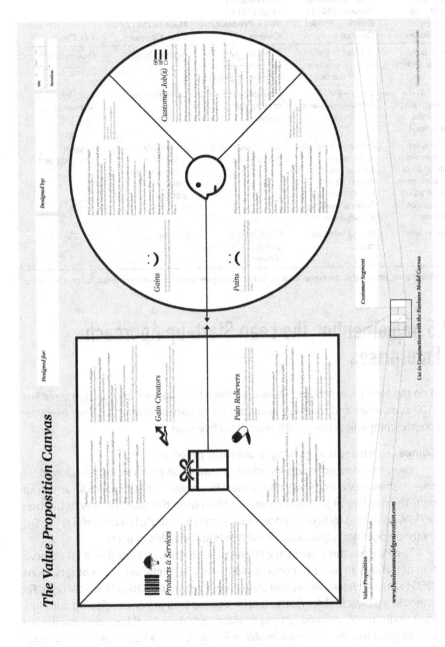

Figure 19.8: Value proposition canvas.
Source: Osterwalder, Pigneur, Bernarda, & Smith (2014).

Table 19.2: Comparison of Human-Centered Design Attributes with the Different Business Model Canvas

Human-Centered Design Attributes	Osterwalder & Pigneur (2010) Business Model Canvas	Osterwalder et al. (2014) Value Proposition Canvas	Maurya (2012) Lean Canvas	FEI Canvas
What is the right problem?	Included in the value proposition part of the canvas	Customer jobs, included as part of customer segments	Problem box	Problem box
Who has the problem (i.e., who is the customer)?	Captured in customer segments	Expanded definition of customer segments	Customer box	Customer Segments box
What is the value to the user to solve the problem (i.e., what are the solution attributes)?	Presumably included in the value proposition box, though it's not exactly clear what "value proposition" encompasses	Gain creators and pain relievers	Presumably included in the value proposition box, though it's not specifically identified as such.	Solution Attributes box
The solution	Missing from the canvas	Highlighted as products and services	Solution box	Solution box

Note: Shaded areas indicate that the canvas has a separate box congruent with the human-centered design attribute.

19.5 Implementing the Lean Start-up Approach in Enterprises

Based on the author's experience implementing a lean start-up approach in three Fortune 100 companies and teaching lean start-up as part of several 14-week executive MBA course, companies consistently stumble in five ways:

1. **Companies struggle at getting to the right problem.**

    Even experienced teams are often unsure what problem they were working on—even as they are typically clear about the unmet customer needs and the solution. The practice of formulating the problem from the point of view of the user, or POV, promoted by IDEO's process (Bootcamp Bootleg[4]), is a powerful reframing methodology that is grounded in the needs and insights of users.

    The POV has three elements: (1) the user, (2) the user's need, and (3) observation of the user in his or her environment and interpretation of the observations. IDEO teams often take weeks and sometimes even months to get the POV right. For example, a typical problem statement for a group working on developing nutritious food might be "A teenage girl needs more nutritious food because vitamins are vital to good health." The same problem formulated as a POV could be "A teenage girl with a bleak outlook needs to feel socially accepted when eating healthy food

[4]http://dschool.stanford.edu/wp-content/uploads/2011/03/BootcampBootleg2010v2SLIM.pdf

because in her group a social risk is more dangerous than a health risk" (Bootcamp Bootleg, 2010, p. 21.) The first formulation is a statement of fact, while the second POV formulation is an actionable description that drives empathy, provides direction for the effort to develop solutions, and serves as a defining vision for the team.

2. **Companies often confuse solution attributes with the solution.**

    It is difficult to separate out solution attributes without falling into the trap of talking about the value of different solutions. The use of a solution attributes map, illustrated in Figure 19.9, can keep teams from falling into this trap. In the example diagram, which offers a hypothetical map for a single-use coffee product, the four key solution attributes are coffee taste, ready to drink time, time to clean, and easy to use. The map illustrates how each competitor measures up on each attribute and assesses the relative importance of each attribute to the user. In the example, the attributes, competitor ratings, and relative importance ratings are all illustrative; in actual use, these factors would be derived from customer feedback.

3. **Teams focus on the wrong customers.**

    In almost all of the projects the author worked with, teams interviewed routine customers rather than lead users or early adopters. Routine customers typically want the same product or service they are currently using with higher performance or at a lower cost; they typically don't see the value of a transformational or disruptive innovation. Steelcase made this error in developing their Aero chair, which eventually turned out to be one of their most successful products. Many of the company's mainstream customers disliked the new chair's design, commenting that it looked like a lawn chair skeleton that was yet to be finished. The chair found an audience among customers who had difficulty being

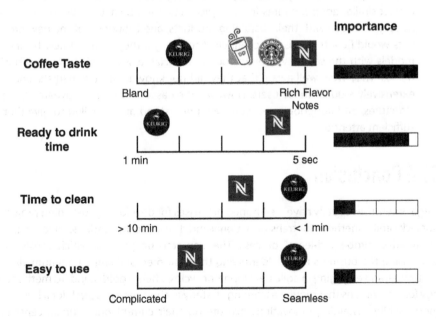

Figure 19.9: Solution attributes map.

comfortable in the existing chairs, some of whom had back problems—in other words, the users with the biggest problems unaddressed by current solutions.

4. **Most teams envision the prototype as a fully featured solution.**

In most cases, team members wanted to show potential customers a fully featured prototype, presumably to avoid embarrassing themselves or offending their users. As one team remarked, "How can we show this very rough prototype to an experienced surgeon? After all, we are a high-quality medical device company." Teams had difficulty understanding that the value of the prototype was to invite conversation and feedback. Proponents of design thinking advocate low-resolution prototypes made up of paper, pipe cleaners, cardboard, and Lego bricks to rapidly depict the solution along a tangible dimension. The objective of the prototype is to test particular solution attributes of the product being developed, *not* to offer a realistic model of the final product.

5. **Teams consistently make incorrect assumptions about channels, cost structure, and adoption rates.**

Based on an in-depth retrospective study of three large enterprises developing business models outside their core, Bertels, Koen, and Elsum (2015) identify three components of the new business model that are most susceptible to false assumptions: channels, cost structures, and product adoption rates. The enterprises had fewer false assumptions in other areas of the canvas, primarily because these changes are relatively easy to identify and firms can, with effort, resolve known uncertainties. For example, one of the new businesses studied involved a large change from the traditional market; the company spent six months conducting sophisticated ethnographic studies to determine the needs of the market. However, companies had ingrained ways of thinking about cost structures, tended to expect similar adoption rates for new products, even breakthrough innovations, as they had seen with their sustaining products, and thought that the new products would fit within existing channels. Accordingly, they adopted new business models with the same overhead structure associated with their sustaining businesses. They were well aware that they did not know their new markets, and so extensively studied those users. However, they assumed channel dynamics, cost structures, and adoption rates were well understood and so failed to give them sufficient attention.

19.6 Conclusion

Large enterprises usually have well-honed processes for developing sustaining projects but lack similar methods for transformational and disruptive innovations, which require an iterative "probe-and-learn" process. The lean start-up process, which consists of developing the business model, identifying the customer, building a minimum viable prototype, and engaging in agile development cycles, offers a gold-standard methodology for innovations that require a learning strategy as they need to search for a business model while sustaining innovations execute on their current one. Human-centered design, which at its root focuses on solving problems by matching needs with what is technologically feasible, moves toward these goals through an iterative approach involving customer empathy and the use of simple prototypes; this iterative approach embodies many of the characteristics of the lean start-up methodology. Just as the lean start-up process focuses on the business model, the human-centered design approach begins with a focus on the problem, building its exploration around four key questions: What is the business problem? Who has the problem? What is the value to the user in solving the problem? What are the attributes of the solution?

The business model canvases used in the lean start-up process accommodate these questions to varying degrees. The original, and very popular, business model canvas (Osterwalder & Pigneur, 2010) does not allow teams to separate out these areas, although the new Osterwalder and colleagues' (2014) value proposition canvas does. The lean canvas (Maurya, 2012), which was developed specifically for start-ups, separates out the first two items, and the FEI canvas, which was developed to support the FEI in large enterprises, offers separate spaces for all of them.

Large enterprises implementing a lean start-up approach struggle in five areas: getting to the right problem; focusing on the right customers; separating solution attributes from the solution; envisioning the minimum viable prototype; and questioning assumptions around channels, cost structure, and adoption rates for the new innovation. The lean start-up process has the potential to become the gold standard project management process for transformational and disruptive innovations in much the same way that the Stage-Gate process is the gold standard process for sustaining innovations.

References

Ajamian, G. M., & Koen, P. A. (2002). Technology Stage-Gate: A structured process for managing high-risk new technology projects. In P. Belliveau, A. Griffin, & S. M. Somermeyer (Eds.), *The PDMA toolbook 1 for new product development* (pp. 267–295). New York, NY: Wiley.

Bellman, R., Clark, C. E., Malcom, D. G., Craft, C. J., & Ricciardi, F. M. (1957). On the construction of a multi-stage, multi-person business game. *Operations Research*, 5(4), 469–503.

Bertels, H., Koen, P. A., & Elsum, I. (2015, March–April). Business models outside the core: Lessons learned from success and failure. *Research-Technology Management* 58(2), 20–29.

Blank, S. (2013a, May). Why the lean startup changes everything. *Harvard Business Review*, 91, 563–572.

Blank, S. (2013b). An MVP is not a cheaper product, it's about smart learning. Retrieved July 22, 2013, from http://steveblank.com/2013/07/22/an-mvp-is-not-a-cheaper-product-its-about-smart-learning/

Bootcamp Bootleg (2010). Retrieved July 22, 2013 from, http://dschool.stanford.edu/wp-content/uploads/2011/03/BootcampBootleg2010v2SLIM.pdf

Brown, B., & Anthony, S, (2011, June). How P&G tripled its innovation success rate. *Harvard Business Review*, 89(6), 64–72.

Christensen, C., & Raynor, M. E. (2003). *The innovator's solution*. Boston, MA: Harvard Business School Press.

Edwards, D. (2012). Innovation adventures beyond on the core. *Research-Technology Management*, 55(6), 33–41.

Koen, P. A., Bertels, H., & Elsum, I. R. (2011). The three faces of business model innovation: Challenges for established firms. *Research-Technology Management*, 55(3), 52–59.

Lynn, G. S., Morone, J. G., & Paulson, A. S. (1996). Marketing and discontinuous innovation: The probe and learn process. *California Management Review*, 38(3), 8–37.

Maurya, A. (2011). Why lean canvas vs business model canvas? Practice trumps theory. Retrieved July 22, 2013, from http://practicetrumpstheory.com/why-lean-canvas/

Maurya, A. (2012). *Running lean*. O'Reilly, Sebastopol, CA.

Osterwalder, A., & Pigneur, Y. (2010). Business model generation. Hoboken, NJ: Wiley.

Osterwalder, A., Pigneur, Y., Bernarda, G., & Smith, A. (2014). Value proposition design: How to create products and services customers want. Hoboken, NJ: Wiley.

Reis, E. (2011). The lean startup. New York, NY: Crown.

Tufte, E. R. (1997). Visual explanations. Cheshire, CT: Graphics Press.

von Hippel, E. (1986). Lead users: A source of novel product concepts. Management Science, 32(7), 791–805.

Wirtz, B. W. (2011). *Business model management*. Gabler Verlag, Springer Fachmedien Wiesbaden GmbH.

About the Author

PETER KOEN is an Associate Professor in the School of Business at Stevens Institute of Technology in Hoboken, New Jersey. He is also currently the Director of the Consortium for Corporate Entrepreneurship (CCE), which he founded in 1998, whose mission is to significantly increase the number, speed, and success probability of highly profitable products and services at the "Front End of Innovation" (www.frontendinnovation.com). Current consortium members include 3M, Corning, ExxonMobil, Goodyear, Intel, P&G, and WL Gore. Peter has also extensively published articles on the front end and founded the popular practitioner front end conference, which is now in its 12th year in the United States, and coined the term FEI. He has 19 years of industrial experience. His academic background includes a BS and MS in mechanical engineering from NYU and a PhD in biomedical engineering from Drexel (peter.koen@stevens.edu).

Part IV

CONSUMER RESPONSES AND VALUES

20

CONSUMER RESPONSE TO PRODUCT FORM[1]

Mariëlle E. H. Creusen
Delft University of Technology

Introduction

The appearance of a product communicates product information, and also has value for consumers in itself (Bloch, 1995). Product form or appearance refers to the visual exterior design of a product and is often the first information that people perceive about a product. The appearance of a product should therefore attract consumers and communicate the right impression about other product attributes. The visual appearance of products can also be used to express the values of the brand. Ideally, the product form should be recognized as an integral aspect of the product and be addressed early on in the development process. A design thinking approach explicitly includes customer needs—including more elusive ones—in the product development process by iteratively converging on concepts with a high likelihood of market success. Prototypes are made and iterated on based on external feedback (e.g., from consumers), including the appearance of the product in the iteration toward a viable product.

This chapter offers insights that help in making strategic decisions about the appearance of a product. In addition, these insights are useful in creating designs to test with customers, and in interpreting customer feedback in the Create and Evaluate modes of the design thinking framework (see Chapter 1). The focus in this chapter will be on tangible goods. The chapter starts with an overview of the ways in which the visual appearance of a product influences consumer product perception and preference. Next, the chapter will shed light on how product form characteristics, such as shape and color, influence consumer perception. This helps in engendering certain impressions or influences of product form. The chapter also covers factors related to the product, consumer, and context that influence the way in which product form impacts consumers. This helps

[1] Acknowledgment: Thanks to Jan Schoormans, Delft University of Technology, for his useful comments on this chapter.

303

managers in determining what a new product should look like. The chapter ends with implications for the practice of new product development.

20.1 How Product Form Influences Consumer Product Evaluation

The different ways in which the appearance of a product can influence consumers are summarized in six "roles of product appearance" (Creusen & Schoormans, 2005). Product form can *provide aesthetic value* to consumers. The appearance of a product may attract consumers or repel them. Second, the appearance of a product can *provide symbolic value* to consumers. People often choose product forms that fit with or express their personality, as products can look serious, playful, or masculine. For example, deodorants for men express masculinity by their dark colors, while deodorants for females often have softer colors. In addition, a unique design can be used to convey social status, such as in the case of an exclusive handbag.

The appearance of a product also communicates impressions about functional types of product value. Such impressions influence feature judgments even after more objective feature information is provided (Hoegg & Alba, 2011). By looking at a product, consumers form an impression about the *ease of use* and *functionalities*. For example, a product with a small number of buttons looks easy to operate, while many buttons seems to indicate many functionalities. In addition, the appearance of a product can communicate *high-performance quality* to consumers. For example, a black coffee maker with metal parts looks of higher quality than a white plastic one. Such impressions about ease of use, functionalities, and quality may or may not be correct. In any case, it is useful for companies to be aware of the inferences people make by just looking at their product.

The exterior design of a product can *draw attention*, for example, in a retail environment. A design that differs from other designs within that product category attracts consumer attention, such as the Philips Alessi coffeemaker did at the time of its introduction (see Figure 20.1). However, when a product looks too different from other products, people may not recognize it for what it is, for example, a coffeemaker. This brings us to the ease with which a product can be identified, and whether it will be categorized into a new subcategory, setting it apart from other products in the category. The *categorization* of a product can be influenced by its visual appearance. For example, the distinctive design of the Dyson bladeless fan promotes that people categorize it as a new kind of fan (see Figure 20.2). And vegetarian alternatives often look similar to meat products so that consumers will consider them as an alternative.

For managers, it is important to know which of these "product appearance roles" will be influential in their market, and some guidelines are given later in this chapter. The next section shows how specific product form characteristics influence the roles of product appearance for consumers. This helps in designing a product form that engenders certain impressions or influences.

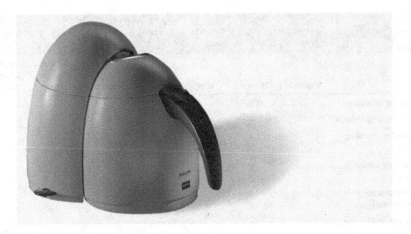

Figure 20.1: Philips Alessi coffeemaker.
©Philips.

Figure 20.2: Dyson Cool™ tower fan.
Courtesy of Dyson.

20.2 Product Form Characteristics and Consumer Perceptions

A lot of research has focused on determining design factors that influence the aesthetic attractiveness of objects and products, such as visual unity, visual complexity, symmetry, visual typicality, size, and color. However, the influence of these design factors on

Table 20.1: The Influence of Product Form Characteristics on the Different Product Appearance Roles

	Typicality	Novelty	Unity	Com-plexity	Symmetry	Good Propor-tion	Size/ Shape/ Color
Drawing attention	−	+					
Ease of categorization	+	−					
Providing aesthetic value	+	+	+	−	+	+	
Providing symbolic value		+					
Communicating functionality		+					
Communicating ease of use	−				−		
Quality impression		+	+	+			

Note: "+" indicates that a positive influence has been found, "−" indicates that a negative influence has been found, and empty shaded cells indicate that the influence depends on the execution of these factors (see the text).

consumer perception of other types of product value, such as performance quality and ease of use, has received far less attention. For managers, it is important to consider the influence of product form characteristics on all relevant types of product value, not only aesthetic appeal. For example, a smaller size and fewer buttons may increase the aesthetic attractiveness of a product but may also decrease its perceived performance and ease of use.

The influence of some important visual design factors on consumer product perceptions is described below (see Table 20.1 for an overview).

Visual Typicality and Novelty

Visual typicality is the similarity to the appearance that most consumers associate with the product category (Garber, 1995). For example, a chair with four legs looks more typical than a one-legged chair does. Visual typicality has been found to positively influence aesthetic preference. *Visual novelty* pertains to the originality of a design. Visual typicality and novelty are negatively correlated, and people tend to prefer designs with an optimal combination of both, such as a table lamp with a typical overall shape but novel material (Hekkert, Snelders, & van Wieringen, 2003). Such products look familiar but also slightly different and thereby interesting.

Although in general visual typicality with a touch of novelty seems to be preferred, for some products or people a more distinctive design is better. This is the case for products for which prestige or exclusiveness is important, such as sports cars. Also, people with a high need to distinguish themselves (i.e., a high "need for uniqueness") tend to prefer atypical designs. An atypical appearance can also help to differentiate products from competitors or from a negative category image, such as for a wheelchair. A distinctive appearance can help in communicating new functional attributes, as in the Dyson bladeless fan (see Figure 20.2). In addition, visual novelty tends to lower perceptions of usability and heighten perceptions of performance quality (Mugge & Schoormans, 2012a, 2012b).

Often to the frustration of designers, consumers tend to dislike novel designs at first sight, and react negatively to those in concept tests. Repeated exposure increases ease of processing and thereby aesthetic liking (Reber, Winkielman, & Schwarz, 1998). Indeed, repeated exposure increases the perceived attractiveness of innovative designs, but not of more typical designs (Carbon & Leder, 2005). So in order to get a valid assessment of consumer liking, companies should use repeated exposure or allow consumers time to get used to novel designs.

Visual Design Principles

General visual design principles such as complexity, symmetry, unity, and proportion influence aesthetic preference. Research shows that, in general, people aesthetically prefer low (but not too low) complexity, high symmetry, high unity, and good proportions. Unity indicates a congruity among the elements in a design. The proportion of the length to width of a product can influence purchase intentions (Raghubir & Greenleaf, 2006). Aesthetic preference for specific proportions (e.g., the "golden ratio") has been proposed, but there is little evidence for this; the proportions that consumers value depend on the kind of product.

Next to aesthetic preference, these visual design principles also influence the perception of other types of product value. For example, higher visual *unity* in a design increases perceived product quality (Veryzer & Hutchinson, 1998), as does higher visual *complexity* (Creusen, Veryzer, & Schoormans, 2010). Visual complexity indicates functional complexity to consumers. In addition, visual complexity lowers perceived ease of use when people want few functionalities (no "bells and whistles"), but increases usability impressions for people desiring many functionalities. The level of visual complexity in a design should be determined with care, as people generally dislike complexity. Lower *symmetry* increases the perceived ease of use of a product, probably because differentiation in button placing, shape, and size—that is, less symmetry—helps the user to distinguish these buttons in use (Creusen et al., 2010).

Size, Shape, and Color

The influence of *shape* on product value perceptions has been shown in the previous section about visual design principles, but there are more ways in which shape influences perceptions. For example, product or package shape influences the perception of stability (Murdoch & Flurscheim, 1983), and thereby perceived ease of use. For instance, a tapered form with a large base looks stable compared to a small and tall product. Although the product is designed to be stable, consumers might conclude it is not after seeing it, and choose another product. Curved products are in general preferred to *angular* ones (e.g., Bar & Neta, 2006), although such preferences may change over time. For example, cars shifted from angular shapes in the 1980s to more organic shapes in the 1990s and beyond. However, square or angular products more easily fit into a corner than rounded forms, and may be preferred for this ease of use-related reason (Creusen & Schoormans, 2005). Also, people attach symbolic associations to certain product shapes (Schmitt & Simonson, 1997). For example, rounded shapes tend to look soft and feminine, while angular and straight forms tend to look dynamic and masculine.

Large shapes are perceived as powerful and strong, while small shapes appear delicate and weak. In addition, bigger objects look heavier (Walker, Francis, & Walker, 2010). The way *size* is evaluated varies strongly with cultural and regional norms (Schmitt & Simonson, 1997). Product size also influences more indirect consequences of use, such as whether the product fits in a drawer, which could be important for consumers (Creusen & Schoormans, 2005). In addition, package container height influences volume perceptions of consumers (e.g., Raghubir & Krishna, 1999).

Color influences aesthetic appreciation, symbolic associations, ease of use, and quality perceptions. In addition, it can grab attention and is used to foster company and brand identity and recognition (Elliot & Maier, 2014; Schmitt & Simonson, 1997). An example of a usage-related perception is that darker objects look heavier than more brightly colored ones (Walker et al., 2010). In addition, buttons that contrast in color from a product's casing make it easier to locate controls, which might, for example, be important for an alarm clock. Bright or colorful packages can imply low quality, whereas the use of low saturated colors suggests higher quality (Scott & Vargas, 2007). The colors of food and their packages establish taste expectations (e.g., Hoegg & Alba, 2007). Some associations with color seem to be relatively constant, although the desirability and meaning of a color depend on the object (e.g., a coffeemaker or a table lamp) and the style of this object (e.g., modern or classic) (e.g., Labrecque, Patrick, & Milne, 2013). The effect of a color depends on the rest of the product, as an aesthetic judgment is found to be holistic. For example, the salmon pink color of the Philips Alessi coffeemaker (see Figure 20.1) suits this product, but not a more typically shaped coffeemaker. Furthermore, there are large differences in the experience of form and color between individuals, cultures, times, and contexts. So consumer associations with color should best be tested in the right context with the actual target group.

20.3 In What Way Will Product Form Impact Consumer Product Evaluation?

Several factors that influence the way in which product appearance influences consumer product evaluation will be treated below.

Product Category-Related Factors

The product aspects that are important to consumers, and thereby the influence of different roles of the product appearance, depend on the type and amount of consumers' purchase motivation and the social significance of the product category (see Table 20.2).

Type of purchase motivation. Two main types of purchase motivation can be distinguished, namely utilitarian and expressive motivation, of which the latter can be subdivided into hedonic and symbolic motivation (e.g., Park & Mittal, 1985). For a product that is mainly bought for *utilitarian* reasons, such as a power drill, functional

Table 20.2: Product Category-Related Factors and the Relative Importance of Different Product Appearance Roles

	Low-Involvement Product	Low Product Knowledge (Newness)	Expressive Purchase Motivation	Utilitarian Purchase Motivation	Socially Significant Products
Drawing attention	▓	▓			
Influencing categorization	▓	▓			
Providing aesthetic value			▓		▓
Providing symbolic value					▓
Communicating functionality				▓	
Communicating ease of use				▓	
Quality impression	▓	▓			

Note: Shaded cells indicate a higher relative influence.

performance is of main importance to the buyer. When *hedonic* motivation is the main reason for purchase, sensory enjoyment is important to consumers, such as when buying ice cream, a DVD, or a nice picture to put on the wall. *Symbolic* purchase motivation indicates the desire to enhance your self-esteem and/or project a desired image to others by means of the product, such as a watch or a handbag. Many products have both utilitarian and expressive significance for consumers. Think of a car: performance aspects, such as fuel consumption, and hedonic and symbolic aspects, such as an attractive styling that fits the kind of person you want to be, both play a role.

Importance of the product. Consumers are more involved in the purchase decision for a car or a pair of shoes than for a stapler or a carton of milk, and therefore will put more effort into making a decision about such a product. Consumer involvement differs between product categories that are low in purchase risk—such as consumer packaged goods—and more expensive or socially significant product categories. When the product is not important to them, consumers want to minimize their effort in making a purchase decision. *Attention drawing* and *ease of categorization* based on the product's appearance will be influential in such a case, as consumers will only look at product alternatives that either draw their attention or are easily identified because of their typical look (see Garber, 1995). People often buy the same brand out of habit, so for well-known brands it is not wise to change their design too much, as people might no longer recognize the product. This could lead to the choice of another brand. In addition, consumers with little interest or product knowledge in a certain product category tend to use easy-to-spot product characteristics, such as price or brand name, as cues for quality. The appearance of a product can be such a quality cue (Dawar & Parker, 1994). This means that product form is more often used as a cue for quality for products in which consumers are less involved or when they have little product knowledge, such as in the case of more radically new products. Because of their lack of interest or knowledge in interpreting more detailed information, the *impression about the product quality* that the appearance of the product gives consumers can be rather influential.

Social significance of products. Both expressive and functional product aspects are found to be more important for socially significant products (Creusen, 2010). These are products that are used in public rather than in private. For example, a car, chair, or coffeemaker can be seen by other people on the street or visiting your home. However, only few people will see your alarm clock, bathroom scale, or shaving device. The importance of *aesthetic and symbolic aspects, functionalities, ease of use,* and *quality* is found to be higher for socially significant products.

Product categories can be classified based on the general level of involvement and the extent to which this involvement is expressive and/or utilitarian (e.g., Ratchford, 1987; Voss, Spangenberg, & Grohmann, 2003). Examples of low-involvement utilitarian products are insecticide and paper towels. Some low-involvement expressive products are pizza and greeting cards. Cameras and washer/dryers are high-involvement utilitarian products, and high-involvement expressive products include sports cars and wallpaper. Although this gives some general idea of how product form will influence consumer evaluation (see Table 20.2), more specific insight is needed for utilitarian aspects. Product appearance can influence the perceived utilitarian product value by showing functional features, ease of use, and performance quality, and the relative importance of these aspects differs between product categories (Creusen, 2010).

The Type of Consumer

Personality characteristics, demographic characteristics, and the amount of product knowledge influence the product aspects that are important to consumers, and thereby the information that the appearance of a product should ideally provide (see Table 20.3). In addition, the type of purchase motivation and the importance of a certain

Table 20.3: Consumer Characteristics and the Relative Importance of Different Product Value Types

	Female	Age	Education	Income	CVPA	Need for Uniqueness
Aesthetic value	+	?			+	+
Symbolic value	+	- SSP only	?	+		+
Functional value	+	+		?		
Ease of use	+	+		+		
Quality		+	+	+		

Note: "+": a positive influence has been found; "-": a negative influence has been found; "?": different results have been found across studies/countries.
CVPA = centrality of visual product aesthetics; SSP = socially significant products.

product (treated in the previous section) also differ between individuals; some people are more involved in the purchase of a computer than others, and some people pay more attention to the aesthetic value of a product than others.

Personality characteristics. People differ in the importance they attach to visual product aesthetics, which can be assessed by the CVPA (centrality of visual product aesthetics) scale (Bloch, Brunel, & Arnold, 2003). People scoring higher on this scale (such as design professionals) attach more importance to the aesthetic value of a product and are more able to evaluate aesthetics (see Chapter 21). Another personality variable that influences the importance of aesthetic and symbolic product value is *need for uniqueness* (Hunt, Radford, & Evans, 2013). Consumers with high uniqueness needs want to feel distinct from others and often prefer distinctive and novel-looking designs (Bloch, 1995), thus paying more attention to the aesthetic and symbolic value of a product.

Demographic characteristics. The importance of several product aspects differs with gender, age, education level, and income.

Expressive product aspects are found to be more important for females (e.g., Williams, 2002). Females indicate a higher importance of the aesthetic attractiveness of products and of the product portraying the correct image to others or themselves (i.e., symbolic value) than males. In addition, females indicate that many functionalities and ease of use are more important than males do (Creusen, 2010).

Younger people are found to attach more importance to expressive product aspects (Henry, 2002), although Creusen (2010) showed that this effect seems to be restricted to the symbolic aspects of socially significant products. As said earlier, portraying the correct image (i.e., symbolic value) is more important for publicly used products, but this appears to be the case for younger people only; the image you portray apparently becomes less important when getting older. Furthermore, in general, many functionalities, ease of use, and high product quality seem to be more important to older people. For ease of use this is obvious, as cognitive and physical abilities diminish with age.

Different effects have been found for education level and relatedly for social class (for which education is a good predictor), which might be due to the different countries investigated. In Australia and the United States, higher social class is found to heighten attention to taste and self-expression in buying products (Henry, 2002; Holt, 1998). However, in the Netherlands, higher-educated people are found to attach less importance to symbolic product value and education does not seem to influence the importance of aesthetic aspects. In addition, higher-educated people indicate product quality to be more important. Also, for income the results differ. In the Netherlands, no effect of income level on the importance of aesthetic aspects was found, while higher income people paid more attention to symbolic product aspects, ease of use, product quality, and whether the product has many functionalities (Creusen, 2010). In the United States, increasing income was found to decrease attention to functional purchase criteria, especially for less socially relevant products (Williams, 2002). So the effects of education level and income are equivocal and may differ between countries/cultures.

Amount of product knowledge. Similar to involvement, a lack of product knowledge promotes the use of cues for quality, as consumers are less able to evaluate all product information. Therefore, the quality impression communicated through the appearance of a product will be more influential when consumers have less product knowledge or are less involved in the product category (see Table 20.2). Product knowledge differs between individual consumers depending on factors such as their interest in the product. In addition, product knowledge may differ between product categories; for example, consumer knowledge of a really new product will in general be low.

Brand Strength and Image

Brand strength is an important cue for product quality. Product appearance has a bigger influence on perceived product quality for a weak as opposed to a strong (i.e., well-known and positively valued) brand, implying that communicating high quality by means of the appearance of products is especially important for a weak as opposed to a strong brand (Page & Herr, 2002).

Product appearance is a powerful communicator of brand image and identity to consumers and can be used for brand identification. For example, many car brands are recognizable from their visual design as they use similar elements over subsequent models. Brands should strategically decide whether the visual appearance of a new product should be similar to other products in their portfolio, and similar to previous products of the brand (Person, Snelders, Karjalainen, & Schoormans, 2007). Creating visual brand recognition is more important for a strong brand than for a weak brand, as a strong brand wants to be easily recognized and to transfer the positive brand associations to new products.

The visual design of products can also be used to express the core values of a brand (e.g., Karjalainen & Snelders, 2010). Orth and Malkewitz (2008) distinguished several key types of holistic package designs that fit certain types of brands. For example, exciting brands should have contrasting designs, while sophisticated brands should have natural or delicate designs.

Phase of the Product Life Cycle

The way in which product form impacts consumers differs with the phase of the product life cycle (PLC) that the product is in (Bloch, 1995; Luh, 1994). During *introduction*, attracting consumer attention by using a fresh form may be essential. However, the appearance should not look too new as this makes it more difficult for consumers to categorize the product, which might have negative effects (Goode, Dahl, & Moreau, 2013). In this stage, the target market often comprises high-income pioneer users, who want the product to be visible and conspicuous to others (Luh, 1994). In addition, the design should communicate superior functionality and safe operation. As many consumers have low product knowledge, product form may have a stronger influence as a cue for quality (see Table 20.2). This means that a novel form will be beneficial in this stage (see Table 20.1). In the *growth* phase, functions become more standardized and criteria such as ease of use or quality may become more important. In addition, product form and styling should be acceptable to more mainstream consumers. In the *maturity* phase, differentiation becomes important. Design may become important in emphasizing performance improvements (Bloch, 1995) and there is an emphasis on intuitively understandable operation or on aesthetic value (Luh, 1994). The needs for self-expression, and thus the symbolic value of the appearance, will be more important. In the *decline* phase, most of the expectations from the maturity phase should be maintained (Luh, 1994).

Table 20.4 provides an overview of ways in which product form is likely to be influential in different stages of the PLC. An empty cell does not mean that this role is not important in that phase; the influence of different product appearance roles also depends on other factors (see the previous sections).

Culture and Time

Culture- and time-related differences have been found in symbolic associations and aesthetic preferences for product designs (Bloch, 1995; Crilly, Moultrie, & Clarkson, 2004; see also Chapter 21). However, the influence of culture and time on the perception of other types of product value is rarely investigated. There are probably greater differences in perceived aesthetic and symbolic product value between cultures and times than in perceptions of functional performance and ease of use, as these are less subjective (Creusen & Schoormans, 2005). For example, most people will agree that larger buttons are easier to operate. But although functional perceptions may be relatively similar between cultures and times, preferences may differ. For example, more buttons on a product tend to make it look more technologically advanced and less easy to use (Norman, 1988). Some cultures may have a greater preference for technologically advanced and complicated products than others, although not much research has been done in this area.

Table 20.4: The Role of Product Form in Different Phases of the Product Life Cycle			
	Introduction Phase	**Growth Phase**	**Maturity and Decline Phases**
Drawing attention	Unique design/fresh form		
Influencing categorization	Novel appearance		
Providing aesthetic value	Attract high-income pioneer users	Attract mainstream consumers	Offer differentiation
Providing symbolic value	Conspicuous consumption (status)		Opportunity for self-expression
Communicating functionality	Communicate superior function		Emphasize performance improvements
Communicating ease of use	Communicate safe operation	Communicate user friendliness	
Quality impression	Communicate quality		

Context Factors

The context of the purchase situation influences the relative importance of certain types of product value. For example, a matching environment can emphasize the aesthetic value of a product (e.g., Bloch, 1995) and heighten its importance for the consumer. Indeed, products are sometimes displayed in a matching environment with other products in the same style or colors, so that they look their best. It is therefore important to take the environment in which the product will be sold, including competitor products, into account in designing a product and its appearance. For example, in general, bright colors draw consumer attention. But when many competitors use bright colors, the use of darker colors might be a better way to draw consumer attention. In addition, categorization of a design can also be influenced by the context of product presentation—either in store or in advertising—as the context can, for example, influence how typical a design looks. An illustration is that typical-looking product designs are perceived as especially typical in an atypical context (Blijlevens, Gemser, & Mugge, 2012). Furthermore, the aesthetic context of one's home may influence the aesthetic value of a product for consumers, as they may want it to fit their home interior (e.g., Bloch, 1995). Someone may like the way a certain product looks, but not buy it because the colors do not fit in their home.

20.4 Practical Implications

The appearance of a product provides value to consumers and influences their perceptions on several product attributes. This chapter provides an overview of different ways in which product form or appearance influences consumers, namely drawing attention, influencing categorization, providing aesthetic value, providing symbolic value, communicating functional value, providing and communicating ease of use, and communicating quality (Bloch, 1995; Creusen & Schoormans, 2005). The influence of several product form characteristics on consumer perception of different types of product value is shown. In addition, an overview of factors that impact how product form influences consumers is provided.

The information presented in this chapter may help in generating a certain impression or attracting a certain target group with a design. For example, an impression of high quality can in general be engendered by a design that has some visual complexity, high unity, and a novel look. However, the influence of product form on consumers is difficult to predict, as the combined influence of the characteristics of a form cannot be foreseen. Designers are trained in creating aesthetics that are appealing and engender certain associations. However, in order to ensure that the appearance of a product has the intended effect on consumers, the appeal of a visual design and the associations and inferences it provokes should be checked with the target group. Influencing the aesthetic and symbolic value of the appearance of a product is especially challenging, as such aspects are more subjective and differ more strongly between cultures and times than the perception of functional and usability aspects does. In addition, aesthetic taste may differ between designers and consumers (see e.g., Crilly et al., 2004). For this reason, it is important for designers to immerse themselves in the context of the target group and its aesthetic and symbolic tastes. Developing personas that vividly describe different kinds of consumers might help in this (see Chapter 3). Furthermore, letting consumers indicate the kinds of products or packages that engender certain associations (e.g., "natural" or "masculine") might help designers in getting a feel for the kind of design that expresses certain associations. For some products, another approach is to rely on mass customization to let consumers themselves to some extent determine how the product looks aesthetically by using an online tool.

Ideally, the visual appearance should be recognized as an integral part of the product and be strategically considered from the start of the product development process. A design thinking approach is suited for this, as possible solutions are tested using consumer feedback and improved in following iterations. The intended market positioning of the product and the type of product value that will be the focus in designing the product and its visual appearance or package should be determined at the start of the development process, based either on consumer needs and preferences or on the strategic decisions of the company and the core values of the brand. In this way, the appearance is likely to fit the other elements of the marketing mix and to communicate the intended impressions to consumers, leading to a fit between the product, the target group and the intended market positioning.

References

Bar, M., & Neta, M. (2006). Humans prefer curved visual objects. *Psychological Science* 17(8), 645–648.

Blijlevens, J., Gemser, G., & Mugge, R. (2012). The importance of being "well-placed": The influence of context on perceived typicality and esthetic appraisal of product appearance. *Acta Psychologica*, 139(1), 178–186.

Bloch, P. H. (1995). Seeking the ideal form: Product design and consumer response. *Journal of Marketing*, 59, 16–29.

Bloch, P. H., Brunel, F. F., & Arnold, T. J. (2003). Individual differences in the centrality of visual product aesthetics: Concept and measurement. *Journal of Consumer Research*, 29(4), 551–565.

Carbon, C. C., & Leder, H. (2005). The repeated evaluation technique (RET). A method to capture dynamic effects of innovativeness and attractiveness. *Applied Cognitive Psychology*, 19, 587–601.

Creusen, M. E. H. (2010). The importance of product aspects in choice: The influence of demographic characteristics. *Journal of Consumer Marketing*, 27(1), 26–34.

Creusen, M. E. H., & Schoormans, J. P. L. (2005). The different roles of product appearance in consumer choice. *Journal of Product Innovation Management*, 22(1), 63–81.

Creusen, M. E. H., Veryzer, R.W., & Schoormans, J. P. L. (2010). Product value importance and consumer preference for visual complexity and symmetry. *European Journal of Marketing*, 44(9/10), 1437–1452.

Crilly, N., Moultrie, J., & Clarkson, P. J. (2004). Seeing things: Consumer response to the visual domain in product design. *Design Studies*, 25, 547–577.

Dawar, N., & Parker, P. (1994). Marketing universals: Consumers' use of brand name, price, physical appearance, and retailer reputation as signals of product quality. *Journal of Marketing*, 58, 81–95.

Elliot, A. J., & Maier, M. A. (2014). Color psychology: Effects of perceiving color on psychological functioning in humans. *Annual Review of Psychology*, 65, 95–120.

Garber, L. L. (1995). The package appearance in choice. *Advances in Consumer Research*, 22(1), 653–660.

Goode, M. R., Dahl, D. W., & Moreau, C. P. (2013). Innovation aesthetics: The relationship between category cues, categorization certainty, and newness perceptions. *Journal of Product Innovation Management*, 30(2), 192–208.

Hekkert, P., Snelders, D., & van Wieringen, P. C. W. (2003). "Most advanced, yet acceptable": Typicality and novelty as joint predictors of aesthetic preference in industrial design. *British Journal of Psychology*, 94, 111–124.

Henry, P. (2002). Systematic variation in purchase orientations across social classes. *Journal of Consumer Marketing*, 19(5), 424–438.

Hoegg, J., & Alba, J. W. (2007). Taste perception: more than meets the tongue. *Journal of Consumer Research*, 33(4), 490–498.

Hoegg, J., & Alba, J. W. (2011). Seeing is believing (too much): The influence of product form on perceptions of functional performance. *Journal of Product Innovation Management*, 28, 346–359.

Holt, D. B. (1998). Does cultural capital structure American consumption? *Journal of Consumer Research*, 25(1), 1–25.

Hunt, D. M., Radford, S. K., & Evans, K. R. (2013). Individual differences in consumer value for mass customized products. *Journal of Consumer Behaviour*, 12(4), 327–336.

Karjalainen, T. M., & Snelders, D. (2010). Designing visual recognition for the brand. *Journal of Product Innovation Management*, 27(1), 6–22.

Labrecque, L. I., Patrick, V. M., & Milne, G. R. (2013). The marketers' prismatic palette: A review of color research and future directions. *Psychology & Marketing*, 30(2), 187–202.

Luh, D. (1994). The development of psychological indexes for product design and the concepts for product phases. *Design Management Journal (Former Series)*, 5(1), 30–39.

Mugge, R., & Schoormans, J. P. L. (2012a). Newer is better! The influence of a novel appearance on the perceived performance quality of products. *Journal of Engineering Design*, 23(6), 469–484.

Mugge, R., & Schoormans, J. P. L. (2012b). Product design and apparent usability. The influence of novelty in product appearance. *Applied Ergonomics*, 43(6), 1081–1088.

Murdoch, P., & Flurscheim, C. H. (1983). Form. In C. H. Flurscheim (Ed.), *Industrial design in engineering* (pp. 105–131). Worcester, England: The Design Council.

Norman, D. A. (1988). *The psychology of everyday things*. New York, NY: Basic Books.

Orth, U. R., & Malkewitz, K. (2008). Holistic package design and consumer brand impressions. *Journal of Marketing*, 72(3), 64–81.

Page, C., & Herr, P. M. (2002). An investigation of the processes by which product design and brand strength interact to determine initial affect and quality judgments. *Journal of Consumer Psychology*, 12(2), 133–147.

Park, C. W., & Mittal, B. (1985). A theory of involvement in consumer behavior: Problems and issues. *Research in Consumer Behavior*, 1, 201–231.

Person, O., Snelders, D., Karjalainen, T. M., & Schoormans, J. P. L. (2007). Complementing intuition: Insights on styling as a strategic tool. *Journal of Marketing Management*, 23(9–10), 901–916.

Raghubir, P., & Greenleaf, E. A. (2006). Ratios in proportion: What should the shape of the package be? *Journal of Marketing*, 70(2), 95–107.

Raghubir, P., & Krishna, A. (1999). Vital dimensions in volume perception: Can the eye fool the stomach? *Journal of Marketing Research*, 36(3), 313–326.

Ratchford, B. T. (1987, August/September). New insights about the FCB grid. *Journal of Advertising Research*, 27(4), 24–38.

Reber, R., Winkielman, P., & Schwarz, N. (1998). Effects of perceptual fluency on affective judgments. *Psychological Science*, 9(1), 45–48.

Schmitt, B. H., & Simonson, A. (1997). *Marketing aesthetics: The strategic management of brands, identity, and image*. New York, NY: Free Press.

Scott, L. M., & Vargas, P. (2007). Writing with pictures: Toward a unifying theory of consumer response to images. *Journal of Consumer Research*, 34(3), 341–356.

Veryzer, R. W. Jr., & Hutchinson, J. W. (1998). The influence of unity and prototypicality on aesthetic responses to new product designs. *Journal of Consumer Research*, 24, 374–394.

Voss, K. E., Spangenberg, E. R., & Grohmann, B. (2003). Measuring the hedonic and utilitarian dimensions of consumer attitude. *Journal of Marketing Research*, 40(3), 310–320.

Walker, P., Francis, B. J., & Walker, L. (2010). The brightness-weight illusion: Darker objects look heavier but feel lighter. *Experimental Psychology*, 57(6), 462–469.

Williams, T. G. (2002). Social class influences on purchase evaluation criteria. *Journal of Consumer Marketing*, 19(3), 249–276.

About the Author

MARIËLLE E. H. CREUSEN is an Assistant Professor of Marketing and Consumer Research in the Faculty of Industrial Design Engineering at Delft University of Technology in the Netherlands. Her research interests include consumer research methods in new product development and consumer response to product design. She has published in journals such as the *International Journal of Research in Marketing, Journal of Product Innovation Management, European Journal of Marketing*, and *International Journal of Design*.

21

DRIVERS OF DIVERSITY IN CONSUMERS' AESTHETIC RESPONSE TO PRODUCT DESIGN

Adèle Gruen

Introduction

When asking around what people think the main difference between art and product design is, one answer that you will often come across is that design is an activity with a purpose, and the purpose of a product is to be used. This basic idea has paved the way to the movement of user centered design, which invites product developers to reflect on who will be using their creation, how will they use it, for what purpose, when, and with whom. Designers are expected to think of how to create positive experiences between the user and the product, not just how to create beautiful products. User knowledge has thus become central in the process of product development. In the field of marketing research, academics have been studying consumers (i.e., product users) for more than a century. Looking at research developments in this discipline is interesting for product developers, as they will find theories, tools, and examples on the understanding of consumer behavior. Looking in the direction of marketing academic research can benefit not only designers, but also design managers, product developers, product managers, or anyone who wants to develop a product for a targeted consumer!

Design thinking theorists encourage new product developers to be concerned about and to focus on user experience. A great part of that experience depends on the aesthetics of the product. Aesthetic considerations therefore have their rightful place in design thinking, especially regarding user research. It is important to identify the variety of consumers and consumption contexts that can impact preferences overall, including aesthetic preferences. Consumer aesthetic preferences are strong determinants of future approach or avoidance behaviors toward the product (Bloch, 1995). That is why when analyzing feedback on a prototype, for instance, we understand how crucial it is to reflect on aesthetic preferences.

In this chapter we look at what marketing can tell us about the diversity of consumers' aesthetic responses. Aesthetics can be defined as "a sensitivity to the beautiful or a branch of philosophy that provides a theory of the beautiful and of the fine arts" (Veryzer, 1993). Aesthetic response refers to the reaction a person has to an object (e. g., product) based on his or her perception of the object (Berlyne, 1974). Veryzer (1993) goes further by stating that aesthetic response can be a reaction to conscious or unconscious aspects (i.e., stimulus, such as color, shape) of the product and that it leads to the registering of affect or pleasure (prior to buying the product, for instance).

Chapter 20 addressed consumer responses to product designs that are shared by most consumers. In this complementary chapter we will try to provide our readership with a (nonexhaustive) list of the drivers of differences in consumer's response, with a focus on aesthetic preferences. Why don't we all like the same designs? The chapter starts with the broad influence of culture (the culture of a country, the consumer culture, and the culture of class), before narrowing down the topic to individualities (personality, taste), and concludes with the importance of situational factors (Figure 21.1).

21.1 Culture

National and Regional Cultures

Culture is a lens through which people view a phenomenon, apprehend and assimilate it (McCracken, 1986). It has been said that beauty is partially in the eyes of the culture

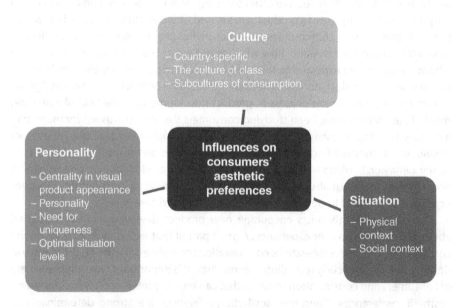

Figure 21.1: Summary of the main influences on consumers' aesthetic preferences.

and not of the individual (Berlyne, 1971). Cultures influence how people see the world and impact how they visually appreciate the design of a product. Thus, the culture of a given individual is likely to significantly guide his or her aesthetic tastes.

Hofstede (1980) proposed a tool to classify cultures according to their values. This helps us to understand how cultures can systematically differ in terms of aesthetic preferences. Below, I present three cultural dimensions that vary across cultures and that may influence an individual's aesthetic response[1] (Table 21.1):

- **Individualism/collectivism.** Western countries are traditionally more individualistic cultures (North America and Europe), whereas Eastern and African countries tend to be more collectivistic.
- **Masculinity/femininity.** Masculine cultures value competitiveness and materialism (Japan is one of the most masculine country). Feminine cultures value quality of life and happiness (the Nordic countries are known to be highly feminine countries in this respect).
- **Long-term orientation/short-term orientation.** This dimension refers to the extent to which a country gives importance to the past and tradition. Countries that score high on this dimension will tend to have a pragmatic approach and encourage modernism. On the opposite countries that score low prefer to maintain traditions and norms.

In Japan and China, designs that foster social harmony will be appreciated more because they fit with the collectivist view of these countries. For instance, symmetric and balanced designs will be favored because these are factors for harmonious design (Henderson, Cote, Leong, & Schmitt, 2003). In individualistic countries such as the United States (which scores 100 out of 100 on this value!), creativity is highly valued. In this context, new, innovative designs will have a better chance of being accepted (Henderson et al., 2003; Schmitt & Simonson, 1997). Time orientation also can influence people's visual tastes. Countries that give a lot of importance to their history and that value their past, such as China, and also Europe ("the old continent"), will place higher value on traditional designs with long life expectancy (think of the longevity and success of the Channel No. 5 fragrance for instance).

Henderson et al. (2003) also found that American consumers tended to favor angular shapes, whereas their Asian counterparts favor rounded shapes. An explanation of this fact can be found in the work of Zhang, Feick, and Price (2006). Zhang et al. found that rounded versus angular shapes preferences could be linked to self-construal. More specifically, an independent self-construal (i.e., the self is defined as independent of others) is associated with confrontation, whereas an interdependent self-construal (i.e., the self is defined as interdependent with others) is associated with conflict avoidance. Independent individuals who value confrontation perceive angular shapes as more attractive and rounded shapes as less attractive than individuals with interdependent self-construal do. Angular shapes reflect conflict and dynamism and are also often associated with masculine cultures. Feminine cultures, however, are often believed to favor more rounded, soft shapes. When thinking about developing a product, the

---

[1] Dimensions and their extensions are fully explained on Hofstede's website: http://geert-hofstede.com/. Country comparisons and specifities are also available.

Table 21.1: Visual Preferences and Cultural Dimensions			
Dimension	**Example of Country**	**Value Associated**	**Visual Preferences**
Collectivist	China Most of Asia and Africa	Social harmony Interdependence within society members	Round Symmetric Balanced
Individualistic	United States	Creativity Independence of society members	New, innovative Disruptive
Short-term oriented	France	Tradition, norms, Past	Vintage, old-fashioned
Long-term oriented	USA	Modernism Future	New Innovative
Masculine	Japan	Materialism, competitiveness, dynamism	Angular
Feminine	Nordic countries	Quality of life Harmony, happiness	Rounded

choice of angular shapes may receive less approval in a feminine culture that values harmony in life than in a masculine culture (Schmitt & Simonson, 1997).

These are just guidelines: what of a country such as Japan, which scores high on masculinity and low on individualism? According to Hofstede, masculinity prevails over collectivism in Japan, yet a careful look at the values and culture needs to be taken before launching a new product in a culture with seemingly contradictory values.

There are also differences among cultures in color associations. For instance, in Western society, which is more individualistic, white stands for purity and brides are mostly dressed in white. In Japan, white is the color used for mourning and grieving (Whitfield & Wiltshire, 1983).

Cultures influence design beyond visual aesthetics by affecting the role of a product. Across countries, different lifestyles impact the designing of products. Bike designs, for example, differ in the United States and in the Netherlands. In the United States, roads are wider and people are used to having cars to travel around. In the United States, bikes are leisure products, used with family or friends on the weekends to ride through the countryside. North American bikes need to be robust to respond to those outdoor activities. In the Netherlands, the roads are quite narrow. The average traveling distance is shorter, and people have developed the habit of traveling by bike, which is also quicker in case of traffic jams. In the Netherlands and most of the Nordic countries, urban residents use their bikes individually as a means of transport. Dutch bikes are lighter, more practical, that is, more adapted to city riding than North American ones. In France and in the United Kingdom, bikes need to answer to both activities at the same time, thus contributing to the success of hybrid bike designs such as the successful Decathlon's B'Twin of France (as shown in Figure 21.2).

American bikes are traditionally designed for outdoor activities, whereas Dutch bikes must respond to urban transportation needs. In France and in the United Kingdom, where both activities are equally important, hybrid bikes are very successful (Figure 21.2).

Figure 21.2: Example of bike designs in (a) the United States; (b) the Netherlands; and (c) France. *Sources:* (a) Wikimedia Commons/Rbv123; (b) Wikimedia Commons/Vijverln; (c) Wikimedia Commons/ Przemysław Jahr.

Cultures cannot be reduced to national or regional boundaries. Not every North American consumer likes innovative product designs, nor does every African like traditional ones. Culture is a complex, multilevel phenomenon. In order to understand a consumer, one must grasp the many influences on which his identity is built, among which (but certainly not exhaustively) are social class and consumer culture.

The Concept of Class

The French sociologist Pierre Bourdieu established in his book *Distinction: A Social Critique of the Judgment of Taste* (1979, English trans. 1984) that thanks to the cultural capital we inherit our tastes are defined by the social class in which we are born. He found that abstract paintings were highly valued by the "petite bourgeoisie," for instance, in order to differentiate their tastes from those of the working class. The working class favored pictures of everyday life such as a first communion or a folklore dance, which they understood and identified with. Though Bourdieu has had many critics, his analysis remains intriguing and relevant. A too visually complex, abstract product will probably not be appealing to the working class. However, individuals who wish to belong to the social elite will probably disregard trivial or very common products and favor distinguished ones. This phenomenon can explain the achievement of Nespresso®, the successful coffee machine and capsule manufacturer. The aesthetic of Nespresso is very peculiar, both product-wise (neat, elegant) and distribution-wise (flagship stores with a doorman). This specific design is associated with upper-class tastes and can be considered snobby by those who do not identify with it. Indeed, owning a Nespresso machine is a sign of social status in Europe today. Bourdieu has shown that taste has a lot to do with the social origins of an individual.

Holbrook (1999, 2005) studied taste and expertise and found a strong correlation between the two. People with expertise have the ability to define what is of good taste and what is not. Taste can be defined as: "An individual's consistent and appropriate response to aesthetic consumption objects through any of the five senses that is highly correlated with some external standard" (Hoyer & Stokburger-Sauer, 2012). The correlation between taste and external standards, defined by experts, helps us understand the work of Bourdieu (1979 [1984]). It is the upper class of society that has the "knowledge" and "expertise" in term of tastes. The lower class, with lower cultural background will, with time, transfer its tastes to the standards established by "experts." However, the standard of beauty of the upper class will by that time already be elsewhere. Thus, there will always be a distinction in taste between people of different social class (Bourdieu, 1979 [1984]).

Subcultures of Consumption

The subculture of consumption is yet another frame affecting visual preferences. People who follow a specific type of consumption are likely to adopt its aesthetics standards (Schouten & McAlexander, 1995). This is the case for Goths, for instance, whose community has its own aesthetics in clothes, music, and way of life (Schilt, 2007). Individuals who identify with this particular culture of consumption will value products with design congruent to the community's standards. In the case of Goths, for instance, there is a preference for the color black for clothing, or dark objects with sharp metal to express dramatization of the self. When designing a store for that target, for instance, designers need to be intentional given that individuals who identify with the Gothic community might feel less at ease in a store playing loud pop music than they would in one playing Lou Reed's albums. Aesthetic standards are present in many subcultures or communities, such as the gay and lesbian community, bikers, rockers, punks, and hipsters, for example.

Culture is a multilevel, complex phenomenon. It is important to always ask yourself a few questions regarding your future user's culture before thinking about developing a product (see Figure 21.3).

21.2 Individual Characteristics

Though the same influences are exerted in a given culture, people within that culture still can have completely opposite aesthetic tastes. Let's take the example of two young girls, Cate and Jenny. They are both 25, British, and living in London. They graduated

Figure 21.3: Nonexhaustive external influences of consumer's aesthetic preferences.

from the same business school and are now dashing, young executives. On paper, a first glimpse at their situation and we could be tempted to "classify" them into the same segment, for instance, "cosmopolitan young adult." Yet, when looking closer, we realize that Cate, who is an analyst in a major bank, comes from a British countryside bourgeois family, has her own flat, likes to go shopping in antique boutiques, and wears vintage clothing. Jenny works in a design consultancy firm. Her family is from London, and she lives with her mother, who is a fashion designer; she never misses any avant-garde exhibition, and makes her own clothes most of the time. Our perspective on their aesthetic preferences has changed. If the challenge is to design a desk for those young women, we might decide on something more traditional, wood-made for Cate, for instance, and perhaps something more modern for Jenny. But how can we know who our consumers are? If the analysis of culture helps new product developers to get a broad sense of the aesthetics standards and expectations of their targeted population, adding individual characteristics into the equation will allow a better understanding of who the consumer is and what he or she desires. For instance, the importance of design in one's life and personality are great influencers of one's aesthetic preferences.

Centrality of Visual Product Aesthetics

Bloch, Brunel, and Arnold (2003) postulated that individual's aesthetic preferences differed depending on the importance of a product's visual appearance in one's life. They named this characteristic the *centrality of visual product aesthetics* and developed a conceptual model and a scale to measure this variable. What the centrality of visual product aesthetics (CVPA) exactly refers to is "the overall level of significance that visual aesthetics hold for a particular consumer in his/her relationships with products" (Bloch et al., 2003). This concept can be explained through three subdimensions: value, acumen, and response (Figure 21.4).

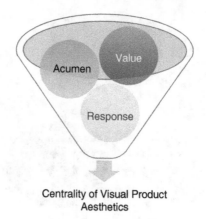

Centrality of Visual Product
Aesthetics

Figure 21.4: The making of CVPA.

Value. This subdimension refers to the importance given by someone to a product's design in general. This is the extent to which people believe design has the ability to enhance "quality of life, both personally and for society in general" (Bloch et al., 2003). For example, do you agree that "beautiful product design make our world a better place"? This subdimension will measure the extent to which individuals believe that product appearance can improve their personal and the society's well-being. Designers, artists, and probably anyone with a sensibility for art are likely to score high on this dimension. In the example we used at the beginning, we can guess that both Cate and Jenny might score high on this dimension, as they are both careful in choosing their clothes and their furniture.

Acumen. This subdimension assesses the ability to evaluate, to identify, and to categorize a product design. Some people are likely to have a higher aesthetic awareness than others. These individuals are expected give greater weight to design elements in product decision making. People who work in a highly aesthetic environment may show more design acumen than others. Jenny works in a fashion consultancy, is surrounded every day by highly aesthetic products and clothes; discussions around her are revolving around artistic or visual issue. We can guess that due to her working environment she will score higher than Cate on this dimension.

Response. This dimension measures the level of aesthetic response to product designs. It consists of both psychological response (enjoyment, affect, etc.) and physical (behavioral) response (approach, willingness to buy, etc.) (Bloch, 1995). Good design is often said to provoke impulse buying. Some individuals will be more prone to act on their feelings toward a product. Those persons often engage more with their possessions and build strong relationships with them. They will not see a product as a mere "dead object" but will grant it a personal intrinsic life. This is perhaps more the case for Cate, who likes vintage clothes and furniture: she gives importance to the historical value and significance of the product.

The authors showed that high-CVPA individuals evaluate highly aesthetic products as more pleasing and, have a more positive attitude toward them and a higher purchase intention for these products. In the world of fashion, these individuals are more likely to be opinion leaders and fashion innovators than low-CVPA individuals, who most likely tend to be fashion followers (Workman & Caldwell, 2007). It is crucial to understand, when developing a product, if the targeted users have higher or lower centrality of visual product aesthetics. Low-CVPA individuals discriminate less strongly between aesthetically pleasing and not pleasing products. They presented no differences in behavioral and purchase intentions between low- and high-aesthetic products with same utility (Bloch et al., 2003). In short, people who have the ability to enjoy more complex designs and for whom aesthetics in life is important will be more difficult to please. Those who do not might look at other criteria as primary influencers such as price or function when buying a product.

Visualizing Tendency

Morris Holbrook (1986) argued for the importance of the role of personality variables in affecting aesthetic response. He identified personality variables that can promote variety in aesthetic judgments such as the tendency to process information visually or verbally, the tendency to be more romanticist or classicist, and gender. His research showed that as personalities differ so do aesthetic judgments. Visualizers, for instance, will give more importance to holistic patterns when processing information (Holbrook, 1986). Holbrook found that these persons were more represented among romanticists, which means that visualizers are more likely to enjoy complex designs than those with verbalizing processing tendencies. Bloch et al. (2003) confirmed this with a study showing higher scores for visualizers on the centrality of visual product aesthetics. Holbrook identified visualizers as also being more strongly represented amongst women. Those visualizing women, for instance, in terms of fashion choices, will prefer "plain jackets and (at most) one isolated or two nonadjacent set(s) of stripes" (Holbrook, 1986). Visualizers have a more holistic view of the outfit and thus give more importance to the combination of elements around the jacket. Verbalizing women, however, who do not have such a holistic view, will prefer "plaid jackets." The difference between visualizers and verbalizers is less important for men; in the end they "play it safe by disliking clashing designs." Finally, and not surprisingly, classicists might prefer products with more neat designs, while romanticists may prefer more stylish, ornate designs. These results are to be taken with care. It would be inaccurate to suggest that all men dislike clashing designs, for instance.

Need for Uniqueness and Optimal Stimulation Level

The need to distinguish oneself from others is a strong determinant of behavior toward objects (Snyder & Fromkin, 1980). The products we own are used every day in the slow construction of identity; they are bought, consumed, displayed, and destroyed in order to build who we are (Belk, 1988). For some individuals who feel a need to be unique, the product they buy will play a much important role. These individuals will put more care into choosing a product: "Is it unique enough for me?" Snyder and Fromkin (1980) showed that these individuals tend to prefer novel or unusual products that reflect their unique personalities. Not surprisingly, Bloch et al. (2003) found that people who score high on the CVPA scale are also the ones who will seek more unique products.

For the majority of people, however, the optimal level of stimulation (OSL) needed to feel arousal is moderate (Berlyne, 1960). For some people, however, this optimal level is greater; they need more stimulation than others to feel stimulated in their everyday lives. These individuals have been identified as "high optimum stimulation levels (OSL) individuals" (Steenkamp & Baumgartner, 1992). The authors found that these consumers will engage in more variety-seeking behavior as they get bored more quickly by a product. They also are more risk-seeking individuals. Raju (1980) studied the relationship between OSL and personality and demographic traits. High-OSL individuals are attracted to new or unusual stimuli or situation. They may, for instance, work in highly stimulating environments such as museums, theatres, and art schools. Drawing their attention requires greater visual stimuli and higher creativity in the designing of a product. In terms of demographics, younger, educated, and employed people seem to have a higher OSL (Raju, 1980). To decorate her home interior, Jenny, who works in a design consultancy, may prefer original and innovative furniture that will offer her psychological benefits such as risk, variety, or novelty.

21.3 Situational Factors

Situational factors are well known in consumer research to be crucial factors influencing behaviors (Belk, 1975). They play a moderating role in the relationship between product design and consumer response (Bloch, 1995), which means that depending on the situation, a person may not have the same response (aesthetic, behavioral) toward a product. We can distinguish between the physical context and the influence of others (social environment).

When talking about context, we mean the context in which the product will be displayed or consumed. The fit of the new product with products already owned by the individual might, for instance, influence its perception. Consumers may like a product's appearance but may not buy it because it does not fit aesthetically with their home interior (Bloch, 1995). If someone owns an old cottage in the countryside decorated with old furniture, he or she may decide not to choose a twenty-first-century-design sofa for the living room. The context can also change a product's function. For instance, many vintage products are used for aesthetic purposes and have lost their utilitarian function (Veryzer, 1995). It is common to find old sewing machines or vintage clocks that have stopped functioning in living rooms for decorating purposes.

The social context also greatly influences the aesthetic response of a consumer to a product. Being alone or being with others will change the way consumers behave toward a product (Belk, 1975). The work of Zhang et al. (2006) offers a great example of that fact. They found that independent persons tended to favor angular shapes. However, that relationship was significant only when in the presence of others. In society, we try to give an image of a confident self, perhaps more confident than in reality. Showing a preference for angular shapes may be believed to reflect a sharp personality. Indeed, angular shapes are associated with masculine features, as we have seen previously.

It is important to reflect on what the product will be used for, where it will be used, and with whom. Will this chair be for the bedroom, where no one except family goes? Or for the living room, which is the place where we can show others who we are (by what we own)?

21.4 Discussion

This chapter provides those who wish to develop new products with a list of factors to pay attention to. This list is not exhaustive. My goal is to encourage product developers to consider the many influences of their consumers' aesthetic preferences. I particularly focused on the influence of cultures, individual characteristics, and situation.

First, we saw the crucial role of culture. Culture acts as a lens through which we see the world (McCracken, 1986), therefore modifying the way consumers approach new products. Chinese consumers may be more sensitive to rounded shapes due to the collectivist nature of their culture. This has an impact on logo design. In China and Singapore it has been found that brand logos that are harmonious and natural create more positive affect towards the brand (Henderson et al., 2003). However, American consumers will probably value innovative design more. Culture can also modify the role of a product as we have seen with bikes, which are used for leisure in North American and for transport in northern Europe. Culture is not only a question of geography, and we have seen that social classes have their own aesthetic standards. When a consumer is involved in a specific consumption activity, it impacts his entire life, including aesthetic preferences like in the case of Goths or bikers.

Second, we looked at personality traits and individual characteristics that may impact the way consumers react to a product. We have seen that for some consumer aesthetics plays a big part in life; they will be more attentive to product design and more demanding as well. People who have greater need to feel unique than average will focus their choices on differentiating products. Others need greater visual stimulation than average to feel stimulated and thus will favor more flashy colors or more unique designs. Even though it is difficult to segment a population according to such criteria, some clues can be taken into account. Holbrook (1986), for instance, found that there were more visualizers among women. Raju (1980) found that young, educated, and employed people were more likely to need novel, varied, and innovative products. The working environment may be a way of knowing if our consumers give greater importance to aesthetics in their lives or if they need more stimulation. People who work in the art world might be more demanding, for instance. The question of whom we are designing the product for is crucial. If the wish is to make a product that will appeal to a broad audience, it is worth noting that a medium level of stimulation will be preferable: too little and the product might not be noticed by consumers; too much and it might be rejected (Berlyne, 1960; Steenkamp & Baumgartner, 1992). However, if the strategy is to target opinion leaders to create an image for the brand or the product, a highly stimulating design with innovative features might a good choice (Workman & Caldwell, 2007).

Finally, it is crucial to look out for the consumption context of the future product. Depending on where and with whom we consume, the product may not have the same appeal. We invite product developers to reflect on the situation in which their consumers will buy, use, display, and consume their product.

21.5 Conclusion

This chapter provides the reader with a nonexhaustive list of forces that sway the aesthetics preferences of consumers. This list is to be used carefully: in the topic of aesthetics there is nothing ever certain and definite. Fashion progresses at a rapid pace and tastes change, evolve, and mutate fast. Also, there are many more factors not mentioned here that consumers consider before choosing a product, like price or functionality. We encourage anyone who wishes to launch a new product to engage into a reflective activity around the targeted consumer. Who do we want our consumers to be, and what do we know about their aesthetic preferences or about the forces that will influence their choices? These are things to consider before thinking of the design itself.

References

Belk, R. W. (1975). Situational variables and consumer behavior. *Journal of Consumer Research*, 2(3), 157–164.

Belk, R. W. (1988). Possessions and the extended self. *Journal of Consumer Research*, 15(2), 139–168.

Berlyne, D. E. (1960). *Conflict, arousal, and curiosity*. New York, NY: McGraw-Hill.

Berlyne, D. E. (1971). *Aesthetics and psychobiology*. New York, NY: Appleton-Century-Crofts.

Berlyne, D. E. (1974). *Studies in the new experimental aesthetics: Steps toward an objective psychology of aesthetic appreciation*. Oxford, England: Hemisphere.

Bloch, P. H. (1995). Seeking the ideal form: Product design and consumer response. *Journal of Marketing*, 59, 16–29.

Bloch, P. H., Brunel, F. F., & Arnold, T. J. (2003). "Individual differences in the centrality of visual product aesthetics: concept and measurement." *Journal of Consumer Research*, 29, 551–565.

Bourdieu, P. (1979; English trans. 1984). *Distinction: A social critique of the judgment of taste*. Boston, MA: Harvard University Press.

Henderson, P. W., Cote, J. A., Leong, S. M., & Schmitt, B. (2003). Building strong brands in Asia: Selecting the visual components of image to maximize brand strength. *International Journal of Research in Marketing*, 20(4), 297–313.

Hofstede, G. (1980). Culture and organizations. *International Studies of Management & Organization*, 10(4), 15–41.

Holbrook, M. B. (1986). Aims, concepts, and methods for the representation of individual differences in aesthetic responses to design features. *Journal of Consumer Research*, 13(3), 337–347.

Holbrook, M. B. (1999). Popular appeal versus expert judgments of motion pictures. *Journal of Consumer Research, 26*(2), 144–155.

Holbrook, M. B. (2005). The role of ordinary evaluations in the market for popular culture: Do consumers have "good taste"? *Marketing Letters, 16*(2), 75–86.

Hoyer, W. D., & Stokburger-Sauer, N. E. (2012). The role of aesthetic taste in consumer behavior. *Journal of the Academy of Marketing Science, 40*(1), 167–180.

McCracken, G. (1986). Culture and consumption: A theoretical account of the structure and movement of the cultural meaning of consumer goods. *Journal of Consumer Research, 13*(1), 71–84.

Raju, P. S. (1980). Optimum stimulation level: its relationship to personality, demographics, and exploratory behavior. *Journal of Consumer Research, 7*(3), 272–282.

Schilt, K. (2007). Queen of the Damned : Women and Girls' Participation in Two Gothic Subcultures, In L. M. Goodlad & M. Bibby (Eds.), *Goth: Undead subculture*. Durham, NC: Duke University Press

Schmitt, B. H., & Simonson, A. (1997). *Marketing aesthetics*. New York, NY: Free Press.

Schouten, J. W., & McAlexander, J. H. (1995). Subcultures of consumption: An ethnography of the new bikers. *Journal of Consumer Research, 22*(1), 43–61.

Snyder, C. R., & Fromkin, H. L. (1980). *Uniqueness: The human pursuit of difference*. New York, NY: Plenum Press.

Steenkamp, J. B. E., & Baumgartner, H. (1992). The role of optimum stimulation level in exploratory consumer behavior. *Journal of Consumer Research, 19*(3), 434–448.

Veryzer, R. W. (1993). Aesthetic response and the influence of design principles on product preferences. *Advances in Consumer Research, 20*(1), 224–228.

Veryzer, R. W. (1995). The place of product design and aesthetics in consumer research. In F. R. Kardes & M. Sujan (Eds.), *Advances in consumer research* (pp. 641–645). Provo, UT: Association for Consumer Research.

Whitfield, A., & Wiltshire, T. (1983). Color. In C. H. Flurscheim (Ed.), *Industrial design in engineering* (pp. 133–157). Worcester, England: The Design Council.

Workman, J. E., & Caldwell, L. F. (2007). Centrality of visual product aesthetics, tactile and uniqueness needs of fashion consumers. *International Journal of Consumer Studies, 31*(6), 589–596.

Zhang, Y., Feick, L., & Price, L. J. (2006). The impact of self-construal on aesthetic preference for angular versus rounded shapes. *Personality and Social Psychology Bulletin, 32*(6), 794–805.

About the Author

ADÈLE GRUEN is a PhD student at the Université-Paris Dauphine in Paris, France. Her doctoral dissertation focuses on the role of design in the appropriation of objects in the context of collaborative consumption. This present work is based on her thesis for the master of research and research methods she attended at that same university.

22

FUTURE-FRIENDLY DESIGN: DESIGNING FOR AND WITH FUTURE CONSUMERS

Andy Hines

University of Houston Hinesight

Introduction

New product ideation and design is aimed at future markets, but the ideas and designs are typically developed using current consumer needs. This chapter offers a framework for understanding long-term values shifts that provide insight into how consumer preferences are changing into the future. Two emerging values types that are driving these changes are introduced and their trajectory over time is described. These shifting values are at the core of five emerging consumer needs, which are illustrated and brought to life with representative future personas. Implications for designers and developers are identified both in terms of adding to a tool kit as well as identifying themes of change cutting across the consumer landscape.

Designers and new product developers (hereafter "developers") looking to develop innovative designs and products are continually challenged to understand how consumer preferences are changing. If only it were as simple as asking them what they will need in the future! The truth is that they don't know either. Thus, one looks for clues to future consumer preferences. My experience as the Global Trends manager with the Kellogg Company back in the 1990s introduced me to the notion that consumer values provided insight to changing consumer preferences, that is, what they might want to buy. More importantly, there were long-term patterns in how these values were changing. I built the initial values framework for Kellogg's and have since used it with dozens of clients over the years (Hines, 2011).

Long-term patterns in values change can provide a useful framework for understanding emerging consumer needs that in turn provide clues to their purchase preferences. They are at the core of five emerging needs that provide insight for designers and developers to produce "future-friendly" designs and products. The focus is on shifting values and emerging needs since the goal is to explore how the future is changing. Focusing on novel values and needs also provides an excellent opportunity for breakthrough innovation. It should be kept in mind, however, that current values and needs are still important and will characterize or drive the majority of offerings.

22.1 A Framework for Understanding Changing Consumer Values

Values are defined as "an individual view about what is most important in life that in turn guides decision-making and behavior" (Hines, 2011, p 9). In essence, they are the priorities consumers use to help them with important decisions, whether it is where to go to college or what kind of car to buy. They also implicitly guide more routine decisions, for example, should I buy the natural product that's a bit more expensive?

A snapshot of values in the present could be useful by itself, but of even greater utility is that the data suggests that they have been changing in a consistent direction over time. The World Values Survey (see www.worldvaluessurvey.org/) identified this pattern and offers a view on values change based on longitudinal data it has been gathering since the 1970s. In addition, research has identified more than two dozen systems relating to values (Hines, 2011). This paper extracts the insights as they relate to change in the future.

Let's begin with the four types of values. The first three types are derived from the World Values Survey data, while the fourth was hypothesized by the Spiral Dynamics system (Beck & Cowan, 1996; Inglehart, 1997).

The long-term shift is from left to right in Figure 22.1: from traditional to modern to postmodern to integral. The percentage estimates below derive from the World Values Survey data and Spiral Dynamics. They have not been updated recently and should be judged with caution.

- Traditional values have been around the longest and were prevalent for a great deal of human history, but are declining in the affluent countries (now at 25 to 30 percent of the population). In design terms, the key task of the designer is to produce a "consistent" design, appropriate to their beliefs, practices, histories, protocols, textbooks, and so on.
- Modern values are peaking in the affluent countries (35 to 40 percent) while they are surging in emerging markets. In design terms, the key task of designers is to produce the "best" design—one that beats the competition is most valuable.
- Postmodern values are growing in the affluent countries (25 to 30 percent). The key task here is a "participatory" design process, one in which everyone is heard and has input, honoring the unique perspectives that individuals bring.
- Integral values are just emerging in the affluent countries (about 2 percent). They key task here is "co-created" design that moves beyond input to direct involvement in the design process, such as open-source approaches.

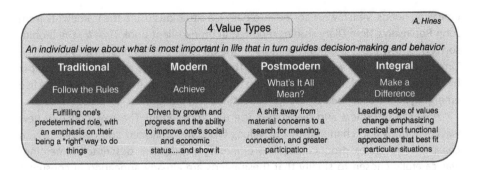

Figure 22.1: Four value types.

The value types highlight patterns in preferences and priorities, but it should be noted that variations are more degrees of emphasis rather than either-or. For instance, postmoderns put a high priority on self-expression, but this doesn't mean that moderns or traditionals will not self-express—it is just less important to them in general.

In affluent countries, traditional and modern values are declining and postmodern and integral are growing. Postmodern values emerged in the late 1960s/early 1970s, and may be approaching critical mass similar to Gladwell's "tipping point," or "magic moment when an idea, trend, or social behavior crosses a threshold, tips, and spreads like wildfire" (Gladwell.com, n.d.). Integral values today are at a similar point to postmodern values in the sixties/seventies in that they just now emerging and, if the model follows, may be poised to reach critical mass in a generation or two.

A key implication is that the divide between designer/developer and customer will get increasingly blurry; "us and them" is evolving toward "we." The "Maker Movement" is one manifestation of this. Designers and developers have a great opportunity to join with these consumers to develop offerings in light with their emerging needs.

22.2 Emerging Consumer Needs

Most product development assumes knowledge of what the consumer wants to do—the problem is at least somewhat defined. Much of working with innovation, development, and design groups, particularly when working with futurists, involves coming up with concepts and ideas aimed at future problems and challenges that are not well defined. A core principle of this future-oriented work is starting with consumers and their needs: combining shifting values with trend identification to forecast the emerging needs to provide a preview of how the future may evolve. Keep in mind that there is continuity along with change and that the postmodern and integral values emphasized here are part of the change while the traditional and modern values are part of the continuity.

Several steps were taken to create the five emerging needs profiled here. Thirty-nine universal needs, 34 postmodern values, 16 integral values, and 13 consumer trends were identified, and then clustered, analyzed, and synthesized into a raw list of 120 potential emerging needs, which were sorted into daily life situations using the framework from John Robinson's Time Diary studies reported in his excellent book *Time for Life* (Robinson & Geoffrey, 1997) and now the basis for the US Bureau of Labor Statistics' American Time Use Survey (see www.bls.gov/tus/). The team looked for patterns in this matrix by combining, consolidating, and pruning the large list. After several iterations, the five emerging needs described below were identified and refined.

How can designers and developers use this knowledge? The emergence of new value priorities and changing mix of values types was analyzed and combined with consumer trends to produce the five emerging needs. The descriptor "emerging" was carefully chosen to signify that these needs are already appearing in consumer life to some degree today and will become increasingly important in the future. In geographies that index high for postmodern and/or integral, for instance, such as Sweden globally, or in California in the United States, the five emerging needs will be more apparent than in places indexing lower. Certainly, traditionals and moderns may also pursue these needs—they are not the exclusively held by postmoderns and integrals, but more prevalent among them. The five emerging consumer needs are:

- *Keeping it real*: Preference for the straight story.
- *The [relentless] pursuit of happiness*: Taking responsibility for one's well-being.
- *Community first*: Preference for things local.
- *We [really] are the world*: Feeling responsible for the well-being of the planet.
- *Glass houses*: Everyone is watching.

These emerging needs are described below with an accompanying representative future persona. Personas are representative characters that fit a profile of someone who has that need or needs. They are emerging today and are forecast to become increasingly evident over the next decade. Future personas can help designers and developers imagine "who" they are designing for. They help to make the abstract needs more concrete. It should be kept in mind that these personas are generalizations. As such, they will miss the diversity that will actually show up in how the emerging needs manifest in the future. For example, the demographics suggesting that Annie is a 35-year-old female does not imply that only 35-year-old females will have this need. It is an attempt to find the center of the need, and some trade-offs have to be made in providing details to help create a mental picture of the persona. It is also important to note that individuals may hold several of these needs. In fact that is very likely to be true. Someone who holds "keeping it real" may also hold "community first." The personas are not exclusive. The intent in providing separate personas is to provide developers and designers with an image of how a particular need might manifest. Indeed, it may be possible to build a composite persona that combines the attributes. The future personas are covered in a consistent format that includes the following:

- **Summary description:** The first few paragraphs characterize the persona in general.
- **Demographics:** A profile of a "typical" persona who embodies the emerging need, including gender, age, household income, education, and life stage.
- **Illustration:** A visual to help facilitate the "real-ness" of the persona.
- **Table:** A summary table that includes the supporting need states of the emerging need, the values this persona would likely embrace, and the related trends supporting the persona's emergence.

- **Committed time activities:** How the persona would likely approach several aspects of committed time—work, household and family care, shopping, and personal/biological necessities.
- **Free time activities:** Description of how the persona would likely approach free time—learning, leisure (entertainment/recreation), affiliation, and communication.
- **Vignette:** A brief day-in-the-life snapshot that provides insight into how the persona might operate in daily life.

These particular ingredients are offered as a basic menu and they may be varied, added to, or subtracted as needed for a particular project—one might include generations, personality or thinking styles, worldviews, and so on.

Keeping It Real

A key word for postmodern and integral consumers is authenticity. It is the core value driving them to this emerging need. They are asking organizations to give it to them straight and trust them to be able to handle the truth. They will reject any paternalistic "for your own good" kind of sugarcoating. Their view is, "Treat me as an adult, as an equal, and as someone with a brain. Don't manage me." Figure 22.2 illustrates "keeping it real" through the "authentic Annie" persona.

Demographics
Female, Age 35
HHI: $125,000
Education: MA, Public Health
Life stage: Mid-career, early parenthood

Figure 22.2: Keeping it real: authentic Annie.

Day in the Life Vignette	Activities	
	Committed Time	**Free time**
Annie's friend used to think she was so trendy. They laugh about it now, but having kids changed her somehow. Now it's more about what's *not* in the things she buys her family than the label or logo that the product carries. Thankfully she found a lot of "mommy blogs," which have given her great tips … it's been years since she read a product review in the mainstream media. Sure, sometimes the stuff she buys costs a little more, but that's okay. She likes to think her choices make a difference, but she also likes the recognition that her choices *are* different. Hmm … maybe she hasn't given up her attachment to labels and logos entirely.	**Work:** P/T at women's health NGO **Household/family care:** At home part-time with kids thanks to job-share **Shopping:** Local and face-to-face for things that matter; online ordering of "staples" **Personal (biological necessities):** Uses Tom's of Maine	**Learning:** Gardening course **Leisure (entertainment & recreation):** Catching up on her latest copy of Dwell or Real Simple **Affiliation:** Sierra Club, local PTA **Communications:** Facebook with friends from the mommy blogs

(continued)

(Continued)

Need States	Values	Trends
■ The Authenticity Premium ■ Au Naturale ■ The Simplicity Premium ■ Less Is More	■ Authenticity ■ Experiences ■ Appropriateness ■ Functionality ■ Design ■ Self-expression ■ Simplicity ■ Cool ■ Sustainability	■ Truth & Truthiness ■ Enoughness ■ Sustainable Consumption ■ Living within Limits ■ Lifeshifting ■ Continuum of Ownership

They are reacting against an overly managed world. "Delighting the customer" has gone to an extreme. As management of consumer experiences has gotten increasingly sophisticated, it has created a situation where every aspect of the experience is micro-managed, and these consumers sense that, and feel that they are constantly being manipulated—and they want it to stop.

Implications for Designers and Developers

■ These consumers will appreciate designers and developers who "do the homework" to find out what authentic is, such as being of aware of history, origins, materials, and handling.

■ "Warts-and-all" offerings may appeal to these consumers by representing authenticity.

■ These consumers will appreciate simplicity that embraces or encompasses a great deal of complexity.

■ Characteristics or keywords: authentic, simple, natural, and "less is more."

The [Relentless] Pursuit of Happiness

The values shifts have a major theme of consumers rethinking the purpose of their lives. The pursuit of happiness is a purpose shared by many. It reflects the growing range of choices enjoyed by postmodern consumers who enjoy relative economic security. Figure 22.3 illustrates "the [relentless] pursuit of happiness" through the "Becky 2.0" persona.

Demographics
Female, Age 19
HHI: n/a
Education: Pursuing BA psychology
Life stage: Student

Figure 22.3: The relentless pursuit of happiness: Becky 2.0.

Day in the Life Vignette	Activities	
The meeting with the dean went better than she expected. She had presented her vision for why the college should grow its own fruits and veggies and give up that portion of its corporate food service. Her premise is that connecting students to their food stream in this very tangible way will lower the schools carbon footprint and provide a psychological and health benefit that will pay off down the road. Thankfully she'd found the CampusGrows network on Facebook. She'd learned so much from other kids who were working on similar plans at their schools ... getting crop/menu ideas, work plans, financial advice to make it self-sustaining, etc. Her motto is "there is always a better way." Next stop ... a meeting with the college president! Rock on!	**Committed Time**	**Free time**
	Work: 5 classes, 2 volunteer gigs, and 12 student activities is work enough ... **Household/family care:** Lives in the dorms; hates her roommate **Shopping:** Clothes at thrift store; food at Whole Foods and farmer's markets **Personal (biological necessities):** All-natural products	**Learning:** Training as DJ on campus radio station **Leisure (entertainment & recreation):** Playing her guitar **Affiliation:** Volunteers at day care for kids of local migrant farm workers **Communications:** Connecting with other students interested in local food via Facebook

Need States	Values	Trends
■ Help Me Help Myself ■ Identity Products, Services, and Experiences ■ Systematic and Consistent ■ Reinventing the Self ■ I'm Not a Consumer ■ Pursuit of Happiness, aka Well-Being	■ Contentment ■ Enjoyment ■ Wellness ■ Self-expression ■ Passion ■ Spirituality ■ Sustainability ■ Interdependence ■ Questioning ■ Discovery ■ Authenticity	■ Enoughness ■ Co-creation ■ Consumer Augmentation ■ Truth & Truthiness ■ Empowered Individual ■ Sustainable Consumption ■ Virtual-Real-Digital Tribes

Traditional values do not put a priority on the pursuit of one's own happiness, as people's roles can be ascribed largely at birth in addition to an emphasis on God or others. The modern values pursuit of happiness tends to focus around economic achievement and material prosperity. The postmodern values holder, with relative economic security, has the freedom to consider a wider range of routes to happiness. Ironically, the modern-to-postmodern transition is often accompanied by a sense of angst. Many have experienced a sense of emptiness from the material prosperity route and call the meaning of their lives into question. The resultant search for meaning in life is not always easy or pleasant. Happiness becomes something that has to be achieved—it does not necessarily arrive on its own for the postmodern consumer.

There is a relentless aspect to this pursuit among some, reflecting a seriousness of purpose: "What makes me happy, and what do I have to do to get there?" This pursuit often involves assisting with others and working to benefit the community. It may also involve faith in a higher power, but this conception is often derived from multiple sources rather than subscribing to a single belief system.

Implications for Designers and Developers

■ These consumers are looking to fit offerings into larger lifestyles, values, and sense of purpose, and evaluate purchase decisions with this fit in mind. These consumers are looking for designers and developers that will "help me help myself." They may be particularly interested in co-creation, where they may be provided with tools, templates, and advice—and handle the rest themselves. For some products and designs, it will be simply about fast, easy, and cheap, but for others, which speak to their "identity," designs and products will increasingly be evaluated in terms of how they influence one's sense of "happiness" or well-being.

■ Characteristics and key words: DIY, happiness, well-being, assistance.

Community First

The emerging values shifts suggest a shift in scale from large to small and in scope from mass to custom. This shows up most strongly in this emerging need. It favors decentralized approaches. It is part of the sense, captured in other need states, that life has gotten too complex, moves too fast, and has become impersonal. It is this depersonalization in particular that drives the move to renewed interest in community, as people seek to reconnect with their life and with one another. In the ascent up the growth curve in modern society, the frenetic pace is seen as worth the trade-off for the economic reward. The postmodern consumer is more aware of the costs, has less need for economic security, and thus begins to reject this trade-off. Figure 22.4 illustrates "community first" through the "good neighbor Bob" persona.

Demographics
Male, Age 28
HHI: $34,000
Education: Some college, self-taught web
  guru
Life stage: Lives in group house with fiancé

Figure 22.4: Community first: good neighbor Bob.

Day in the Life Vignette	Activities	
Bob remembered how proud he was of himself when he bought his first pair of pants from American Apparel ... it was a start but even that doesn't seem quite local enough for his taste anymore. That's why for the past 2 years, Bob's been on a mission to connect artists and craftspeople with people in the neighborhood through his new Locals ONLY iPhone app. Part eBay ... and part ePinions, the content is all local. And why not? Brooklyn has everything to offer whether you're looking for artisan bread, an oil painting, or a handmade refurb'd bicycle. And the social aspects of the app take the guesswork out of who you're buying from ... Bob's next challenge—take his LocalsONLY movement to other cities....	**Committed Time**	**Free time**
	Work: Studio engineer in a Brooklyn recording studio	**Learning:** Takes classes at local community college
	Household/family care: Lives in group house; engaged but not quite ready for it	**Leisure (entertainment and recreation):** Fixed gear bike guru; playing in his band
	Shopping: Buys local! If he can't buy local, next goes to Freecycle.org	**Affiliation:** On the Board of Neighborhood Association
	Personal (biological necessities): Trades time in his studio to local herbalist for homemade toothpaste, deodorant	**Communications:** Writes and blogs for local alternative newspaper

Need States	Values	Trends
Local Preferences	Community	Relocalization
Community Support	Connectivity	Living within Limits
Trust the Network	Appropriateness	Sustainable Consumption
	Influential	Empowered Individuals
	Interdependent	Virtual-Real-Digital Tribes
	Collaboration	Emerging Markets Arise
	Sustainability	
	Questioning	
	Skepticism	
	Tolerance	

This desire for connection manifests in both the physical and the virtual worlds. These consumers question why they don't know their neighbors or even the mayor. They are looking for ways to get involved with what's going on directly around them, as this helps to provide an anchor or security in what is seen as an increasingly chaotic world. The explosion of Facebook and other social networking sites is evidence of how the virtual world can serve as a mechanism for connection.

Implications for Designers and Developers

- These consumers value locally produced offerings as way to support their local community. They will also tend to favor small, local producers in other jurisdiction if they are competing against a big, multinational competitor.
- These consumers could be key drivers of a move to require designers and developers provide some kind of local benefit.
- These consumers place trust in their physical as well as virtual networks and may rely on them for advice, referrals, or even to co-create via crowdsourcing approaches, which will continue to grow stronger.
- Characteristics and keywords: local, community, network, crowdsourcing.

We [Really] Are the World

The title of this emerging need plays on the 1985 song "We Are the World," which was recorded to support charitable causes in Africa. That effort spurred some short-term attention, and while things soon returned to business as usual, the song lived on; the thought apparently touched something in these consumers that is now coming back to life, thus the "really" in parentheses. This time, the feeling of global responsibility or planetary consciousness is emerging as a stronger and more genuine force. Figure 22.5 illustrates "we [really] are the world" through the "Stewart'ship'" persona.

Demographics	
Male, Age 58 HHI: Living off nest egg Education: BA, Yale, MA, Columbia Life stage: Launching his "encore" career	 Figure 22.5: We really are the world: Stewart"ship".

Day in the Life Vignette	Activities	
Stewart couldn't wait to get off the plane and hit the ground running. He'd heard from his team that the villages where they were going to launch the microfinance pilot program were really excited by the possibilities. This sure was gonna be different than doing a deal on Wall Street, but he was glad he'd left that all behind. When he thought about it, his transformation probably started sometime after his church's mission trip to Haiti. He didn't go soft or anything ... if anything, it reinforced his belief that free markets and commerce were the only answer. What it did do was make him realize that he could make a real difference. So he took his the nest egg he'd made in 20 years as a VP on Wall Street, set up a little foundation, and was going to do his part to bring people into the fold of the global economy and fight the powers that were driving people to extremism one microloan at a time.	**Committed Time**	**Free time**
	Work: Living off Wall Street nest egg; starting microfinance foundation	**Learning:** Language classes, so he can connect with his foundation staff and clients
	Household/family care: Family living on a real budget now	**Leisure (entertainment and recreation):** Trying to stay connected to the Yankees
	Shopping: Supports small businesses in emerging markets, e.g., fair trade coffee, TenThousandVillages.com	**Affiliation:** Unitarian church, Optimists International
	Personal (biological necessities): Seeks to be footprint-neutral in choices	**Communications:** Avid blogger and offers free local personal finance workshop

Need States	Values	Trends
■ Global Citizens ■ Making a Difference	■ Thoughtfulness ■ Influential ■ Integration ■ Appropriateness ■ Interdependence ■ Transcendence ■ Self-Expression	■ Sustainable Consumption ■ Empowered Individuals ■ Emerging Markets Arise ■ Living within Limits ■ Relocalization ■ Consumer Augmentation

What has changed alongside the strengthening of the supporting values is the "flattening" (Friedman, 2005) of the world that enables easily accessible and real-time information about any event or situation almost anywhere in the world. Few geographies are beyond the reach of global media and communications. The connection to distant problems is more easily maintained and the options for action have increased as well. It has become much easier to act on these values now than it was back in 1985. So, while the values supporting this emerging need may well have been present 25 years ago, the supporting infrastructure was not—but it is now and increasingly so in the future.

Implications for Designers and Developers

- These consumers think of themselves as global citizens and will think through the ramifications of designs and products that go beyond national borders, with a genuine concern for planetary welfare and a willingness to act on that.
- These consumers are for ways to make a tangible difference in the pursuit of idealistic grand schemes that suggest designs and products that blend vision and practicality.
- A sense of global social responsibility will be an added criterion for designs and products that appeal to these consumers; they will be inclined to ask what it adds the common welfare?
- Characteristics and keywords: global, sustainable, vision, difference.

Glass Houses

These consumers are the activists and many will have an aggressive orientation. They are intolerant of behavior they deem wrong and are not afraid to let the offender, or any interested party, know about it. They feel they are not to be trifled with and that their values and beliefs are important and need to be respected. Figure 22.6 illustrates "glass houses" through the "high-tech Tina" persona.

Demographics
Female, Age 61
HHI: $65,000
Education: BA
Life stage: Empty-nester, husband
    retiring

Figure 22.6: Glass houses: high-tech Tina.

Day in the Life Vignette	Activities	
It was fitting she thought that they launched their NGO on the 40th anniversary of Woodstock. It was where she and James met. They were so young then … but man, if they'd had the technology they have now back then. The NGO—called The Watchtower Group is going to build a web tool to help individuals track the social performance of their investments in real time. Users enter their stock and mutual fund holdings and pick from a list of 50+ issues that they care about—such as the company's stance on fair trade, treatment of employees, environmental record, local vs. global sourcing, etc. They get a baseline report as well as real-time alerts. They'd also get the option to ping the investor relations departments to voice their support … or displeasure.	**Committed Time**	**Free time**
	Work: Community college professor; husband retiring. Non-work hours spent on Watchtower Group	**Learning:** Her son's teaching her mash-ups with Google Maps
	Household/family care: Enjoying the empty nest	**Leisure (entertainment and recreation):** Kayaking, hiking, genealogy
	Shopping: Local and face-to-face for things that matter; online ordering of "staples"	**Affiliation:** Friends Church (Quaker), socially responsible investing club
	Personal (biological necessities): Cross-compares and checks up on all companies she does business with	**Communications:** Careful about what she shares publicly
Need States	**Values**	**Trends**
■ Trusted Partners for the New Insecurity ■ The Truth, Whole Truth, and Nothing but the Truth ■ Expanding Accountabilities	■ Sustainability ■ Community ■ Assistance ■ Commitment ■ Authenticity ■ Questioning ■ Skepticism ■ Integration	■ Sustainable Consumption ■ Truth & Truthiness ■ Empowered Individuals ■ Living within Limits ■ Emerging Markets Arise ■ Relocalization ■ Continuum of Ownership

These consumers are watching, often all the time. They are often savvy users of technology and expert in the world of information, and they use that to support their cause. Accountability is the buzzword; it won't always be pleasant; and it won't always be fair. The best an organization can do is stay consistent and true—or, closing the circle back to our first emerging need, be authentic. "Spin" and message control and such tools will only get organizations into trouble. Telling the truth will, eventually at least, earn respect and credibility that will be appreciated and rewarded over the long haul.

Implications for Designers and Developers

■ These consumers will be inclined to transparent and open approaches; they will want to know how a design or product was derived, or at least know that they can have access to that information.

■ These consumers are looking for trusted partners to help them navigate through what they see as a complicated and even insecure future. They will appreciate partners who are willing to admit faults and mistakes, seeing it as a sign of good faith.

■ These consumers are likely to embrace a collaborative approach to design and development.

■ Characteristics and keywords: open, transparent, participation, collaboration.

22.3 Going Forward

The emerging needs are offered to designers and developers as a means to develop future designs and products. The personas help to illustrate and create a mental picture or image of what those consumers might be like. Two principal ways to use the ideas in this chapter going forward are suggested:

■ The first and perhaps more valuable long-term applications are the potential additions to the design and development tool kit.
■ The second involves using the specific implications accompanying each need and persona and a set of key themes derived from them that will be identified below.

The Tool Kit

The values framework provides a foundation for understanding and insight into future consumer preferences, thus enabling designers and developers to align their work with the future. Consumer expectations of designers and developer are growing—they will expect to be understood. Insight into values provides a basic framework for understanding what is important to them.

Designers and developers are, of course, busy, thus the tool kit provided here can be "laddered up" depending on how much time is available for understanding consumer preferences:

1. The four value types provide a "quick-and-dirty" framework.
2. The values, when combined with consumer trends provide additional understanding.
3. Combining the values into five emerging consumer needs adds another layer of understanding.
4. Studying the personas that accompany the emerging consumer needs helps bring those needs alive.
5. Finally, the personas can be customized and enhanced to provide a multilayered approach to understanding the consumer targets.

The future personas are intended to serve as targets for designers and developers by providing a visual picture of consumers who embody the emerging needs. They present a means to address the challenge mentioned at the beginning of this piece to design and develop products based on a view of how the future will be different, rather than assuming that the present situation will continue. The personas can be used during ideation sessions both to generate ideas—what types of designs or products will consumers with needs want. The personas can also help refine concepts by helping designers and developers tailor them with a more informed sense of what motivates these consumers.

Some Key Themes

This chapter suggests that the consumer landscape is changing in ways that can be understood. It is important to understand the values as they provide the "why" behind the needs. Consumers will expect designers and developers to be aligned with them, to understand them, and to relate to them. Each of the five emerging needs and personas were accompanied by implications for specifically appealing to them. We'll conclude this chapter with some themes or big ideas that apply in general to the two values types—the postmoderns and integrals—at the leading edge of changing consumer preferences. It is not intended to suggest that these needs are inevitable. Futurists recognize that for every trend there is a potential countertrend, and thus it is important to monitor the future as it unfolds for any changes in direction.

- Consumer preferences are bifurcating into those offerings that are generic or commodity-like where they seek fast, cheap, and easy solutions, and those where the product or design means something to them, that is, it speaks to their identity and their values. For the latter, consumers will be more concerned and involved with the design and development process.
- There will be a great desire for more open approaches that encourage participation and co-creation, especially for those offerings that appeal to their identity.
- For these offerings to appeal to one's identity, the story behind the design and development will be a key ingredient driving the purchase decision.
- In cases where consumers want to be involved, deeper association with designers and developers who provide advice, tools, and templates in more of a coaching role to offer an environment within which consumers can create.

It should be kept in mind that the emphasis of this chapter has been on the future and on emerging needs. Current needs will not only remain with us but will characterize the majority of offerings. Along those lines, even the emerging needs are indeed already with us to a degree. While at the leading edge of change today, the research suggests that these emerging needs will increasingly join the mainstream over the next decade. Understanding and embracing their emergence will provide designers and developers insight upon which to develop their designs.

References

Beck, D., & Cowan, C. (1996). *Spiral dynamics: Mastering values, leadership, and change.* Malden, MA: Blackwell.

Friedman, T. (2005). *The world is flat: A brief history of the twenty-first century.* NY: Farrar, Straus and Giroux.

Gladwell.com. (n.d.). The tipping point. Retrieved October 23, 2014, from http://gladwell.com/the-tipping-point/

Hines, A. (2011). *ConsumerShift: How changing values are reshaping the consumer landscape.* Tucson, AZ: No Limits.

Inglehart, R. (1997). *Modernization and postmodernization: Cultural, economic, and political change in 43 societies.* Princeton, NJ: Princeton University Press.

Robinson, J. P., & Geoffrey, G. (1997). *Time for life: The surprising ways Americans use their time.* University Park, PA: Pennsylvania State University Press.

About the Author

DR. ANDY HINES is Assistant Professor and Program Coordinator for the University of Houston's Graduate Program in Foresight and is also speaking, workshopping, and consulting through his firm *Hinesight*. His 24 years of professional futurist experience includes a decade's experience working inside first the Kellogg Company and later Dow Chemical, and consulting work with Coates & Jarratt, Inc. and Social Technologies/Innovaro. His books include *Teaching about the Future, ConsumerShift: How Changing Values Are Reshaping the Consumer Landscape, Thinking about the Future, and 2025: Science and Technology Reshapes US and Global Society*. His dissertation was "The Role of an Organizational Futurist in Integrating Foresight into Organizations." Dr. Hines can be contacted about this chapter at ahines@uh.edu or 832.367.5575.

Part V
SPECIAL TOPICS IN DESIGN THINKING

23

FACE AND INTERFACE: RICHER PRODUCT EXPERIENCES THROUGH INTEGRATED USER INTERFACE AND INDUSTRIAL DESIGN[1]

Keith S. Karn

Bresslergroup

Introduction

When users flip a light switch, turn a knob to adjust the volume on a radio, or swipe their fingers across a touch screen, they are interacting with the product's user interface (UI). The user interface encompasses the physical and digital components that allow a user to communicate with a machine or device. The devices we use in everyday life are constantly evolving, and they have morphed drastically in the past 50, and even 20, years. Likewise, user interfaces and the UI design process have changed considerably.

This chapter begins with a call for reintegration of hardware and software UI development, which have evolved into separate silos within many organizations. Next, I provide an overview of emerging UI technologies in new product development (NPD) that are providing designers with opportunities to expand products beyond the limits of

[1] Acknowledgement: I borrowed the phrase "Face and Interface" from Richard Rohr.

physical controls and screens. In the last half of the chapter, I suggest methods for teams who are ready to dig in and develop UI and industrial design (ID) in parallel. (This chapter assumes that teams have already discovered and defined the problems to be solved by the product and are at the threshold of creating and evaluating concepts.) And as no two projects progress identically—and as no process informed by design thinking proceeds in a wholly linear fashion—I conclude with seven questions to ask yourself along the way in order to guide your particular, sometimes unavoidably meandering, but hopefully more focused path.

Defining Terms

In this chapter, I use the term *digital* in reference to digital visual displays, often paired with a touchscreen for user input. There is a tendency today to give every new product a touch screen, but that may not always be for the best—more on that later.

In the product development world, there is some confusion around the term user interface (UI) and other similar terms, including user experience (UX) and interaction design (IxD). Because the disciplines are relatively new, they are still being defined. For the purposes of this chapter, UI refers to both the physical and digital (on-screen) interactions between a human user or operator and a device or piece of equipment. I like to think of the user interface as the means of communication between human and machine.

A person who designs UIs is an interaction designer. IxD refers to the art and science of user interface design, but people often misuse it to indicate only the digital portion of the interface. The term *UX* is frequently misused as the exclusive domain of website design, but it actually denotes a broader, more holistic human-product experience. It takes into account considerations such as the purchasing process, the maintenance of a product, how it will be stored, customer support, and activities all the way through to end of life.

While UI encompasses both the physical and digital components of a product's controls and displays, this chapter will occasionally call out physical versus digital features since the goal is to clarify the process of developing each.

23.1 Divergent Paths: User Interface in Physical and Digital Products

Separate Development Paths

Prior to 1980, user interface design fell under the domain of industrial design (ID) and mechanical engineering because it was so physical, and it was primarily driven by the selection of appropriate controls such as buttons, switches, and knobs. That changed with the advent of the Age of Computers, and hardware and UI software development processes have evolved separately, even within companies, ever since. Hardware generally takes longer to design, build, and test and typically follows a more linear Stage-Gate (or *phase-gate*) development process. The Stage-Gate model divides the process into a series of tasks (stages) and decision points (gates) that a team advances through sequentially. This development process is not as fluid or flexible as the software development process. Much like building a house, you have to put down a foundation and have a well-defined architectural plan before you start building product hardware.

Software development, however, is generally more flexible and typically follows an agile development process. The concept of agile development was introduced in the early 2000s. It emphasizes adaptive planning, an iterative approach/and rapid, flexible response. It is characterized by lots of loops, short sprints with working software output, and minimal documentation. Because of this, software goes through more and shorter cycles during development. Rather than producing a physical prototype, like their hardware development counterparts, to test and refine, software developers relatively quickly write, test, and rewrite code.

Even though 3D printing and other technology is making it easier to prototype physical products, developing a physical device to production-ready status still takes longer than software development. The result is that we often develop hardware first because of this longer lead time, and software is brought in later. When the hardware team specifies the control and display elements of the UI before the interaction design work has even begun for the UI software, a suboptimal user experience is almost certain. Thus, this separation of hardware and software development—both organizationally and temporally—results in a lower-quality product.

A Call for Reintegration

Much of the UI work that was happening in the 1980s was for the screen, with the keyboard and mouse controlling the action. Gamers were the first to realize the limits of this paradigm. When everything was on a computer, most users took the mouse and keyboard for granted. To most users, those were the only input devices they knew. Video gamers refused to settle for these input devices, though, and they recognized the limitations of the typical computer output devices (displays with a small color gamut and crude audio systems). As a result, the game industry was born and began to develop new tools, such as joysticks and handheld controllers to better simulate natural user inputs. Soon the industry was pushing the envelope on higher-resolution displays and developing its own game processors like the Xbox and PlayStation for handling higher-quality animation and better audio.

These trends—developing processors for specific purposes (rather than general-purpose computers) and designing user inputs that go beyond the mouse and keyboard—have continued to grow. Today, countless products center around dedicated microprocessors and are restoring physical controls, like buttons and switches. The "Internet of Things" is populated by devices (aka "things") that can connect to the Internet. That connection allows devices to communicate with each other, without physical input or assistance from users. This will surely lead to product innovations and new business models and processes, and has the potential to spur efficiency and reduce costs and risks. Product developers need to consider how these inputs will contribute to the use of the product. ID considerations must, once again, be reassessed.

At the same time, consumers and product developers are realizing that not everything should be controlled by a touch screen—sometimes physical buttons or knobs are best. Ideally, touch screens enable compact and multipurpose devices because the screen allows a small surface to be reconfigured for a variety of tasks. A smartphone, for instance, can be a phone, camera, and web browser all in one. But for many devices, a combination of hardware controls (physical buttons and switches) and touch interfaces is ideal. To get the best mix of hardware and onscreen/digital controls, product developers need to reunite UI design with engineering and industrial design processes—ideally within the Design Thinking framework.

23.2 Emerging User Interface Technologies

Today, the interaction designers' toolbox is growing. As emerging technologies such as advanced audio technology, haptic or tactile feedback technologies, and gestural interfaces travel down the learning and cost curves, designers have more opportunities than ever to expand products beyond the limits of visual screens. More designers are realizing that, when used appropriately, these new tools can have a positive impact on usability, user experience, and brand recognition.

Auditory Feedback

UI design is, for the most part, missing out on the auditory dimension. Many product design firms do not even have dedicated sound designers. It is hard to imagine Apple's iPhone without the "swoosh" that signals a sent email or the chime to indicate a phone is connected to its charger. Very few products popular today have that sort of auditory element. Compared to products developed pre-1980 (when physical buttons, knobs, and switches provided their own inherent sounds), products today are way behind in terms of auditory feedback.

Imagine a wearable, mountable camera similar to the GoPro Hero camera. This camera is going to spend most of its time bolted to a helmet or the tip of a kayak or surfboard—somewhere out of reach or out of sight. The user is going to control the camera with a remote or by pressing buttons that are out of sight. For such a product, auditory feedback is essential since the user will not be able to see the camera during use. The user will rather rely on tactile cues to locate and identify the correct controls and rely on auditory cues to provide the feedback that the correct function is selected.

Haptic Technology

Haptic technology allows systems to stimulate the user's sense of touch by applying force, vibrations, or motion to the user. This is typically targeted to the users' hands or feet but can be applied to any body surface. For instance, instead of seeing or hearing a device power on, the user might feel it vibrate as it turns on. Haptic technology appeals to the user's tactile senses and kinesthetics. Tactile senses are those associated with skin contact (or, more technically, *cutaneous stimulation*) and include feelings such as temperature, pain, vibration, and pressure. The kinesthetic sense refers to the perception of movement and position of our limbs based on muscle forces exerted.

Most people are familiar with the haptic technology of a smartphone vibrating, but designers are only scratching the surface of what they can do with haptics. Today a vibration is analogous to a simple beep in the auditory domain or a single indicator light in the visual domain. It alerts the user but provides limited information. Imagine the difference in the audio domain between using a simple beep and conveying information to a user through speech or a beautiful song. Today, the haptic domain is using vibrations as "beeps" when they could be creating the equivalent of beautiful songs. The problem is that more sophisticated force feedback is still complex and costly.

Gestural Interfaces

Gestural interfaces allow computers to interpret human gestures via algorithms. For users, this means the ability to issue commands to a computer without making physical contact. Gestural interfaces are evolving from simple presence detection, though. Inexpensive cameras and high-speed processing are making camera inputs more common and resulting in higher performance. The Amazon Fire phone, for instance, has four front-facing cameras that enable gestural control, through which a user can control the phone's screen with the nod of a head or the wave of a hand. Similarly, the Leap Motion Controller is a device that allows users to manipulate a desktop computer's screen using hand and finger motions similar to a mouse, but does not require hand contact or touching. In other words, a user can control a computer with his or her fingers without ever making physical contact.

Augmented Reality

Since the 1980s, screens have diminished the use of auditory and tactile feedback. Visual feedback has gotten better, but there is always room for improvement. Augmented reality has the potential to take visual feedback to the next level. Augmented reality allows computer-generated sensory inputs like sound, video, or graphics to be overlaid on top of a real-world view. The user sees both the real-world view and the virtual images at the same time. Advances in augmented reality hardware and three-dimensional modeling are making this technology more viable from a user's standpoint.

23.3 New Technology Demands a New Development Process

As discussed earlier in this chapter, hardware and UI development have become separated over time. Today, the physical hardware components of a product are often designed by ID teams closely coupled with mechanical engineers. The digital interfaces are often designed separately by UI designers, who work closely with software engineers. More often than not, though, designing the hardware and digital interface components separately detracts from the final product. To build better products, development teams should integrate the hardware and software development processes. The design thinking mind-set could provide the means to enable this reunification.

Merge Development Timelines

Hardware development typically happens on a linearly constrained timeline. First there is a long design and development process. Then production, assuming tooling lead times, testing, compliance, and so on, can take upward of six months before finished goods are ready to ship. This forces the ID process into a strict timeline, and industrial designers must lock down the design as quickly as possible. In contrast, UI design operates on a much more flexible, faster development process, which is typically nonlinear. Interaction designers can be iterating the software weeks before launch, and even post launch in the form of software updates. Since the development timelines do not naturally line up, the ID and UI are typically developed separately and brought together at the end. As mentioned previously, that disconnect can diminish the final product.

To marry the UI and ID development processes, the development timelines can be merged at certain key points (Figure 23.1). Some elements of UI and ID are the same. For instance, both require the up-front user research necessary for developing empathy with the intended end user, discovering unmet needs, and distilling these down into key customer insights. Later, both require some form of prototyping, though digital UI (screen) prototypes tend more often to be the lower fidelity prototypes so helpful in early, iterative evaluation. Both hardware and software will need visual brand language development. So, the product development processes can be pulled together at those common points. For example, the software and hardware evaluation (typically in the form of usability or concept testing), can be done at the same time.

The UI development process, which evolved from the software development process, is typically broken into segments called sprints. The idea is that development happens in a series of sprints that are limited in time, typically a few weeks. In each sprint, designers tackle as much as they can within the timeframe and are required to present functional deliverables at the end of each sprint. This process is flexible and open to change, which is much different than the ID development process. Hardware designers start wide, with a number of options, and then funnel their options until they lock down a final concept that can be refined until it is ready for production.

Bridging the agile development process of UI and the linear Stage-Gate process of hardware design is challenging, but it can be done by forcing a more iterative process. Product developers can commit to a series of sprints that are set in time (Figure 23.2). In each sprint, the hardware team must deliver a physical prototype, but the expectations for those prototypes can be lowered. They do not need to be perfect because their primary function is to test hardware and software together and to generate user feedback. This approach to problem solving is, of course, in line with the notion of the design thinking approach in the context of NPD: teams develop simple prototypes, then, armed with feedback, iterate further. The result is a better, faster, more efficient process and a more cohesive final product.

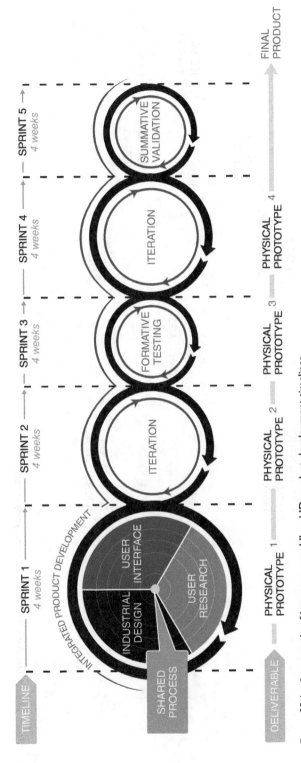

Figure 23.1: An example of how to merge UI and ID product development timelines.

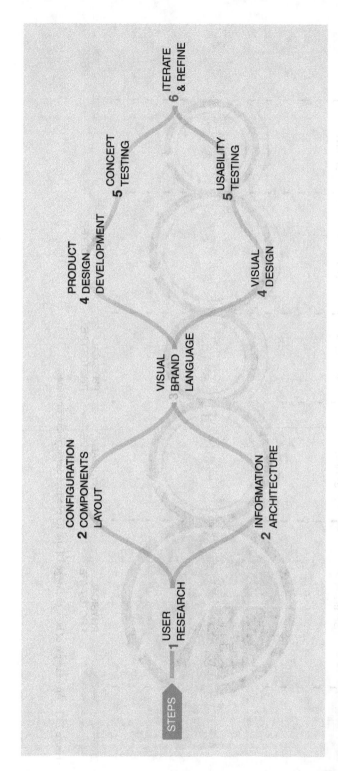

Figure 23.2: A blended process plan for prototyping in parallel.

Prototype in Parallel

Early on in the process, the UI software will likely be created with wire frames and work-flow diagrams, which can be printed out and turned into paper prototypes. Then, when initial form-factor physical prototypes are ready, the paper screens can be stuck on the physical prototype to represent the screen designs in the context of the physical device.

At higher levels of fidelity, the development team might run the UI prototypes on the screen hardware of the intended final display. The screen might not yet be connected to the hardware, but it can still be driven by a computer concealed "behind the curtain" to get a sense of the UI interactivity on the real display and a higher fidelity look and feel of the micro-interactions. It is important to integrate the software and hardware during the early, iterative prototyping to really understand the capabilities and limitations of particular display hardware. Do the colors display correctly? Can the screen adjust to ambient lighting conditions? Do elements of the UI need to be altered to interact more coherently with the industrial design, or vice versa? These are the types of questions a holistic development process will draw out.

23.4 Seven Questions to Guide the Integration of Industrial Design with User Interface Design

For managers looking to integrate hardware and UI development, these seven questions or decision points can help guide a project team. They are drawn from the collective experience of myself and my colleagues, with the caveat that each project differs based on goals and circumstances.

Who Is Leading the Process?

This question refers to both organizational leadership and which aspects of the project are taking the lead simply by starting first. When it comes to organizational leadership, a collaborative process that includes both hardware and software teams is best. Typically, a mechanical or electrical engineer will lead the hardware development team and a software engineer will lead the software team.

Ideally, each project will have its own multidisciplinary project team. The project team members will vary depending on the product you are developing, but the project team may include interaction designers, software engineers, industrial designers, mechanical engineers, interaction designers, product planners, user research experts, and marketing experts. The idea is to have one unified team working together on a product rather than separate hardware and software teams working independently of one another. It is all about getting the right people to the table and getting them there early.

In most cases, it makes sense for hardware to lead—to a certain extent. It takes more time to produce the hardware, and the hardware specs generally inform the software. (The development team cannot develop code to run on a processor that has not yet been selected or might not even exist yet, and if the product has a screen, the interaction designers will need to know the specifications of that screen.) The problems that teams run into start when the software designers arrive too late in the game—after the hardware engineers have already determined critical features based on cost constraints, size, and weight.

For instance, a major camera company was developing a camera with a touch screen that had menu-scrolling capabilities similar to the scrolling function of the iPhone's contacts list. Unfortunately, the UI team was brought into the process too late to achieve this. The hardware team had already chosen a processor that could not meet the demands of a fast-scrolling function. The menu scrolled, but the processor was so busy scrolling, it could not receive the next command to stop the scrolling and select a menu item. To solve the problem, the software team put more "friction" in the system (a deceleration function) and essentially slowed the scroll function so the processor could keep up. The final product would have been better if UI considerations had been taken into account earlier on, when the engineering team was specifying the hardware.

Managing the sometimes conflicting needs of UI and ID teams is a balancing act. Most UI teams would love to have high-resolution, capacitive touch screens on every product, but each project has its own constraints. The ID team might specify, for instance, that the final product cannot be large enough for a six-inch screen. It is important to keep product planning and marketing teams involved, too. Marketing team members would supply the "voice of the customer" and say, "Yes, a large touch screen would be great but the final product has to sell for less than $200."

What Are the User's Tasks and Needs?

Just as it is important to know how much a user will pay for the product, it is critical to understand who the user is and how we expect the user to interact with the product. Once the project team is in place, the next step is to define the system requirements. Human Factors 101 teaches the importance of the relationship between users, their tasks, the products, and the environments in which the products will be used—and, finally, the demands they place on the user (Figure 23.3).

The takeaway for a development team is to apply these questions to their product: What is the machine going to do? What is the human going to do? Once we know what we are asking the human to do, the question becomes: what information has to be communicated from the human to the machine in terms of inputs to the device, and what information needs to be communicated from the machine to the human in terms of outputs from the device?

Environment also plays a role. Is the product going to be used in a public space or a private space? Will one person use the device, or will it be used by multiple people, either individually or as a group? Will it operate in bright sunlight? A backlit LCD screen used outdoors or in a bright environment will need to crank up the brightness to overcome ambient light. As soon as the user has to lift a hand to shield the screen from light, you have lost the usability battle. That said, in a typical domestic environment, standard LCD screens with LED backlights do work well, as do LED-based displays like the seven segment modules on stoves and microwaves.

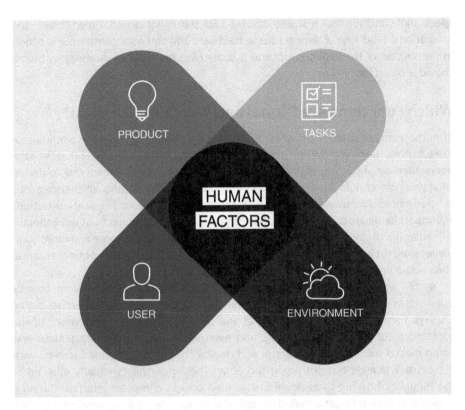

Figure 23.3: Human factors sits in the intersection of products, tasks, users, and environment.

The user's tasks are important, too. More tasks mean more complexity, and complexity will influence UI design. How many functions do there need to be and how complex is their presentation? If a device is going to display temperature, can the display show just three general ranges (hot, medium, cold)? Or does it need to display temperature readings to the precision of a tenth of a degree? Do all of the functions appear on the display at once? It might be beneficial to hide extraneous information in order to make the product appear less complex, but developers should not try to diminish complexity by making the screen smaller than necessary or burying functions behind too many layers of interface.

Industrial designers might ask if the user is going to hold the device, if it is going to be mounted on a wall or another device, or if it will be freestanding. This might influence the shape and weight of the final product. Is it being used outdoors and does it need to be waterproofed? What kind of physical use and abuse will it need to withstand?

All of these questions will direct the work of a project team. The answers will lead to different takeaways for interaction designers and industrial designers, so it is important that everyone involved in product development take these questions into consideration. Build a matrix that lists possible hardware and software UI elements and the resulting impact of each on the user experience. A sample checklist of hardware specs might include touch screen display, LED status lights (red, green, blue), and navigational hard keys. Balancing these hardware and software constraints is critical to the success of the final product and is something the product development team should account for early on.

Which Functions Are Digital and Which Are Physical?

After the team has determined what information the user and device will communicate to each other, the next question is: how will they communicate with each other? Will the interactions be physical or digital? This is where the development team should decide what types of technologies and interfaces the product will feature and which commands will be dedicated hardware controls versus on-screen, soft buttons. Physical interactions will impact hardware capabilities and industrial design. Digital interactions will generally fall under the UI domain and require software capabilities. Today, more and more companies want their products to have screens, but these features should not be gratuitous. Their suitability needs to be thought out carefully.

In the early 1980s, military aircraft cockpits were transitioning from literally hundreds of electromechanical displays and controls to integrated, interactive, digital displays. I worked on the F/A-18 Hornet and the A/V-8B Harrier where much of the navigation, communication, sensing, and even weapons system management was being moved from dedicated control and display "heads" to integrated screens with dynamically labeled buttons along the edge. The fighter pilots initially rebelled at the thought of having to navigate menus on a cockpit computer interface during a dogfight or critical target acquisition task. The design team had to consider carefully which functions still needed separate, dedicated controls (and sometimes dedicated displays) and which could be integrated into a more central system. Only the controls that made sense as part of a layered, visual presentation were moved to the screen to make the most of limited cockpit space. The rest of the interactions remained physical.

Today's product developers are making similar decisions. Some functions still need physical controls, but many others can be adapted for digital screens. It is important to envision in advance exactly how your device will work. Imagine if Apple had designed the physical shape and layout of the iPhone before deciding a swipe of a finger on a touch screen or a fingerprint ID on a home button would activate the device.

No matter what the product, interaction design needs to be thought of holistically. A lot of people think of interaction design as pretty pictures on-screen, but it also encompasses all the ways users physically interact with their products. Determine early in the design process what will be physical and what will be digital.

What Are the Hardware Characteristics? Define the Display and Other UI Elements

Companies should aim for a unified look and feel for their product lines—all the products in a line or family characterized by a cohesive brand language—translated to the physical design as well as the UI design. Users do not want to see dials on one product and a toggle switch on another for the same function in the same product line. Transfer of learning carries over from one device to the next and adds value. Clients often ask us to develop products in a line at different price points typically resulting in screens of differing sizes or other hardware differences. It is important during the design process to consider how UI elements might translate to a display on, for instance, a small tablet versus a larger monitor.

Because humans are such visual animals, design teams sometimes overlook speakers and audio capability. With some products, due to cost constraints, the team will choose a tone generator that can only beep or buzz instead of a more sophisticated audio system. That can constrain the overall design. Sometimes a killer tone for the power on and power off functions can become a strong signature for a brand or product. It is important to expand your definition of UI and interaction design to include auditory elements, along with planning for these elements early on in the design process.

How Can UI and Industrial Designers Best Work Together?

Industrial designers are not experts in interaction design, and UI or interaction designers are not experts in industrial design. To help bring the two together, co-location, team-building activities, and cross-training gets everyone in the same space and on the same page. Another strategy is to divide teams by product lines rather than by discipline. For instance, a printer and office supply company might divide its product developers into a Home Products Group, an Office Products Group, and so on. Each group would consist of industrial designers, interaction designers, and graphic designers who all report to one team manager.

Storyboarding

Another way to bridge the divide is to storyboard product development so each discipline can envision how UI and ID contribute to a functioning final product. A storyboard might show how a user would interact with the product and how the product would look at different points in the process (Figure 23.4).

To storyboard, it is best if the interaction and industrial design teams work together to generate the storyboard concept and determine user inputs and system outputs. (For more on storyboarding, see the chapter, "Visual Storytelling" in Bill Buxton's *Sketching User Experiences*.) The collaborative team should discuss the flow of elements and best-case scenario for the user-product interaction. The interaction design team can map out how the UI architecture contributes to the process and the industrial design team can outline the role of the hardware. The storyboard helps bring those two elements together to show how UI and industrial design will work in tandem. Storyboarding is especially helpful when co-location is not an option and team members must work remotely.

Sprints

A best practice model for designing in parallel is one that views the entire process as a series of sprints. For instance, if a team has 12 weeks to design a product, they could divide the process into three "sprints." In the first sprint, industrial designers, interaction

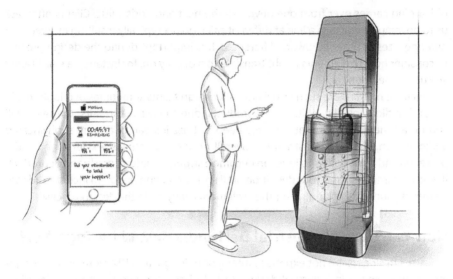

Figure 23.4: A slide from a storyboard mapping out a typical day in the life of an automated homebrew device with mobile app.

designers, and engineers would work together to develop the first physical prototype. In the second sprint, the same team would iterate the process and advance a second physical prototype. In the third four-week sprint, the cycle would repeat and produce a third and final product. This helps marry the faster, agile development process of software development to the more rigid Stage-Gate process of hardware development.

What Kind of Prototyping Does This Product Need?

Determine the goals of prototyping at each point in development, and base the fidelity of the prototype—whether it needs to function, and how closely it needs to mirror the final product in terms of appearance and behavior—on these goals. At times, a paper prototype will suffice. At other times a three-dimensional print will work. As you advance, the prototypes may need to become indistinguishable from the final product.

Often, the industrial design and interaction design prototypes are developed separately, which feeds a disconnect that expands as you move through the design process. For example, when interaction designers know they will run the product on a touch screen, it can be tempting to test the UI on an iPad or in Flash. Meanwhile, the industrial design team will be working on a beautiful physical model. When the two eventually come together, they may not mesh.

Even if both the ID and UI prototypes have to be slightly dumbed down in terms of their level of fidelity, a more integrated approach to prototyping will ultimately end up saving time. It can also help avoid having to force incompatible UI and industrial designs together at the end of the development process. The UI and industrial design are much more powerful when they are co-prototyped.

Test Early and Often

To make sure your product is on the right track, it is best to test early and often in an iterative cycle. Design, prototype, test, evaluate. Rinse and repeat. Even at the paper prototype stage, it is not too early to do usability testing. Early testing can help determine if users understand the proposed information architecture, even if the physical interactions are not quite there yet.

The earlier you start testing, the better. Product development can be like hardening concrete. At first it is easy to redirect, shape, and form, but as time goes on, the concrete begins to set and is very difficult to change. The longer you wait, the more your prototype is going to represent the final product and the harder it will be to change. Companies developing new products often ask: Are we ready? Should we cancel or postpone this test? Almost always, it is best to continue with a test. You can learn a lot with what you have, and it is better to learn more, earlier than later.

How Will You Specify the Integrated Design?

When it comes to creating product specs, it is not about integration between the software or UI team and the hardware or ID team. Here, integration between the user interface design and industrial design teams and their respective execution or development teams is more important. The on-screen portions of the UI design need to be coordinated with a software or firmware development team. The physical portions of the UI and the industrial design need to be coordinated with engineering, manufacturing, and purchasing teams.

23.5 Practice Makes Perfect

Keep in mind that a book like this is full of "how-tos" describing perfect situations that never really exist. So do not panic if your project is not tracking along in a picture-perfect way. Remember that you are not going to change corporate culture overnight—especially if you work in a large organization. Take small, steady steps in the direction to which we are pointing in this chapter and toward the overall design thinking mind-set presented throughout this book. And keep your eyes open for rare opportunities for sudden leaps forward, such as when your CEO gets converted and wants to elevate your design function to the C-suite level.

By answering the questions in the preceding section, the manager of the development team should have a good sense of how well the UI and ID teams are working together. Furthermore, discussing these questions among the team should provide some thoughts on how to increase collaboration and the quality of the resulting design.

All of this may seem like an excessive amount of stopping to strategize and evaluate, but merging the UI and ID processes introduces enough variables to disqualify anything resembling a "standard," one-size-fits-all template. For the successful reintegration of two disciplines that have—to the detriment of the quality of our end products—become siloed, we need to become comfortable with a looser, more flexible paradigm. Speediness comes with practice, and after working through a few integrated process projects, teams will begin to accumulate repeatable processes and best practices.

About the Author

KEITH S. KARN, the Director of User Research and Human Factors at Bresslergroup, leads contextual inquiry and usability testing to add the rigor required to create successful user interfaces and intuitive product experiences. He speaks and publishes frequently on usability evaluation methods, improving customer acceptance through user-centered product design and user-interface design for emerging technologies. He earned his PhD in experimental psychology at the University of Rochester and degrees in ergonomics and industrial engineering at North Carolina State University and Penn State. Keith also taught human factors and human-computer interaction for many years at the Rochester Institute of Technology and the University of Rochester.

25
DESIGN THINKING FOR SUSTAINABILITY

Rosanna Garcia, PhD,
North Carolina State University

Scott Dacko, PhD,
University of Warwick

Introduction

The focus in this chapter is on how a sustainability approach can be merged with design thinking to develop socially responsible and environmentally sustainable products. Design thinking brings a human-centered approach to designing for sustainability by combining empathy for the people impacted by the service/product being designed with creativity in developing radical solutions, and rationality to analyze what is feasible in the given context. As such, design thinkers have the potential to slow down environmental and social degradations more so than economists, engineers, or even governmental agencies because they create products and services that incorporate empathy for the person-and-problem situation into product and service design. When designs inspire individual consumers/end users to change their behaviors and act in a more sustainable manner, environmental longevity and social benefit will be more likely to ensue (Young, 2010). Just one example of design thinking and sustainability being in disharmony is the single-use coffee capsules (such as Keurig K-cups). In 2013, Keurig Green Mountain produced 8.3 billion K-cups—enough to circle the Earth 10.5 times.[1] Although convenient, K-cups are not environmentally friendly.

In this chapter, design thinking is merged with design for sustainability insights to provide a means whereby consumers become inseparable partners in ensuring the longevity of our natural, social, and economic environments. There is a growing recognition and acknowledgement that designers and manufacturers are substantially responsible for many of the man-made stresses imposed on and throughout our planet

[1] www.motherjones.com/blue-marble/2014/03/coffee-k-cups-green-mountain-polystyrene-plastic

as 80 percent of products are discarded after a single use and 99 percent of materials are discarded in the first six weeks of use (Shot in the Dark, 2000). Clearly, a sustainable perspective to design thinking approach is necessary if environmental and related social and economic issues are to be targeted and addressed effectively.

25.1 Design for "X"?

Assimilating sustainability into design thinking first requires a definition for sustainability. We describe it as a three-legged stool: the integration of environmental, economic, *and* social issues (Makower, 2014) or in the systemic viewpoint of biologist Barry Commoner (1971): "everything is connected to everything else." An introduction to and evaluation of a range of design for "X" sustainable strategies establishes a foundation for further discussion. Over the years, numerous product design criteria have been introduced to manufacturers in varying degrees as a way of thinking about the impact of new products and services on the environment (see Table 25.1). Such design for "X" strategies focus on specific engineering/research and development (R&D) issues and typically view consumers passively—as opposed to partners—in the design criteria.

As a means to evaluate more critically the various design for "X" strategies, *Design for Sustainability* (DfS) (also referred to as Design for Efficiency), *Design for Effectiveness* (DfEffv) and *Design for Environment* (DfEnv) are presented as the three overarching approaches that encompass most of these more specific design strategies.

Design for sustainability/efficiency. In 1992, McDonough and Braungart penned the Hannover Principles to insist on the rights of humanity and nature to co-exist;

Table 25.1: A Range of Design for "X" Strategies		
Design for Reuse and Recovery[e]	Design for Durability[b]	Design for Benign Waste Disposition[c]
Design for Abundance (Upcycling)[a]	Design for Disassembly[b,d]	Design for Hazard Reduction[c]
Design for Materials Optimization[e]	Design for Repair & Upgrade[b]	Design for Manufacturability[c]
Design for Waste Efficient Procurement[e]	Design for Dematerialization[b]	Design for Maintainability[c]
Design for Deconstruction and Flexibility[e]	Design for Servicization/Servitization[c]	Design for Human Safety[c]
Design for Energy and Material Conservation[c]	Design for Revalorization[c]	Design for Human Capital[c]
Design for Natural Capital[c]	Design for Economic Capital[c]	Design for Product Recovery[c]
Design for Product Disassembly[c]	Design for Recyclability[c,d]	Design for Release Reduction[c]

Sources:
[a]McDonough & Braungart (2002)
[b]Autodesk (2014)
[c]Fiksel (2011)
[d]White, St. Pierre, and Belletire (2013)
[e]WRAP (2015)

accepting responsibility for the consequences of design; creating safe objects of long-term value; and eliminating the concept of waste. A central tenet of sustainable design is *eco-efficiency*. As defined by the World Business Council for Sustainable Development (WBCSD, 2000), "eco-efficiency is achieved by the delivery of competitively priced goods and services that satisfy human needs and bring quality of life, while progressively reducing ecological impacts and resource intensity throughout the life cycle to a level at least in line with the Earth's estimated carrying capacity." In short, it is concerned with creating more value with less impact.[2]

Although eco-efficiency has become more prevalent in these and other firms' design strategies in the past 20 years, critics of eco-efficiency argue that design for sustainability/efficiency solutions are likely to result in short-term cost savings and efficiency gains with "low-hanging fruits." Laszlo and Zhexembayeva (2011) describe this as "bolted-on" sustainability versus "embedded" sustainability. Bolted-on sustainability is sometimes seen as "green washing" when sustainability efforts are conducted primarily as marketing measures, whereas embedded sustainability is seen as developing a "green gestalt" within the company that drives the firm's strategy. Embedded sustainability embraces not just "less harm" but "zero harm" and seeks positive environmental benefits as core business activities (see Table 25.2).

Table 25.2: Bolted-on versus Embedded Sustainability Design Strategies (Laszlo & Zhexembayeva, 2011)		
	Bolted-on Sustainability	**Embedded Sustainability**
Goal	Pursue shareholder value.	Pursue sustainable value.
Scope	Add symbolic wins at the margin.	Transform core business activities.
Customer	Offer "green" and "socially responsible" products at premium prices or with diminished quality.	Offer "smarter" solutions with no trade-off in quality and no social or green premium.
Value capture	Focus on risk mitigation and improved efficiency.	Reach across all levels of sustainable value creation.
Value chain	Manage company's own activities.	Manage across the product or service life-cycle value chain.
Relationship	Leverage transactional relationship. Stakeholder such as customers, employees, and suppliers are resources to be managed and sources of input.	Build transformative relationships. Co-develop solutions with all key stakeholders including NGOs and regulators to build system-level change.
Competitor	Operate only in win–lose mode in which any gain is competitor's loss.	Add cooperation with competitors as potential source of gain.
Organization	Create a "scapegoat" department of sustainability.	Make sustainability everyone's job.
Competencies	Focus on data analysis, planning, and project management skills.	Add new competencies in design, inquiry, appreciation, and wholeness.
Visibility	Make green and social responsibility highly visible and try to manage the resulting skepticism and confusion.	Make sustainability performance largely invisible but capable of aligning and motivating everyone.

Source: Laszlo and Zhexembayeva (2011).

[2] www.wbcsd.org/pages/EDocument/EDocumentDetails.aspx?ID=13593

Design for effectiveness. Following on the embedded sustainability perspective, McDonough and Braungart (2002) introduced *Design for Effectiveness*. DfEffv argues that a product or service could meet criteria for *eco-efficiency* but not be eco-effective, which refers to not just minimizing a negative footprint, but also having a positive footprint through sustainable growth. For example, some opponents to electric vehicles (EVs) argue that the additional resources needed to manufacture the more expensive EV, the use of nonrenewable energy to power the vehicle and the disposal of toxic batteries does more harm to the environment than good. EVs may be eco-*efficient* but some critics doubt if they are eco-*effective*. Eco-effectiveness takes a cradle-to-cradle approach to the product/service life cycle where products are not taken to the 'grave' but are upcycled back into the system. In Cradle to Cradle[3] design (also referred to as C2C, cradle 2 cradle, and regenerative design), at a product's end-of-life, all materials used in a product or service become either biological nutrients (organic materials) or technical nutrients (nontoxic inorganic or synthetic materials) (McDonough & Braungart, 2002); see Figure 25.1. By choosing to adopt fully the DFEffv approach for one line of its products, sport

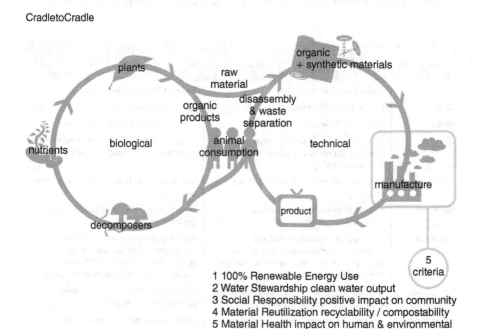

CradletoCradle

1 100% Renewable Energy Use
2 Water Stewardship clean water output
3 Social Responsibility positive impact on community
4 Material Reutilization recyclability / compostability
5 Material Health impact on human & environmental

5 criteria

Figure 25.1: Cradle-to-cradle approach to design for effectiveness.[4]

[3]The term *Cradle to Cradle* is a registered trademark of McDonough Braungart Design Chemistry consultancy.
[4]"Biological and technical nutrients (C2C)" by Zhiying.lim, licensed under Creative Commons Attribution-Share Alike 3.0 via Wikimedia Commons. http://commons.wikimedia.org/wiki/File: Biological_and_technical_nutrients_(C2C).jpg#mediaviewer/File:Biological_and_technical_nutrients _(C2C).jpg

lifestyle company PUMA has designed the InCycle collection, a range of footwear and apparel that are specially labeled as being made from materials that are able to be relatively easily turned into both biological and technical nutrients at the end of their useful lives and where in-store bins make it convenient for consumers to return the used products.

In sum, DfEffv's emphasis on regeneration from both technical and biological perspectives clearly presents new product and service developers with ambitious aims, a broad scope and multiple challenges spanning potentially long periods of time. The design for environment approach, discussed next, is seen as consistent with the DfEffv approach but the emphasis is on pursuing an integrated set of *specific* design means by which such system-wide and longer-term sustainability goals can be achieved.

Design for environment. This approach is "the systematic consideration of design performance with respect to environmental, health, safety, and sustainability objectives over the full product and process life cycle" (Fiksel, 2011, p. 6). DfEnv merges sustainability strategies with the new product development (NPD) process and takes into detailed consideration how the needs and expectations of stakeholders can be achieved in the most environmentally benign and socially and economically sustainable manner. Specifically, the design principles of DfEnv are embedded within four sustainability strategies (see Figure 25.2): design for *dematerialization,* design for *detoxification,* design for *revalorization,* and design for *capital protection and renewal.* As these approaches are important in establishing a sustainable innovation design thinking perspective, a quick summary of each is provided next.

Design for dematerialization focuses on the reduction of material, as well as the corresponding energy requirements, for a product and its associated processes

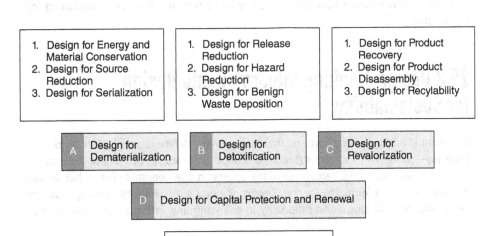

Figure 25.2: Four major strategies of design for environment.

throughout the life cycle. An example is when Procter & Gamble reduced the amount of water of their Tide laundry detergent by compacting two or three times as much cleaning power into the same amount of liquid detergent. This significantly reduced water content, package size, transportation weight, and shelf space.

Design for detoxification focuses on reducing or eliminating the toxic, hazardous, or otherwise harmful characteristics of a product and its associated processes, including waste streams that may adversely affect humans or the environment. For example, in 2013, Unilever started to use smaller, compressed aerosol cans for their Sure, Dove, and Vaseline brands. The new cans use on average 25 percent less aluminum and, due to the smaller size, reduce the overall carbon footprint of the product by an average of 25 percent per can.[5]

Design for revalorization focuses on recovering, recycling, or otherwise reusing the residual materials and energy that are generated at each stage of the product life cycle, thus eliminating waste and reducing virgin resource requirements. For example, European Union Member States are required to establish collection systems for end-of-life autos and ensure that all vehicles are transferred to authorized treatment facilities through a system of vehicle deregistration based on a certificate of destruction.[6]

Design for capital protection and renewal focuses on ensuring the safety, integrity, vitality, productivity, and continuity of the human, natural, and economic resources that are needed to sustain the product life cycle. Supporting both the continuity and renewal of natural resources, UK specialty paper manufacturer James Cropper PLC developed breakthrough processes for recycling cocoa husk waste into paper and for recycling both the paper and plastic components of disposable coffee cups.[7]

In the next section, the DfEnv approach is overlaid on the design thinking philosophy to demonstrate how the consumer can become a partner in designing for sustainability.

25.2 Design Thinking Integrated into Design for Sustainability

As exemplified in earlier chapters, there are several different approaches to design thinking—some with six stages (Stanford school[8]), five stages (IDEO[9]), or four stages (Liedtka & Ogilvie, 2011). Regardless of the approach, the general belief is that design thinking combines "empathy for the context of a problem, creativity in the generation of insights and solutions, and rationality in analyzing and fitting various solutions to

---

[5] www.sustainablebrands.com/news_and_views/articles/unilever-redesigns-aerosol-deodorants-smaller-environmental-impact

[6] www.epa.gov/oswer/international/factsheets/200811_elv_directive.htm

[7] www.cropper.com/news/post.php?s=2014-07-08-sdra-and-james-cropper-present-durapulp

[8] https://dschool.stanford.edu/groups/k12/wiki/332ff/Curriculum_Home_Page.html

[9] http://socialinnovationmn.com/design-thinking-toolkit-for-educators/

the problem context" (Kelley & Kelley, 2013), by inviting the end user/consumer to be a part of the innovation process (Liedtka & Ogilvie, 2011).

Toward achieving an integration of design thinking with sustainability, we overlay the design thinking approach of Liedtka and Ogilvie (2011) with the design for environment framework of Fiksel (2011) expanded for further emphasis on social sustainability to develop a method for bringing human empathy into the design process for sustainable innovations. While Liedtka and Ogilvie's (2011) approach may appear simplistic with its four steps of *What is, What if, What wows,* and *What works,* it aligns solidly with the concepts of design for sustainability. When merging DTh with DfEnv, *What is* includes *analysis methods* such as sustainable life-cycle assessments; *What if* and *What wows* both include *rules and guidelines,* such as design for revalorization; and *What works* includes *sustainability indicators and metrics* (Figure 25.3). This is discussed in detail as design thinking for sustainability (DThfS).

What Is?

The goal of the *What is* phase of discovery is to frame sustainability problems from the consumers'/end-users' perspective. Typical questions posed are: What are the customers' sustainability needs and wants? What sustainability problems do end users see from product use? What recycling programs are in place for recovery and reuse? How do consumers participate? During the *What is* phase, a deep understanding of customers' habits, routines, and customs is uncovered. Methods to use in the *What is* phase include:

Visualizing sustainability. "Conjuring up visual depictions of customers and their experiences [to] make them human and real" (Liedtka & Ogilvie, 2011, p. 49),

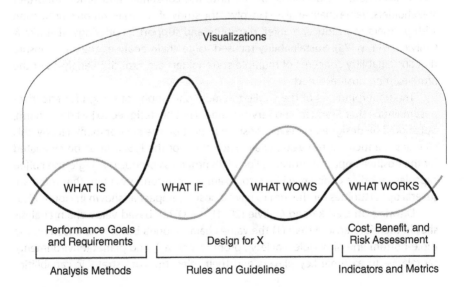

Figure 25.3: Design thinking for sustainability foundation.

Figure 25.4: Photo stories.

visualization makes sustainability less abstract in the design thinking process. Importantly, visualization helps to "match" mental models of stakeholders prior to committing significant resources to the endeavor. Sustainability visualization tasks should be kept simple; drawing and photographs (Figure 25.4) are good methods for making sustainability real and for coming to an agreement on what is being evaluated, for example, landfilling versus recycling component materials.

Sustainability journey mapping. By creating representations of a customer's sustainability-related experiences in flowchart or other graphic format as he or she interacts with a product or service (Liedtka & Ogilvie, 2011, p. 61), sustainability concerns can be assessed. Captured through focus groups, surveys, or other means, they establish the nature of customer involvement (e.g., concerns, joys, disgusts) at each stage of the product/service life cycle (P/SLC).

Sustainability value chain analysis. The primary goal of the *What is* stage is to search where sustainable value exists for the firm, the consumer, and other important stakeholders. Value chain analyses involve "the study of an organization's interaction with partners to produce, market, distribute and support its offerings" (Liedtka & Ogilvie, 2011, p. 75). Sustainability-focused value chain analyses therefore ensure that sustainability concerns of multiple stakeholders are explicit in analyses of the firm's existing business model.

The underpinnings of the cradle-to-cradle philosophy of design for effectiveness maintain that products and services should not be designed to just do no harm, but should be designed to have a positive impact on the environment; clearly, this changes the focus of the value chain. Each stage of the P/SLC must be evaluated for the positive impact it provides. For a B2B firm mass producing single-use coffee capsules for office coffee machines, for example, it becomes evident that there are several opportunities for the firm to have a positive impact as shown in Figure 25.5.

Drawing on Liedtka and Ogilvie (2011),[10] a DThfS-based value chain analysis should therefore ensure that: (1) the value chain for one's business is based on the triple-bottom line of people (society), profit (economy), and planet (environment); (2) relative to external key players and their roles, the core strategic capabilities

[10] See Chapter 4 of Liedtka and Ogilvie (2011) for more details.

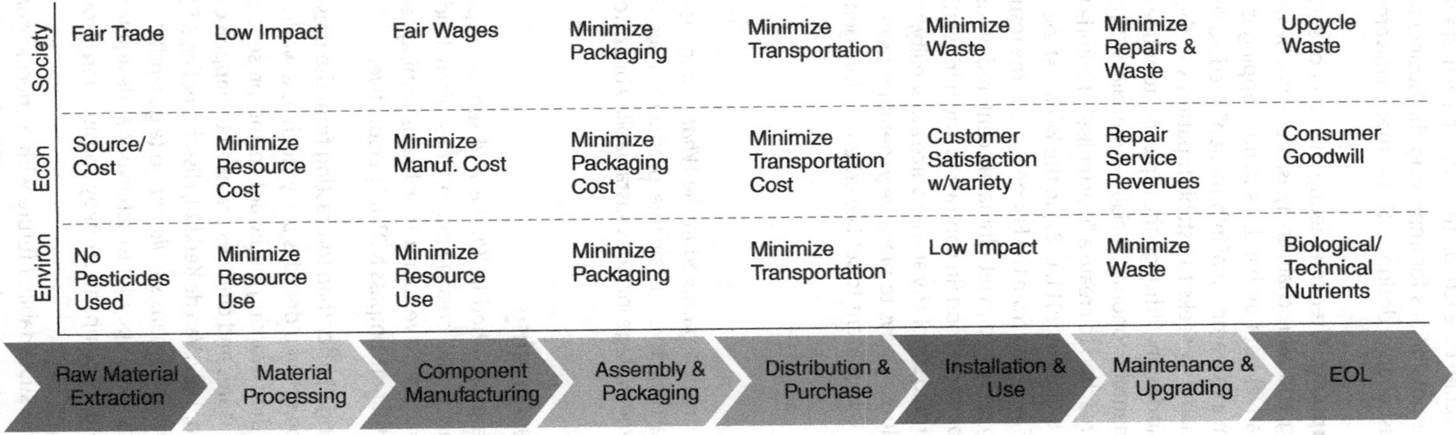

	Raw Material Extraction	Material Processing	Component Manufacturing	Assembly & Packaging	Distribution & Purchase	Installation & Use	Maintenance & Upgrading	EOL
Society	Fair Trade	Low Impact	Fair Wages	Minimize Packaging	Minimize Transportation	Minimize Waste	Minimize Repairs & Waste	Upcycle Waste
Econ	Source/Cost	Minimize Resource Cost	Minimize Manuf. Cost	Minimize Packaging Cost	Minimize Transportation Cost	Customer Satisfaction w/variety	Repair Service Revenues	Consumer Goodwill
Environ	No Pesticides Used	Minimize Resource Use	Minimize Resource Use	Minimize Packaging	Minimize Transportation	Low Impact	Minimize Waste	Biological/Technical Nutrients

Figure 25.5: Value chain analysis in sustainable design thinking.

needed to produce value to each of the 3Ps at each stage of the P/SLC are identified; and (3) the possibilities for improving the business model and product offerings through sustainable design with partner involvement are established and analyzed.

Sustainability mind mapping. After completing the tasks of sustainability visualization, journey mapping, and value chain analysis, there may be much qualitative (and maybe some quantitative) data on hand. As mind mapping is the process for "extracting meaning from a vast amount of information" (Liedtka & Ogilvie, 2011, p. 81), sustainability mind mapping seeks to establish patterns in the data that provides direction in sustainable design. Although the triple–bottom-line elements of sustainable design (environmental, economic, social) may conflict with each other, a general consensus should be reached to create a "master list" of criteria that an ideal design should meet (Liedtka & Ogilvie, 2011, p. 87) in the *What if* stage. For one sustainable design initiative, the Save Food from the Fridge Project, the resulting criteria for the ideal nonrefrigerating design to meet ultimately centered on using traditional and natural processes to preserve food (in contrast to any refrigeration). By factoring in user perceptions regarding simplicity and resource availability, the Project has developed innovations including ways to preserve vegetables using damp sand! As this example illustrates, it is important to set key criteria before generating solutions in the *What if* stage.

What If?

All the sustainability information gathered in the *What is* stage is now utilized in the *What if* stage to create a definitive sustainable product/service design using brainstorming exercises. The *What if* stage includes sustainability brainstorming and concept development.

Sustainability Brainstorming

Taking the design criteria established in *What is* and setting a specific challenge for designers can help to break the creative boundaries often inadvertently imposed by profit-maximizing firms. We have found four brainstorming techniques useful to generate creative energy that encompass a DThfS perspective:

Sustainability Backcasting. Rather than forecasting from the present, starting with a desirable future end-point even decades away (such as a world without single-use coffee capsules) and then working backward to the present can be valuable to assess what policy measures and changes are need to happen over time to achieve the outcome (Robinson, 1982; Van de Kerkhof, Hisschemöller, & Spanjersberg, 2002). It draws attention to what humans *should not* do (e.g., landfill materials) as well as what humans *should do* (e.g., upcycle) to achieve that desirable future. Holmberg's (1998) work on long-term company strategy formulation provides a framework to follow:

1. Identify a long-term sustainability future with a specific goal, such as a world without toothpaste tubes abandoned to landfills. Analyze the present situation.

2. Develop a creative design for what the sustainable product/service might look like that can deliver this future and the structure of the firm that could deliver that offering.

3. Using 5-to-10-year increments, identify and discuss the milestones that need to be achieved (e.g., 90 percent recycled and recyclable material used) to deliver the sustainable product/service.

Sustainability via emotionally durable attachment. Jonathan Chapman (2005) suggests that products should be designed for dependency where a consumer becomes emotionally attached to the products they own (termed *emotional durability*). When emotions to a product are durable, and empathy for the product is sustained from the point of purchase through continued delight and even surprise, consumption and waste are minimized as owners become reluctant to dispose of "prized" possessions.

As a brainstorming exercise, select a product or service to be designed for emotionally durable attachment. Take the product of interest (e.g., a refillable ink pen) and using multiple participants work to build a multilayered narrative on how the product is to become an integral part of the consumer's life. In the narrative, specifically: (1) anticipate the aging process of the product (e.g., pen surface wear) and how it develops character over time, (2) deliberate on how the product can become more self-sustaining (e.g., ink not easily drying out), (3) consider what product attributes (e.g., unique engravings, unique stand) are needed for an *emotional* connection with the consumer, and (4) incorporate both the physical and emotional connections into the product design and reflect on how well the product can continue to delight the consumer over time.

Sustainability via Eternally Yours designs. Similar to "emotionally durable" products, the Netherland Design Institute developed the concept of "Eternally Yours" in 1997 to promote the design of products that increase their "psychological life span": the time products are able to be perceived as worthy objects. Aimed squarely at early product waste simply due to boredom, perceived or actual technological obsolescence or irrepairability, products are designed to age in a dignified way (van Hinte, 1997). A sustainability brainstorming exercise to design products for a longer life (White et al., 2013, p. 16) is as follows:

1. Taking the product under consideration (e.g., a wristwatch), envision how it might provide usage over a 100-year span. Develop a concept scenario and create a storyboard.

2. Design a support system including services (e.g., upgrading) for this product's viability.

3. Develop an advertising campaign to entice someone to buy such a long-lasting product.

4. Considering your support and service system (e.g., for cleaning, repairing, upgrading), reanalyze and rediscuss your design and advertising campaigns.

Product-service replacement. A strategy for innovation that widens the scope for sustainable designs and their development is that of a product-service system (PSS), or the coordination of a set of products and services into a united offering (Goedkoop, van Halen, te Riele, & Rommens, 1999). The social, economic, and environmental impacts of a PSS approach become inherent as the viewpoint forces a shift from production and consumption as two separate entities to a systems perspective of a single closed-loop product life cycle (Mont, 2002). Further, the PSS approach forces more attention to the use phase of the P/SLC as opposed to the product design in R&D. A PSS therefore has the potential to reduce waste by creating alternative scenarios of product use, for example, sharing/renting/leasing schemes to consumers and producer take-back programs for refurbishment.

As shown in Figure 25.6, there are three different approaches to a PSS, each of which has different sustainability implications. In a product-oriented PSS (PO-PSS), where a consumer owns the product (e.g., a home photocopier), the firm can provide additional services to assure the durability of the product. In a use-oriented PSS (UO-PSS), the service provider owns the product, making it easier for the product to be eventually refurbished and reused, and sells only the "function" to the customer through a service contract. In a results-oriented PSS, the customer buys only results and is not concerned how the firm delivers those results. Accordingly, the firm's own photocopiers could be run at off-peak times and ad hoc copy services provided as needed.

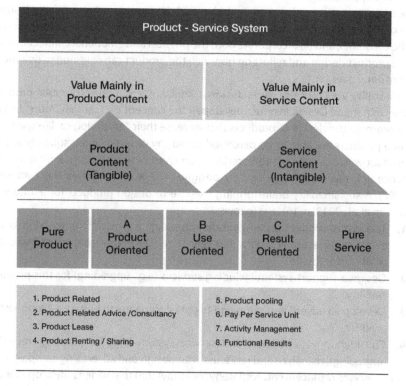

Figure 25.6: Product-service system categories.[11]

[11]www.d4s-sbs.org/MD.pdf; http://www.d4s-de.org/

As a brainstorming exercise, direct the team to think beyond pure product offerings by designing a more-sustainable results oriented PSS:

1. Take the product under consideration, for example, a home music system, and envision how it might be designed as a pure service. The idea may be outlandish but that is acceptable in brainstorming exercises.
2. Design a support system for this service. What type of business model could be viable?
3. Considering the service life cycle, what resources are needed to provide the service in a sustainable manner? Relative to a pure product offering, what value-add would the consumer, the company, and the natural and social environments each receive by consumers subscribing to this service?

After sustainability brainstorming, concept development comes next in the design thinking for sustainability process.

Sustainability in Concept Development

In the design thinking for sustainability process, concept development involves reaching a consensus on a product/service concept that meets the requirements of a sustainable design brief.[12] Such briefs should effectively include the project description, the scope of the project, key assumptions about the product design, target users, expected outcome of the project, success metrics, project timeline, and resources needed (Liedtka & Ogilvie, 2011). For example, the brief for a recyclable *and* upcyclable office chair would give the project team a framework from which to begin, benchmarks by which they can measure progress (e.g., against Steelcase's 99 percent recyclable Think® chair), and a set of objectives to be realized—such as price point, available technology, and market segment (Brown & Wyatt, 2010). Sustainable design criteria, which describe the attributes, constraints, and user perceptions of the product /service to be designed, are also delineated during the concept development stage. These two documents, the design brief and the design criteria, are used as inputs to guide in the *What wows* stage of design thinking for sustainability. To ensure comprehensiveness, the nine guidelines from the Hannover Principles[13] for sustainable design (DfS) and the seven principles of DfEnv (Fiksel, 2011) should be reflected in both concept development documents.

The United Nations Environment Programme's (UNEP) Design for Sustainability workbook (UNEP, 2005) advocates performing a life-cycle analysis (LCA) for sustainable concept development as a means to evaluate the impact of a product or service in its PLC. The analysis can be as simple as stakeholders spending an afternoon discussing their sustainability perspectives on each stage of the P/SLC or as sophisticated as complex software-driven analyses conducted over several weeks.[14] Figure 25.7 shows

---

[12]Design briefs, discussed in detail in Liedtka and Ogilvie (2011), are used in the *What Is?* stage. For DThfS, we have moved them to the *What If?* stage. At either stage, the goal is consistent—to reach a consensus on the product or service that will be designed.

[13]These principles were developed for EXPO 2000, which was held in Hannover Germany. www.mcdonough.com/wp-content/uploads/2013/03/Hannover-Principles-1992.pdf

[14]See Curran (1996) and Keoleian and Spitzley (2006) for a more thorough overview of LCA methods and applications.

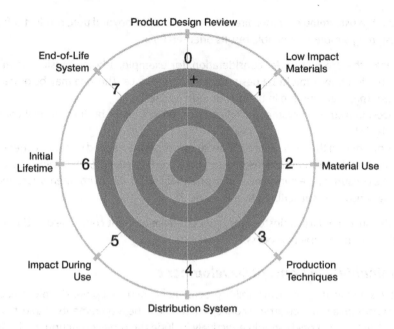

Figure 25.7: UNEP's design for sustainability strategy wheel.

UNEP's eight-spoke strategy wheel that parallels the stages in the PLC, where the process starts at product design review for new products and continues to end-of-life. While the product design should be developed from the perspective of each spoke, typically two to three of the spokes will arise as being more important than the others and there will be sustainability trade-offs at each step as well.

What Wows?

The goal of *What wows* is to determine if the sustainable product/service designed in *What if* actually enchants the consumer as intended. There are two steps in *What wows*: sustainability assumption testing and rapid prototyping. In both steps, physical experiments are conducted with potential users to determine if the product/service meets the sustainable design criteria and "wows" the consumer.

Sustainability Assumption Testing[15]

Sustainability assumption testing takes the design brief and the design criteria and subjects them to the mandates set by the four DfEnv strategies of Fiksel (2011): design for dematerialization, design for detoxification, design for revalorization, and design for capital protection and renewal. The DThfS approach centers on design for revalorization with its focus on the consumer as a partner covering all touch points that the consumer

---

[15] Liedtka and Ogilvie (2011) indicate that assumption testing should be a thought experiment. In design thinking for sustainability, it is a physical experiment in order to test the design for sustainability strategy.

has with the firm and the product, including packaging, supplementary products (utensils, disposable coffee cups, etc.), and end-of-life (EOL) waste. With design for revalorization, rather than becoming waste, products and materials are diverted to economically viable reuse, for example, by upcycling the product for new use or remanufacturing the products and/or its components at the end of their useful life. Safmarine Shipping, for example, turns its old shipping containers into classrooms for South African children.

Revalorization is also coupled with dematerialization since recycled materials reduce the need for new materials. At Xerox in 1991, comprehensive processes were established for taking back end-of-life products, designing for ease of disassembly and recovery as well as for remanufacture, reuse, and recycling.[16] The result was a diversion of billions of pounds of copier and printer waste from landfills as well as savings of billions of dollars to the firm.

In designing for revalorization, assumption testing requires inputs from consumers, which may be established as follows:

1. Presenting the product/service idea to the representative target market customers through prototypes, drawings, or other representations.
2. Using the chosen revalorization strategies, asking the customer to process the product as if it were at the end of its life and observing customer's actions to test the following:
   a. The user can easily process the product for revalorization (at the EOL of a product, instructions manuals have typically been discarded, thus, it may be necessary to include instructions on the product itself).
   b. No more than one readily available tool is required for revalorization.
   c. Components resulting from revalorization can be reused in a meaningful way. For example, although nonrechargeable batteries can be easily removed from an electronic product, they cannot be reused.
   d. The user knows where to properly dispose of waste (i.e., take-back programs).
   e. Above and beyond value-add to the firm and the environment, the consumer *wants* to revalorize because it provides value to *them*.
3. Sending the product/service back to the concept development stage for a rework of the design brief and design criteria if the above assumptions are not confirmed.

Sustainability via Rapid Prototyping

Sustainability by rapid prototyping—creating "visual (and sometimes experiential) manifestations of a concept" (Liedtka & Ogilvie, 2011, p. 141)—completes the *What wows* evaluation by obtaining customer feedback to demonstrate a sustainable product that will "wow" in the marketplace. The process is "rapid" because low-fidelity prototypes are presented to customers for their feedback and then reformulated based on this feedback, with the process sometimes taking just minutes. Some low-fidelity prototypes can be made quickly by 3D printers to convey the product/service concept, yet storyboards or illustrations can also be utilized. The intent of the basic representation and

---

[16]http://resilience.osu.edu/CFR-site/dfe_advisor.htm

rapid redesign is to fail early and often in the search for the product/service that wows. Engaging consumers in the process allows misdirected sustainable designs to be identified quickly and improved on immediately.

Drawing on Liedtka and Ogilvie's (2011) research, guidelines for sustainability via rapid prototyping are:

1. **Start small and simple.** Early feedback invites users to contribute as co-creators of sustainable solutions and builds empathy between users and product.
2. **Figure out the story you want and show it, don't tell it.** Visualize sustainability concepts in pictures and few words. Then make prototypes seem real through imagery, artifacts, and experiences and to create empathy by bringing the user into the sustainability concept.
3. **Visualize multiple options.** Create choices to be made by your audience.
4. **Play with prototypes, don't defend them.** Prototypes are about testing sustainability assumptions that are to be validated or knocked down in order to build a better design.
5. **Test EOL revalorization steps.** Such steps should be part of the rapid prototyping process.

What Works?

There are two steps in *What works*: sustainability via customer co-creation and learning launch. The first step allows designers to gauge reactions and impressions of the customer and to understand how the sustainable product delivers on its intended value. The second step allows designers to capture consumers' *revealed* behaviors regarding sustainability and not just their *stated* behaviors that were captured in previous design steps. Sustainability performance indicators and metrics can then be evaluated to determine if the resulting design indeed meets the design goals.

Sustainability via customer co-creation. Involving customers in the NPD process allows identifying which attributes they value most about the product/service, and is vital to understanding how sustainability improvements might influence attribute values. With co-creation companies may be able to design products in such a way that they may be easily integrated into users' habits and everyday lives (Heiskanen, Kasanen, and Timonen, 2005), while also encouraging behavioral changes that are more socially, economically, and environmentally sustainable (Young, 2010).

A sustainability learning launch. In sustainable design, a learning launch is a selling experiment for the planned sustainable offering that is conducted quickly and inexpensively in a marketplace. The goal of the learning launch is not the successful launch of the new offering, but knowledge about how to improve its success. It is a "dress rehearsal" and an important step in determining the future success of the sustainable design in the real marketplace.

Sustainability metrics. Upon the completion of the design thinking process, managers will want to know if they have been meeting their goals around sustainability measures. Two major trends have been noted in enterprises regarding sustainability metrics (Fiksel, 2011, p. 98): (1) the integration of environmental performance metrics and assessment methods into engineering practices and (2) accounting systems that now recognize and track environmental costs and benefits. Similarly, sustainability goals are being woven into the existing key performance indicators (KPIs) that define what should be measured and metrics defining how KPIs will be measured. Fiksel (2011) provides a list of environmental metrics that can be used to establish product or process design objectives[17]:

- **Energy usage metrics:** Total energy consumed during the product-service life cycle (P/SLC), renewable energy consumed, power used during operation
- **Water usage metrics:** Total fresh water consumed during P/SLC
- **Material burden metrics:** Toxic or hazardous materials used in production, total industrial waste generated during production, greenhouse gases released during the P/SLC
- **Recovery and reuse materials:** Product disassembly and recovery time, purity of recyclable materials recovered, percent of recycled materials used as input to the product
- **Source volume metrics:** Product mass, useful operating life, percentage of packaging recycled
- **Economic metrics:** Average life-cycle costs incurred by manufacturer, purchase and operating costs incurred by the customer, cost of revalorization for the customer
- **Value creation metrics:** Utilization of renewable resources, avoidance of pollutants, human health and safety improvement, enhancement in community quality of life, improvement in customer environmental performance

In the long run, as sustainability becomes an integral part of a company's operations, design thinking for sustainability will just be design thinking, and sustainability measures will be standard indicators all firms track.

25.3 Conclusion

Table 25.3 provides a summary of the four design for sustainable innovation approaches examined in this chapter. The first three are sustainability approaches typically used by engineers in product/service development, while the fourth, the design thinking for sustainability (DTfS) approach as presented in this chapter draws on the strengths of the first three and, further, proactively includes the consumer as a co-development partner. Based on the premise that a lack of empathy in the use of the innovation can derail the sustainability goals of even the best designed product, a DThfS approach has much greater potential to have a significant impact by creating products and services that integrate empathy for the environment and society into product and service designs.

---

[17] For further details on metrics and indicators, see Fiksel (2011), Chapters 7 and 9.

Table 25.3: Design Strategies for Sustainable Products and Services

Strategy	Definition	Product Life Cycle Focus	End-of-Life Goal	Focus
Design for sustainability/ efficiency (Birkeland, 2002)	Delivery of competitively priced goods and services that satisfy human needs and bring quality of life, while progressively reducing ecological impacts and resource intensity throughout the life cycle to a level at least in line with the Earth's estimated carrying capacity.	Cradle-to-grave	Downcycle	Ecological
Design for effectiveness (McDonough & Braungart, 2002)	Systems that emulate nature so that the waste of the production process and the EOL product waste itself are raw material inputs of a new product or service.	Cradle-to-cradle	Upcycle	Industrial
Design for environment (Fiksel, 2011)	The systematic consideration of design performance with respect to environmental, health, safety, and sustainability objectives over the full product and process life cycle.	Cradle-to-gate; cradle-to-grave	Downcycle	Technological
Design thinking for sustainability (this chapter)	Including emphatic design involving the end user in the systematic consideration of design performance with respect to environmental, health, safety, and sustainability objectives over the full product and process life cycle.	Cradle-to-grave; cradle-to-cradle	Both up- and downcycle	Consumer/End user; ecological; industrial; technological

References

Autodesk (2014). *Autodesk sustainability workshop.* Retrieved January 31, 2015, from http://academy.autodesk.com/sustainable-design

Birkeland, J. (2002). *Design for sustainability: A sourcebook of integrated, eco-logical solutions.* London, England: Earthscan.

Brown, T., & Wyatt, J. (2010). Design thinking for social innovation. *Stanford Social Innovation Review,* 8(1), 30–35.

Chapman, J. (2005). *Emotionally durable design: Objects, experiences and empathy,* London, England: Earthscan.

Commoner, B. (1971). *The closing circle: Nature, man, and technology.* New York, NY: Random House, Knopf.

Curran, M. A. (1996). *Environmental life-cycle assessment.* New York, NY: McGraw-Hill.

Fiksel, J. (2011). *Design for environment: A guide to sustainable product development*, 2nd edition. New York: McGraw-Hill.

Goedkoop, M. J., van Halen, C. J. G., te Riele, H. R. M., & Rommens, P. J. M. (1999). *Product service systems, ecological and economic basics*. The Hague, Den Bosch & Amersfoort: Pi. MC, Stoorm CS & PRé Consultants.

Heiskanen, E., Kasanen, P., & Timonen, P. (2005). Consumer participation in sustainable technology development. *International Journal of Consumer Studies*, 29(2), 98–107.

Holmberg, J. (1998). Backcasting: A natural step in operationalising sustainable development. *Greener Management International*, 23, 23–30.

Kelley, T., & Kelley, D. (2013). *Creative confidence: Unleashing the creative potential within us all* (pp. 19–20). New York, NY: Crown Business.

Keoleian, G. A., & Spitzley, D. V. (2006). Life cycle based sustainability metrics. *Sustainability Science and Engineering: Defining Principles*, 1, 127–159.

Laszlo, C., & Zhexembayeva, N. (2011). *Embedded sustainability*. Stanford, CA: Stanford Business Books.

Liedtka, J., & Ogilvie, T. (2011). *Designing for growth: A design thinking tool kit for managers*. New York, NY: Columbia University Press.

Makower, J. (2014). Retrieved March 30, 2015, from http://www.greenbiz.com/blog/2014/02/10/state-green-business-people-side-sustainability-csr-starbucks-levis

Mont, O. K. (2002). Clarifying the concept of product–service system. *Journal of Cleaner Production*, 10(3), 237–245.

McDonough, W., & Braungart, M. (2002). *Remaking the way we make things: Cradle to cradle*. New York, NY: North Point Press.

McDonough, W., & Braungart, M. (1992). *The Hannover Principles: Design for sustainability*. Charlottsville, VA: William McDonough and Partners.

Robinson, J. (1982). Energy backcasting: A proposed method of policy analysis. *Energy Policy*, 10(4), 337–344.

Shot in the Dark. (2000). *Design on the Environment: Ecodesign for Business*. Sheffield, England: Author.

UNEP (2005). *Design for Sustainability: A Practical Approach for Developing Economies*. Paris, France: UNEP.

Van de Kerkhof, M., Hisschemöller, M., & Spanjersberg, M. (2002). Shaping diversity in participatory foresight studies: Experiences with interactive backcasting in a stakeholder dialogue on long–term climate policy in the Netherlands. *Greener Management International*, 37(1), 85–99.

van Hinte, E. (Ed.) (1997). *Eternally Yours: Visions on product endurance*. Rotterdam, Netherlands: 010 Publishers.

White, P., St. Pierre, L., & Belletire, S. (2013). *Okala course guide*. Dulles, VA: Industrial Designers Society of America.

WBCSD. (2000). Eco-efficiency: Creating more value with less impact. Eco-efficiency Learning Module. Geneva, Switzerland: World Business Council for Sustainable Development, Five Winds International.

WRAP. (2015). Design for deconstruction and flexibility. Retrieved January 31, 2015, from http://www.wrap.org.uk/content/design-deconstruction-and-flexibility

Young, G. (2010). Design thinking and sustainability. Retrieved January 31, 2015, from http://zum.io/wp-content/uploads/2010/06/Design-thinking-and-sustainability.pdf

About the Authors

ROSANNA GARCIA is Associate Professor of Marketing at North Carolina State University, where she is a Chancellor's Faculty of Excellence in Innovation+Design. Her role within the Innovation+Design Cluster is to develop curriculum and student-focused programs that help develop and launch technologies germinating from student ideas and faculty research, especially those around environmental sustainability. As an expert in the diffusion of innovations, Dr. Garcia's research has focused on how to introduce "resistant" innovations to a reluctant marketplace. Her recent research and teaching centers on sustainable innovations and the role of trust in collaborative consumption communities. Dr. Garcia's teaching focuses on entrepreneurial marketing and new product launch. She is author of the textbook *Creating and Marketing New Products & Services*. Correspondence regarding this chapter can be directed to S.G.Dacko@warwick.ac.uk.

SCOTT DACKO is Associate Professor, Marketing and Strategic Management, at Warwick Business School, University of Warwick. He has ten years' new product development, management, and marketing experience in large and small companies. He holds a mechanical engineering degree from the University of Minnesota, an MBA from the University of Minnesota, and a PhD in business administration from the University of Illinois at Urbana-Champaign. His research interests include sustainable service innovation and the role of timing in product and service marketing strategies. He is the author of *The Advanced Dictionary of Marketing: Putting Theory to Use* (Oxford University Press, 2008). His publications include articles in *Economics of Innovation and New Technology, Technological Forecasting and Social Change, Journal of Advertising Research, Industrial Marketing Management, Marketing Intelligence & Planning, Benchmarking: An International Journal, International Journal of New Product Development and Innovation Management*, and the *Journal of Marketing Management*.

INDEX